Aspects of symmetry

Selected Erice lectures of

SIDNEY COLEMAN

Donner Professor of Science, Harvard University

CAMBRIDGE
UNIVERSITY PRESS

CAMBRIDGE UNIVERSITY PRESS
Cambridge, New York, Melbourne, Madrid, Cape Town, Singapore,
São Paulo, Delhi, Dubai, Tokyo, Mexico City

Cambridge University Press
The Edinburgh Building, Cambridge CB2 8RU, UK

Published in the United States of America by
Cambridge University Press, New York

www.cambridge.org
Information on this title: www.cambridge.org/9780521318273

© Cambridge University Press 1985

First published 1985
First paperback edition 1988
Reprinted 1989, 1990, 1993, 1995

A catalogue record for this publication is available from the British Library

Library of Congress catalogue card number: 84–21364

ISBN 978-0-521-26706-9 Hardback
ISBN 978-0-521-31827-3 Paperback

To my mother

Contents

Contents

Contents

Contents

Preface

I first came to Erice in 1966, to lecture at the fourth of the annual schools on subnuclear physics organized by Nino Zichichi. I was charmed by the beauty of Erice, fascinated by the thick layers of Sicilian culture and history, and terrified by the iron rule with which Nino kept students and faculty in line. In a word, I was won over, and I returned to Erice every year or two thereafter, to talk of what was past, or passing, or to come, at least insofar as it touched on subnuclear theory. Eight of these lectures, or more properly lecture series, are collected here.

No attempt has been made to bring the lectures up to date. Typographical errors, when spotted, have been corrected, and references to works to be published have been changed to references to published works. (I thank Hugh Osborne for taking on this dull task.) Otherwise, these are unaltered reprints of the original publications.

Numerous debts are acknowledged in the individual lectures, but there is one overriding debt that must be acknowledged here. None of this would have existed were it not for Nino Zichichi. Of course, he is the creator and director of the subnuclear school, and of the International Center for Scientific Culture 'Ettore Majorana' which encompasses it, but, more than that, he is personally responsible for each and every one of these lectures. The lecture notes would never have been written were it not for his blandishments and threats, transmitted in a fusillade of urgent cablegrams and transatlantic phone calls at odd hours of the morning. This book may be the least of his many accomplishments, but one of his accomplishments it is, and it should be counted as such.

Finally: These lectures span fourteen years, from 1966 to 1979. This was a great time to be a high-energy theorist, the period of the famous triumph of quantum field theory. And what a triumph it was, in the old sense of the word: a glorious victory parade, full of wonderful things brought back from far places to make the spectator gasp with awe and laugh with joy. I hope some of that awe and joy has been captured here.

Harvard 1984   Sidney Coleman

Acknowledgements

An Introduction to Unitary Symmetry (1966) was originally published in *Strong and Weak Interactions – Present Problems* (Academic Press, New York and London, 1966) © Academic Press, and is reproduced with their permission.

Soft Pions (1967) was originally published in *Hadrons and their Interactions* (Academic Press, New York and London, 1968) © Academic Press 1968 and is reproduced with their permission.

Dilatations (1971) and *Renormalization and Symmetry* (1971) were originally published in *Properties of the Fundamental Interactions* (Editrice Compositori, Bologna, 1973) © Editrice Compositori 1973, and are reproduced with their permission. The work was supported in part by Air Force Office of Scientific Research under contract no. F44620-70-C-0030 and the National Science Foundation under grant no. GP-30819X.

Secret Symmetry (1973) was originally published in *Laws of Hadronic Matter* (Academic Press, London and New York, 1975) © Ettore Majorana Centre and is reproduced with permission of the copyright holder and the publisher. The work was supported in part by the National Science Foundation under grant no. GP-30819X.

Classical Lumps and their Quantum Descendants (1975) was originally published in *New Phenomena in Subnuclear Physics* (Plenum Publishing Co., New York, 1976) © S. Coleman 1976 and is reproduced with permission of the copyright holder and the publisher. The work was supported in part by the National Science Foundation under grant no. MPS75-20427.

The Uses of Instantons (1977) was originally published in *The Whys of Subnuclear Physics* (Plenum Publishing Co., New York, 1979) © Plenum Press 1979 and is reproduced with their permission.

1/N (1979) was originally published in *Pointlike Structures Inside and Outside Hadrons* (Plenum Publishing Co., New York, 1982) © Plenum Press 1982 and is reproduced with their permission. The work was supported in part by the Department of Energy under contract DE-AC03-76SF00515 and by the National Science Foundation under grant no. PHY77-22864.

1

An introduction to unitary symmetry
(1966)

1   The search for higher symmetries

1.1   *The eight-baryon puzzle*

Let us begin with a very simple observation: there are eight baryons. By this I mean there are eight positive-parity particles with spin one-half and nucleon number one (the nucleons, the Λ, the Σs, and the Ξs), and that the masses of these particles are close together, all lying within 20% of their common mean mass. There are no other particles with the same parity, spin, and nucleon number which lie at all close in mass to these eight. Now, eight is a disquietingly large number of particles; although there is no fundamental reason why we should not have eight independent particles in a field theory, with eight arbitrary masses and eight sets of arbitrary coupling constants, still, it would be very pleasant if there were some way to reduce the number of entities. Two methods immediately suggest themselves:

(i) Perhaps some of the baryons are fundamental and the others are composite meson–baryon states. This was the central idea of the original Sakata model, where p, n, and Λ were taken as fundamental. Then the approximate mass degeneracy would arise because the masses of the mesons are small compared to the masses of the fundamental baryons. This idea has not met with much success.

(ii) Perhaps the strong interactions are approximately symmetric under a group larger than the ordinary isospin–hypercharge group. The eight baryons form a basis for an irreducible representation of this 'higher symmetry group'. If the symmetry were exact, the masses would be degenerate; because the symmetry is only approximate, the masses are split. (The situation envisioned here is something like that which prevails for isospin symmetry; if we turned off the weak interactions and electromagnetism, the world would be isospin symmetric, and all the particles in an isotopic multiplet would be degenerate; in fact, the world is not isospin symmetric, and the masses are slightly different.) This is sometimes phrased by saying that the strong-interaction Hamiltonian is a sum of two terms: one due to 'very strong interactions', which is symmetric under the higher symmetry group, and one due to 'medium-strong interactions', which is not.

The second viewpoint is one which has met with considerable success, and we intend to explore it in more detail here.

Now, what do we know about the structure of such a 'higher symmetry group'? Well, in the first place, there are several assumptions we can make, not because they are forced upon us by the problem, but because they are straightforward generalizations of what we know about the isospin group, and because they simplify the problem considerably. We assume, in the limit of exact symmetry, that:

(i) The group commutes with the Poincaré group.

(ii) The group acts on the Hilbert space of the states of the world as a group of unitary operators. (This is connected with conservation of probability.)

(iii) The one-particle states form an invariant subspace under the action of the group. The many-particle asymptotic states transform like tensor products of one-particle states.

(iv) The group commutes with the S-matrix.

A group which satisfies these conditions is usually called 'an internal symmetry group'. It is worth noting that in a field theory any group whose generators are obtained by integrating, over all space, the time components of a set of conserved Hermitian currents is an internal symmetry group.

In addition to these assumptions, there are several conditions forced on the group by the nature of the problem.

(i) The group turns baryon states at rest, with spin up, into baryon states at rest, with spin up. Therefore, it must have an eight-dimensional representation.

(ii) Since the whole point of the hypothesis is to explain the approximate equality of the baryon masses, this representation must be irreducible.

(iii) (This is a rather fine technical point, but it is necessary in order to avoid some pathological cases.) The eight-dimensional representation should be closed, in the sense that any matrix which is the limit of a sequence of representation matrices should be a representation matrix itself. Since the S-matrix is continuous, if we are given a representation that is not closed, we can always close it, and thus obtain a representation of a larger group which is still an internal symmetry group.

(iv) The representation must be faithful. All observed particles can, in principle, be constructed out of the eight baryons and their antiparticles. Thus, any group element which acts trivially on the baryons will act trivially on everything known, and might as well be forgotten.

These conditions are sufficient to ensure that the higher symmetry group must be a compact Lie group. This is a class of mathematical objects which has been thoroughly investigated; all the compact Lie groups have been classified, and so have all of their representations. So finding all the compact Lie groups with eight-dimensional representations is merely a question of knowing what mathematics books to look in.

However, there is another condition, which arises because we want our higher symmetry group to contain the isospin–hypercharge group, the group of the old familiar symmetries of strong-interaction physics. This is:

(v) The group should contain a subgroup isomorphic to the isospin–hypercharge group. Furthermore, when we restrict ourselves to this subgroup, the eight-dimensional representation should decompose into an isodoublet of hypercharge one (N), an isosinglet of hypercharge zero (Λ), an isotriplet of hypercharge zero (Σ), and an isodoublet of hypercharge minus one (Ξ).

To determine the groups which satisfy this last condition, as well as all the others, requires some independent effort on the part of the investigator; however, the work is straightforward and can be done in a few hours. At the end of all this one has a long, and unenlightening, list. However, it has a surprising property; every group on it contains either SU(3), the group of all unimodular unitary transformations on a three-dimensional complex vector space, or G_0, a group called 'the connected part of the minimal global symmetry group', whose structure I will explain below.

These two groups are minimal; if their predictions are wrong, the predictions of any other higher symmetry group must be wrong, and the whole idea of higher symmetry must be abandoned. We will begin with G_0, since that turns out to be wrong; SU(3), of course, turns out to be right – otherwise I would not be giving these lectures.

1.2 The elimination of G_0

G_0 is the direct product of three factors of SU(2),

$$G_0 \cong SU(2) \otimes SU(2) \otimes SU(2).$$

Thus, every element of G_0 can be written as a triplet of elements of SU(2),

$$(g_1, g_2, g_3),$$

the generators of G_0 are three commuting 'angular-momentum vectors',

$$\mathbf{I}^{(1)}, \mathbf{I}^{(2)}, \mathbf{I}^{(3)},$$

and the irreducible representations of G_0 are labeled by three 'spins',

$$(s_1, s_2, s_3).$$

Isospin and hypercharge are imbedded in G_0 in the following way:

$$\mathbf{I} = \mathbf{I}^{(1)} + \mathbf{I}^{(2)},$$

and $Y = 2I_z^{(3)}$.

The eight baryons transform according to the representation

$$(\tfrac{1}{2}, 0, \tfrac{1}{2}) \oplus (\tfrac{1}{2}, \tfrac{1}{2}, 0).$$

(Thus, we have the desired isospin–hypercharge decomposition $I = \tfrac{1}{2}$, $Y = \pm 1$, and $I = 0$, $Y = 0$.) This is a reducible representation of G_0; therefore G_0 does not meet our conditions – though it is contained in many groups which do. The easiest way to display a group which does meet the conditions and which contains G_0 is to add to G_0 a discrete element which has the effect of interchanging $\mathbf{I}^{(2)}$ and $\mathbf{I}^{(3)}$. The representation above then becomes an irreducible representation of the enlarged group.

One particular element of G_0 is

$$R = e^{i\pi I_y^{(3)}}.$$

R is a hypercharge reflection operator,

$$RYR^{-1} = -Y \quad R\mathbf{I}R^{-1} = \mathbf{I}.$$

Simply from R invariance one can deduce an almost endless list of contradictions with experiment. I will give three here.

(i) For every particle or resonance there must be another particle or resonance, of the same spin and nucleon number, with opposite hypercharge, and with approximately the same mass. Thus, there must be a low-lying $\tfrac{3}{2}^+$ resonance with hypercharge -1. Such a resonance does not exist.

(ii) It is easy to show from R invariance, isospin invariance, and the assumption that the electric current transforms like the electric charge (i.e. like $I_z + Y/2$), that the electromagnetic self-energy of the Σ^+ should be the same as that of the Σ^-. In fact, they are separated by 13 MeV.

(iii) From the same assumptions, the magnetic moment of the Λ should be zero. In fact, it is of the order of magnitude of the nucleon moments.

Thus, G_0 must be rejected. All that is left, our last hope, is SU(3).

2   SU(3) **and its representations**

In this section we will develop some of the properties of SU(3) and its representations, and also develop some methods for doing simple SU(3) calculations. We will begin by making some remarks about SU(n).

2.1   *The representations of* SU(n)

SU(n) is defined as the group of all unimodular unitary $n \times n$ complex matrices. This definition immediately tells us one representation of the group. If we let x^i be a complex n-vector, on which the group acts in the following manner

$$U: x^i \to U^i_j x^j,$$

then it is clear that the space of all x^i forms a basis for a representation of U(n). A basis for another representation is formed by a set of vectors y_i, which transform according to

$$U: y^i \to U^i_j y_j,$$

where

$$U^j_i = \bar{U}^j_i.$$

(We use an overbar throughout to indicate complex conjugation.)

The notation we have used, with its upper and lower indices, mimics that of ordinary tensor analysis. This mimicry is not deceptive, for, as a consequence of the unitarity of U,

$$U: x_i y^i \to x_i y^i,$$

and thus, just as in ordinary tensor analysis, the summation of upper and lower indices is an invariant operation.

By the usual method – taking direct products – we can form, from these primitive objects, the spaces of all tensors of rank $n+m$, with n upper and m lower indices. Each of these spaces clearly forms the basis for a representation of U(n).

These representations are, however, not necessarily irreducible. If a_{ij} is a tensor of rank two, we may divide it into the sum of a symmetric and an antisymmetric part

$$a_{ij} = a_{(ij)} + a_{[ij]}.$$

This separation is invariant under the action of the group; thus we have divided the original representation space (of dimension n^2) into two invariant subspaces (of dimension $n(n+1)/2$ and $n(n-1)/2$, respectively). Likewise, if a^i_j is a mixed tensor, we may divide it into two parts,

$$a^i_j = \frac{1}{n} \delta^i_j a^k_k + \hat{a}^i_j,$$

where

$$\hat{a}^i_i = 0,$$

and again we have two invariant subspaces, in this case of dimension 1 and $n^2 - 1$. (Note that in this case we cannot make a further reduction by symmetrizing and antisymmetrizing, because the indices transform differently under the action of the group.)

One of the problems we set in the introduction was finding the irreducible representations of SU(3). Our general method of attack will be to

(i) construct all tensors with a given number of upper and lower indices;

(ii) divide them invariantly into as many parts as we can;

(iii) discard the parts which we can show lead to representations equivalent to those obtained from tensors of lower rank; and

(iv) identify the remaining parts with (hopefully) new irreducible representations.

This method is, in principle, capable of generating all representations of SU(n), for any n. However, the combinatorics required to keep track of the irreducible tensors becomes formidable for n greater than 3, and other methods seem to be more efficient.

We will first tackle SU(2), to warm up.

2.2   *The representations of* SU(2)

It is a great convenience to introduce the antisymmetric two-index tensor ε^{ij}. Under the action of the group

$$U: \varepsilon^{ij} \to (\det U)\varepsilon^{ij};$$

but since U is unimodular, ε^{ij} is invariant. So is ε_{ij}, the corresponding entity with two lower indices. Given any tensor, we can use the ε tensors to raise and lower indices, just as the metric tensor is used in ordinary tensor analysis. In particular, we may write any tensor invariantly in terms of a tensor with all lower indices, and thus there is no representation of SU(2) induced on a general tensor that is not equivalent to one induced on a tensor with only lower indices. Let

$$a_{i_1 \ldots i_n}$$

be a tensor of this kind. We may divide a into two parts, which are respectively symmetric and antisymmetric under interchange of the first two indices. With the aid of ε_{ij} we may write the antisymmetric part in terms of a tensor of lower rank:

$$a_{[i_1, i_2]\ldots} = \varepsilon_{i_1 i_2} b_{i_3 \ldots}.$$

Thus, we need only consider completely symmetric tensors. Thus we have (hopefully) an inequivalent irreducible representation of SU(2) associated with every space of completely symmetric tensors of a given rank, with all indices lower. (We adhere to an ancient convention, and call this rank $2s$.) We will call the representation $D^{(s)}$, or (s) for short. The dimension of (s) is the number of linearly independent tensors of the proper sort. Since the tensors are completely symmetric, and since the indices are allowed to assume only two values, this is the same as the number of ways $2s$ objects can be divided into two sets. That is to say,

$$\dim (s) = 2s + 1,$$

a result which should be familiar.

2.3   *The representations of* SU(3)

Now let us apply the same techniques to SU(3). The invariant ε tensors now have three indices, and thus cannot be used to raise and lower indices. Thus we have to work with tensors that have both lower and upper indices:

$$a^{i_1 \ldots i_n}_{j_1 \ldots j_n}.$$

However, we may still use the ε tensors to write the antisymmetric part of any tensor in terms of a tensor of lower rank

$$a^{[i_1, i_2] \cdots}_{\cdots} = \varepsilon^{i_1 i_2 \, k} \, b^{\cdots}_{k \cdots},$$

and therefore we need only consider tensors completely symmetric in both their upper and lower indices. By similar reasoning, we need only consider traceless tensors.

We will define the representation induced on the space of all traceless, completely symmetric tensors with n upper indices and m lower indices as $D^{(n,m)}$, or simply (n, m) for short. Eventually, we will show that this family of representations forms a complete set of inequivalent irreducible representations. However, it is more expedient to first extract some properties of these representations that are of practical importance, and only afterwards to prove their completeness, inequivalence, and irreducibility. Therefore, for the time being, we will simply call them IRs. If you want to think of 'IR' as an acronym for 'irreducible representation', you are welcome (indeed encouraged) to do so, but, in fact, we shall not use this property until we have proved it.

Note that a simple consequence of our definitions is that

$$\overline{(n, m)} = (m, n),$$

where the overbar indices complex conjugation.

2.4   *Dimensions of the IRs*

To calculate the dimension of an IR is a straightforward exercise in combinatorics. The space of all completely symmetric tensors with n upper indices and m lower indices can be decomposed into the space of all symmetric tensors and to a space of tensors equivalent to the traces. (By 'the traces' we mean those tensors which are obtained by summing one upper index with one lower index.) The space of the traces is equivalent to the space of all completely symmetric tensors with $(n-1)$ upper indices and $(m-1)$ lower indices. In representation language,

$$(n, 0) \otimes (0, m) = (n, m) \oplus [(n-1, 0) \otimes (0, m-1)].$$

Taking the dimensions of both sides

$$\dim (n, m) = \dim (n, 0) \times \dim (0, m) - [\dim (n-1, 0)$$
$$\times \dim (0, m-1)].$$

By arguments similar to those we used in the discussion of SU(2),

$$\dim (n, 0) = \tfrac{1}{2}(n+2)(n+1) = \dim (0, n),$$

and therefore,

$$\dim (n, m) = \tfrac{1}{2}(n+1)(m+1)(n+m+2).$$

There is an alternative method of designating representations, much used in the literature, in which a representation is labeled by its dimension. Since (n, m) and (m, n) have the same dimension, they are distinguished by labeling a representation by its dimension if n is greater than m, and by its dimension with an overbar if m is greater than n. Thus

(1, 0) is often called   **3**,
(0, 1) " " "   **$\bar{3}$**,
(1, 1) " " "   **8**,
(3, 0) " " "   **10**,
(2, 2) " " "   **27**, etc

(This is still not totally unambiguous, since dim $(4, 0)$ = dim $(2, 1)$ = 15, but it suffices for practical purposes.)

2.5 *Isospin and hypercharge*

Merely to state that the very strong interactions are invariant under a given higher symmetry group is not sufficient information to construct a physical theory. We also have to know how isospin rotations and hypercharge rotations, the old familiar symmetries of strong-interaction physics, are imbedded in the group. The most convenient way of specifying this, for SU(3), is by giving the isospin and hypercharge transformation properties of the fundamental representation $(1, 0)$.

$(1, 0)$ is a three-dimensional representation. Therefore, when we restrict SU(3) to the isospin group SU(2), it can decompose in only three ways: into the sum of three isosinglets, into an isosinglet and an isodoublet, or into an isotriplet. The first case is mathematically impossible, for it implies that all the elements of SU(2) are inside the identity element of SU(3). The third case is mathematically possible, but physically uninteresting: if the fundamental triplet contains only integral isospin, then all the IRs (which are made from direct products of fundamental triplets) would contain only integral isospins, which is not very satisfactory for explaining nature.

Thus, only the second possibility remains. We will choose our basis in unitary space so that the $(1, 0)$ representation looks like

$$\begin{pmatrix} q^0 \\ q^{+1/2} \\ q^{-1/2} \end{pmatrix},$$

where we have labeled the basis vectors by the appropriate eigenvalue of I_z. All that remains is to assign the hypercharge.

In order that the hypercharge differences between observed particles be integers, it is necessary that the hypercharge difference between the singlet and the doublet be of magnitude one. We choose the hypercharge of the doublet to be the greater. (This is just a matter of convention, although it does not appear to be so at first glance. If we were to choose the opposite assignment, the representation $(0, 1)$ would have our original assignment. Thus, the structure of all calculations would be the same, except that (n, m) would everywhere be replaced by (m, n). This only has to do with our conventions about how we write things on paper, not with what goes on in the world. Formally, this degree of freedom corresponds to the existence of an outer automorphism of SU(3).)

Thus, if the hypercharge of the singlet is y, the hypercharge of the doublet is $1 + y$. Now, if hypercharge rotations are to be a subgroup of SU(3), the hypercharge itself must be a generator of SU(3). The generators of unimodular matrices are traceless and, therefore, if Y is the 3×3 hypercharge matrix, then

$$\text{Tr } Y = 3y + 2 = 0$$

which means

$$y = -\tfrac{2}{3}.$$

(If we had assigned a different value to y, we could not have put the hypercharge rotations inside SU(3). However, we could still have put them inside U(3), which is just as good a symmetry group for the purposes of physics. Unlike the choice connected with the previous degree of freedom, a different choice for y would make a real difference in the physics of the world. However, it would not effect what we plan to calculate here, the isospin and hypercharge assignments of the observed particles. The reason is that all observed particles may, in principle, be constructed from baryons and antibaryons. The baryons are assigned to the representation (1, 1), and the hypercharge assignments within this representation do not depend on y.)

2.6   *Isospin–hypercharge decompositions*

We want to determine how an IR decomposes into irreducible representations of the isospin–hypercharge subgroup when we restrict SU(3) to that subgroup. We will label representations of the subgroup by their isospin and hypercharge, thus

$$(i)^Y.$$

We already know one decomposition from the preceding section:

$$(1, 0) \rightarrow \left(\frac{1}{2}\right)^{1/3} \oplus (0)^{-2/3}.$$

From this formula it is trivial to calculate the decomposition of $(n, 0)$ since this is constructed by forming the completely symmetric product of n factors of (1, 0). Thus,

$$(n, 0) \rightarrow \left(\frac{n}{2}\right)^{n/3} \oplus \left(\frac{n-1}{2}\right)^{n/3-1} \oplus \ldots (0)^{-2n/3}.$$

Likewise,

$$(0, m) \rightarrow \left(\frac{m}{2}\right)^{-m/3} \oplus \left(\frac{m-1}{2}\right)^{-m/3+1} \oplus \ldots (0)^{2m/3}.$$

Table 1. *Graphical representation of the decomposition of*
$(n, 0)\otimes(0, m)$

	$\left(\frac{n}{2}\right)^{\frac{n}{3}}$	$\left(\frac{n-1}{2}\right)^{\frac{n}{3}-1}$	$\cdots$
$\left(\frac{m}{2}\right)^{-\frac{m}{3}}$			
$\left(\frac{m-1}{2}\right)^{-\frac{m}{3}+1}$		/////	/////
		/////	/////
		/////	/////

From these formulae we may graphically (see Table 1) represent the decomposition of $(n, 0)\otimes(0, m)$. We form an $m+1$ by $n+1$ rectangular array, and label each column by a term in the first decomposition and each row by a term in the second. Inside each box we put the product of the factor associated with the row and the factor associated with the column. (To calculate this requires only the ability to add hypercharges and to multiply representations of the rotation group.) The total content of the array is then the desired isospin–hypercharge decomposition.

The table shows such an array. (The shading will be explained momentarily.)

Likewise, in the same manner, we may decompose $(n-1, 0)\otimes(0, m-1)$. The relevant primary decompositions are

$$(n-1, 0)\to\left(\frac{n-1}{2}\right)^{n-1/3}\oplus\dots,$$

and

$$(0, m+1)\to\left(\frac{m-1}{2}\right)^{-m-1/3}\oplus\dots.$$

However, as far as calculating the product goes, we could with no error subtract $\frac{2}{3}$ from all the hypercharges occurring in the first series and add the same amount to all the hypercharges in the second series. But then the decomposition of this product is revealed as just the content of the shaded portion of the original array.

But since

$$(n, 0)\otimes(0, m)=(n, m)\oplus[(n-1, 0)\otimes(0, m-1)],$$

3	*An introduction to unitary symmetry*

Table 2. *Decomposition of* $(2, 2)$

	$(1)^{\frac{1}{3}}$	$(1/2)^{-\frac{2}{3}}$	$(0)^{-\frac{3}{3}}$
$(1)^{-\frac{1}{3}}$	$(2)^0 \oplus (1)^0 \oplus (0)^0$	$(3/2)^{-1} \oplus (1/2)^{-1}$	$(1)^{-2}$
$(1/2)^{\frac{2}{3}}$	$(3/2)^1 \oplus (1/2)^1$		
$(0)^{\frac{5}{3}}$	$(1)^2$		

this means that the decomposition of (n, m) is nothing but the content of the unshaded portion of the array, the border. In Table 2, we have calculated the decomposition of $(2, 2)$ by this method.

A common (although extremely awkward) method of displaying the isospin–hypercharge decomposition of a representation of SU(3) is to plot a graph, in which the vertical axis is Y and the horizontal axis is I_z. If a state occurs in a representation with a given (Y, I_z) assignment, a dot is placed on the graph in the appropriate location. If two such states occur, the dot is circled. If three, the dot is circled twice, etc. Such a graph is called a weight diagram. Our result may be used to construct the weight diagram of $(2, 2)$ shown in Fig. 1.

It is a good exercise to use this method to construct the weight diagrams for $(1, 1)$, $(3, 0)$ and $(4, 1)$, all of which are of physical interest. ($(1, 1)$ is usually identified with the baryons and mesons: $(3, 0)$ with the $\frac{3}{2}^+$ resonances in baryon–meson scattering. It has recently been proposed that some of the higher resonances may be part of $(4, 1)$.)

2.7 The Clebsch–Gordan series

I now want to present a method for decomposing the direct product of two IRs into a direct sum of IRs. The method proceeds in two steps: first we decompose the direct product of two IRs into a direct sum of certain special reducible representations, which will be defined

Fig. 1

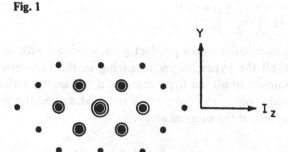

below. Then we decompose the special reducible representations into a direct sum of IRs.

We shall denote the special reducible representations by $D^{(n,n',m,m')}$ – or for brevity, simply by $(n, n'; m, m')$. The representation $(n, n'; m, m')$ is defined as that representation which has for its basis the set of all tensors with $n+n'$ upper indices and $m+m'$ lower indices, that are completely symmetric among the first n upper indices, completely symmetric among the last n' upper indices, completely symmetric among the first m lower indices, completely symmetric among the last m' lower indices, and traceless. Roughly speaking, $(n, n'; m, m')$ may be thought of as the direct product of (n, m) and (n', m') with all traces removed, but without any symmetrization.

It is a simple matter to decompose the direct product of IRs into special reducible representations. We merely separate out all tensors that can be obtained by contracting, in all possible ways, indices from the set of n with indices from the set of m', and indices from the set of n' with indices from the set of m. That is to say,

$$(n, m)\otimes(n', m')=(n, n'; m, m')\oplus(n-1, n'; m, m'-1)$$
$$\oplus(n, n'-1; m-1, m')$$
$$\oplus(n-1, n'-1; m-1, m'-1)\oplus \ldots.$$

The process terminates whenever we run out of indices to contract; that is, whenever a zero appears in the series on the right. In more compact form,

$$(n, m)\otimes(n', m')= \sum_{i=0}^{\min(n,m')} \sum_{j=0}^{\min(n'm)} (n-i, n'-j; m-j, m'-i),$$

where the summation sign indicates a direct sum.

We now wish to decompose one of our special reducible representations into direct sums of IRs. In the language of tensors, we want to decompose an arbitrary tensor from the basis of $(n, n'; m, m')$ into a sum of linear combinations of completely symmetric traceless tensors. Let us begin with the upper indices. Let

$$T^{i_1 \cdots i_n i_{n+1} \cdots i_{n+n'}}_{j_1 \cdots j_m j_{m+1} \cdots j_{m+m'}},$$

be an arbitrary tensor of the type under discussion. Let us choose a pair of upper indices; with no loss of generality they may be i_1 and i_{n+1}. We may write the tensor as the sum of two tensors, one of which is symmetric under interchange of these indices, and the other of which is antisymmetric. Using the ε tensor, we may write the antisymmetric part in terms of a tensor of lower rank,

$$S^{i_2 \cdots i_n i_{n+2}}_{kj_1 \cdots} = \varepsilon_{ki_1 i_{n+1}} T^{i_1 \cdots}_{j_1 \cdots}.$$

The surprising fact, which enormously simplifies the whole reduction, is that this tensor is already completely symmetric in its lower indices.

Proof. For example, let us take the indices j_1 and j_{m+1}. We prove the tensor is symmetric under interchange of these indices by showing that their contraction with the ε tensor vanishes,

$$\varepsilon^{rj_1 j_m +1} S_k^{i_2 \cdots} = \varepsilon^{rj_1 j_m +1} \varepsilon_{kl_1 l_{n+1}} T_{j_1}^{l_1 \cdots}$$
$$= (\delta_k^r \delta_{i_1}^{j_1} \delta_{i_{n+1}}^{j_m} {}_{+1} - \delta_k^{j_1} \delta_{i_1}^r \delta_{i_{n+1}}^{j_m} {}_{+1} + \text{cyclic perms.})$$
$$\times T_{j_1 \cdots}^{i_1 \cdots}.$$

But, by the tracelessness of T, the right-hand side of this equation is zero. Similar arguments work for any pair of indices.

Thus, the symmetrization is very simple. We may remove pairs of upper indices, adding a lower index whenever we do so; or, alternatively, we may remove pairs of lower indices, adding an upper index whenever we do so – but we can never remove both a pair of upper indices and a pair of lower indices, for once we have removed a pair of upper(lower) indices, the tensor is already completely symmetric in its lower (upper) indices. The process terminates when we run out of indices. Returning from the basis space to the representation, we may write the decomposition in compact form:

$$(n, n'; m, m') = (n+n', m+m')$$
$$\oplus \sum_{i=1}^{\min(n,n')} (n+n'-2i, m+m'+i)$$
$$\oplus \sum_{j=1}^{\min(m,m')} (n+n'+j, m+m'-2j),$$

where the summation sign again represents the direct sum.

To demonstrate the efficiency of this method, we conclude with two examples. All arithmetic is shown.

Example 1: $(1, 1) \otimes (1, 1)$

$(1, 1) \otimes (1, 1) = (1, 1; 1, 1) \oplus (1, 0; 0, 1) \oplus (0, 1; 1, 0) \oplus (0, 0; 0, 0).$

$(1, 1; 1, 1) = (2, 2) \oplus (0, 3) \oplus (3, 0),$

$(1, 0; 0, 1) = (1, 1),$

$(0, 1; 1, 0) = (1, 1),$

$(0, 0; 0, 0) = (0, 0).$

The desired decomposition is the sum of all the terms on the right. If we write this in terms of the notation in which representations are labelled by their dimensions, we find

$$8 \otimes 8 = 27 \oplus 10 \oplus \overline{10} \oplus 8 \oplus 8 \oplus 1.$$

This is a familiar (and useful) result. It tells us the number of independent amplitudes for the scattering of two octets into two octets (eight, if time reversal imposes no further restrictions), the number of independent Yukawa couplings for antibaryon–baryon–pseudoscalar-meson (two), the number of amplitudes for the decay of a $\frac{3}{2}^+$ resonance into baryon and pseudoscalar meson (one), etc.

Example 2: $(2, 2) \otimes (3, 0)$

$$(2, 2) \otimes (3, 0) = (2, 3; 2, 0) \oplus (2, 2; 1, 0) \oplus (2, 1; 0, 0).$$
$$(2, 3; 2, 0) = (5, 2) \oplus (3, 3) \oplus (1, 4),$$
$$(2, 2; 1, 0) = (4, 1) \oplus (2, 2) \oplus (0, 3),$$
$$(2, 1; 0, 0) = (3, 0) \oplus (1, 1).$$

In the alternative notation,

$$27 \otimes 10 = 81 \oplus 64 \oplus 35 \oplus \overline{35} \oplus 27 \oplus 10 \oplus \overline{10} \oplus 8.$$

2.8 *Some theorems*

I now wish to show that the IRs we have been discussing do indeed form a complete set of inequivalent irreducible representations of SU(3). In order to do this, we need a simple theorem from group theory. Let G be any compact group, and let the irreducible representations of G be arranged in a series, $D^{(0)}, D^{(1)}, \ldots$, with $D^{(0)}$ the trivial representation. Then any representation of G may be decomposed into a direct sum of irreducible representations. We will write this in the following way

$$D = \oplus \sum n_i D^{(i)},$$

where n_i is the number of times $D^{(i)}$ occurs in the sum. The theorem we need is this:

Theorem. In the decomposition of $\bar{D}^{(i)} \otimes D^{(j)}$, $n_0 = \delta_{ij}$.

Corollary. If, in the decomposition of $\bar{D} \otimes D$, $n_0 = 1$, D is irreducible.

Now, we know how to decompose the direct product of any two IRs into a sum of IRs, from the algorithms of the preceding section. Thus, if we know which IRs contain $D^{(0)}$, we can use the corollary to check if the IRs are irreducible, and the theorem to check if they are inequivalent.

We will now prove the following:

Lemma. Of all the IRs, only $(0, 0)$ contains $D^{(0)}$ in its decomposition, and it contains it once.

Proof. Let us assume an IR contains $D^{(0)}$. Then it must contain a hyper-charge-zero isosinglet. It is a trivial consequence* of our algorithm for the hypercharge–isospin decomposition of an IR that only (m, m) contains a hypercharge-zero isosinglet, and further, that it only contains one such state. Thus, any tensor component of such an IR which transforms like a hypercharge-zero isosinglet must be pure $D^{(0)}$. Such a component is

$$a_{11...1}^{11...1}.$$

But this cannot be pure $D^{(0)}$, for there are evidently transformations of the group (for example, those which interchange the first and second axes and leave the third invariant) under which it is not invariant. The only exception to this argument occurs for $(0, 0)$.

Thus, to find how many times $D^{(0)}$ occurs in the decomposition of $\overline{(n, m)} \otimes (n, m) = (m, n) \otimes (n, m)$, we need only find how many times $(0, 0)$ occurs. But it is a trivial consequence* of our algorithm for the Clebsch–Gordan series that this only occurs once. Thus we have the following.

Theorem. (n, m) is an irreducible representation of SU(3).

By the same token, to prove the inequivalence of the IRs we need only see how many times $(0, 0)$ occurs in the decomposition of $\overline{(n, m)} \otimes (n', m')$ $= (m, n) \otimes (n', m')$. But it follows trivially* from the Clebsch–Gordan algorithm that $(0, 0)$ only occurs in this series if $m = m'$ and $n = n'$. Thus we have the following.

Theorem. (n, m) and (n', m') are equivalent only if $n = n'$ and $m = m'$.

To prove the completeness of the IRs, we need some more information from group theory. For any compact group it is possible to define an integral over group space in such a way that

$$\int \bar{D}_{\alpha\beta}^{(i)}(g) D_{\gamma\delta}^{(j)}(g) \mathrm{d}g = 0 \quad (i \neq j),$$

where the Greek subscripts indicate matrix elements. Now, let us assume that there exists an irreducible representation of SU(3) which is not equivalent to any (n, m). Let us call this representation $D^{(?)}$. Then, by assumption

$$\int \bar{D}_{\alpha\beta}^{(?)}(g) D_{\gamma\delta}^{(n, m)}(g) \mathrm{d}g = 0.$$

The representation $(1, 0)$ has eight independent matrix elements. Let me call these z_i ($i = 1 \cdots 8$). They form a set of coordinates in group space. The matrix elements of $(0, 1)$ are the $\bar{z}_i$. The equation above says that the matrix elements of $D^{(?)}$ are orthogonal to z_i and $\bar{z}_i$, that is to say, to all

---

* Prove it.

linear functions of z_i and $\bar{z}_i$. It also says that the matrix elements of $D^{(?)}$ are orthogonal to all the matrix elements of $(1, 0)\otimes(1, 0)$, $(1, 0)\otimes(0, 1)$, and $(0, 1)\otimes(0, 1)$, that is to say, to $z_i z_j$, $z_i \bar{z}_j$, and $\bar{z}_i \bar{z}_j$. In other words, these matrix elements are orthogonal to all polynomials of order two in the coordinates. In fact, by similar arguments, we can show that they are orthogonal to all polynomials in the coordinates and, therefore,

$$D_{\alpha\beta}^{(?)}(g) = 0 \quad \text{for all } g.$$

But this is a contradiction, because for any representation $D_{\alpha\beta}(1) = \delta_{\alpha\beta}$. Thus, we have our final

Theorem. The IRs form a complete set of inequivalent irreducible representations of SU(3).

2.9 *Invariant couplings*

The last general subject I am going to talk about is the construction of invariant couplings. (For trilinear interactions, this is equivalent to the problem of constructing the Clebsch–Gordan coefficients.) I know of no methods here as simple and powerful as the ones discussed above. However, there are some special tricks based on tensorial methods which are useful for a very restricted class of problems. (Which, through the grace of God, turns out to include many cases of physical interest.) My main aim here will be to discuss these; however, I would like to begin with a general discussion, to place these methods within a wider framework.

2.10 *The problem of Cartesian components*

Let us suppose we wish to couple three octets of fields. Just to be definite, let us make them baryon (denoted by ψ), antibaryon ($\bar{\psi}$) and meson (ϕ). The Clebsch–Gordan algorithm tells us that there are only two invariant couplings, and a moment's thought shows what they must be;

$$\bar{\psi}_k^i \psi_i^j \phi_j^k \quad \text{and} \quad \bar{\psi}_k^i \psi_j^k \phi_i^j.$$

(Of course, if these are really baryon and meson fields, ψ is an octet of Dirac bispinors, and there should be a γ_5 in the above expressions to conserve parity. However, we will ignore degrees of freedom associated with space-time transformation properties here, and treat all fields as if they were scalars, for the sake of simplicity.) In a sense, this expression solves the problem completely, for it explicitly gives the invariant coupling in terms of the fields. Unfortunately, it gives the coupling in terms of the Cartesian components of the fields (ϕ_1^1, ϕ_2^1, etc.), and for practical applications, we need the coupling in terms of isospin–hypercharge eigenstates (π^+, K^0, etc.).

Thus, in this formulation, the whole problem of constructing invariant couplings reduces to the problem of expressing the Cartesian components of SU(3) tensors in terms of hypercharge–isospin eigenstates. If I could present a simple algorithm for constructing such expressions, the problem would be completely solved. Regrettably, I know of no such algorithm. However, we can always construct such expressions by what are essentially cut-and-try methods; the main portion of the subsequent exposition will be devoted to an explanation of these techniques. They are definitely not of the back of the envelope class; to do calculations, you need tables, which is unfortunate. However, with this way of looking at the problem you only need one table per representation, while with the more usual technique, you need one table per triplet of representations. So there is some gain.

As before, we will begin by discussing SU(2), where things are simple.

2.11   SU(2) *again*

Let us begin with a fundamental doublet of fields, which we will call p and n, and which we will write as a column vector,

$$N = \begin{pmatrix} p \\ n \end{pmatrix}.$$

The conjugate doublet of antiparticle fields transforms like a row vector

$$\bar{N} = (\bar{p}\,\bar{n}).$$

But, of course, with the aid of the ε tensor, we may write this as a column vector also:

$$\begin{pmatrix} \bar{n} \\ -\bar{p} \end{pmatrix}.$$

(The minus sign appears because of the antisymmetry of the ε tensor.)

Next, let us consider a triplet of fields (the pions, for example). We may represent this triplet in three equivalent ways, as tensors of the three forms

$$\phi_{ij}, \phi^{ij}, \phi^i_j.$$

We will choose the third way, and write the components in the form of a 2×2 matrix:

$$\begin{pmatrix} \phi^1_1 & \phi^1_2 \\ \phi^2_1 & \phi^2_2 \end{pmatrix}.$$

Antisymmetry for one form implies tracelessness for the other:

$$0 = \varepsilon^{ij} \phi_{ij} = \phi^i_i.$$

Thus

$$\phi_1^1 = -\phi_2^2.$$

It is clearly an invariant condition to demand that the matrix ϕ be Hermitian, and we shall impose such a condition here. (There is nothing wrong with non-Hermitian matrices, but they would describe sets of six real fields – perfectly suitable for the Σ hyperons, but not for the pions.) The component ϕ_2^1 carries charge $+1$, and we will normalize the matrix by demanding that it be the π^+ field. Then Hermiticity and tracelessness determine everything to within one free real parameter α:

$$\phi = \begin{pmatrix} \alpha\pi^0 & \pi^+ \\ \pi^- & -\alpha\pi^0 \end{pmatrix}.$$

This is determined by looking at the invariant-mass term in the Lagrangian. This is $\phi_j^i\phi_i^j$, and we want to normalize our fields so that it will be the sum of three squares with coefficients one*. Thus,

$$\frac{1}{2}\phi_j^i\phi_i^j = \pi^+\pi^- + \alpha^2\pi^0\pi^0 = \pi^+\pi^- + \frac{1}{2}\pi^0\pi^0,$$

which means

$$\alpha = \pm\sqrt{\left(\frac{1}{2}\right)}.$$

The sign ambiguity is just a matter of phase conventions. We choose the plus sign.

One advantage of writing these fields as row vectors, column vectors, and matrices, is that the ordinary operations of matrix multiplication (which just involve, with these conventions, the summation of upper and lower indices) are invariant under the action of the group. Thus, the usual pion–nucleon interaction may be written as

$$\mathcal{L} = \frac{1}{2}\operatorname{Tr}[\partial_\mu\phi\partial^\mu\phi - \mu^2\phi^2] + \bar{N}(i\partial_\mu\gamma^\mu - m)N + g\bar{N}\gamma_5\phi N + \lambda\operatorname{Tr}\phi^4;$$

where Tr represents the ordinary matrix trace.

2.12   SU(3) *octets: trilinear couplings*

By exactly the same arguments as we used above, we can write the $(1, 1)$ representation of SU(3) as a 3×3 matrix. To identify this matrix with the physical fields, all we need to do is study its transformation

---

* We do not have to do this of course; we could choose α to be any non-zero real number, and it would all cancel out of any calculation when we were finished with the renormalizations. This is, however, the traditional and convenient choice; it makes the symmetry between the three pions manifest.

properties under the SU(2) subgroup. (I remind you that this is the set of all transformations that leave the first index invariant.) This determines everything to within normalization factors, which are determined by the same arguments we used in the SU(2) case. Thus, we find for the octet of baryon fields,

$$\psi = \begin{pmatrix} \sqrt{(\tfrac{2}{3})}\Lambda & \Xi^- & -\Xi^0 \\ p & -\sqrt{(\tfrac{1}{6})}\Lambda + \sqrt{(\tfrac{1}{2})}\Sigma^0 & \Sigma^+ \\ n & \Sigma^- & -\sqrt{(\tfrac{1}{6})}\Lambda - \sqrt{(\tfrac{1}{2})}\Sigma^0 \end{pmatrix}$$

The minus sign for the Ξ doublet arises because we are writing it as a row vector. (See the discussion for the nucleon doublet above.) Likewise, we may write a matrix for the pseudoscalar meson octet

$$\phi = \begin{pmatrix} \sqrt{(\tfrac{2}{3})}\eta & K^- & \bar{K}^0 \\ K^+ & -\sqrt{(\tfrac{1}{6})}\eta + \sqrt{(\tfrac{1}{2})}\pi^0 & \pi^+ \\ K^0 & \pi^- & \sqrt{(\tfrac{1}{6})}\eta - \sqrt{(\tfrac{1}{2})}\pi^0 \end{pmatrix}$$

Note that ϕ is Hermitian. The corresponding minus sign does not arise here. This is because the usual convention is to define $(K^-, \bar{K}^0)$ as the anti-particles of (K^+, K^0); the particles which form an isotopic doublet with proper phase relations are $(\bar{K}^0, -K^-)$. The two minus signs cancel.

Just as before, the ordinary operations of matrix algebra are invariant under the action of the group. Thus, the two invariant couplings discussed in 2.10 may be written as

$$\text{Tr } \bar{\psi}\psi\phi \quad \text{and} \quad \text{Tr } \bar{\psi}\phi\psi.$$

Actually, it is conventional to consider not these two couplings, but the two linear combinations

$$\text{Tr } \bar{\psi}\{\psi, \phi\} \quad \text{and} \quad \text{Tr } \bar{\psi}[\phi, \psi].$$

These are called d-type and f-type couplings, respectively.

2.13   SU(3) *octets: quadrilinear couplings*

We may also use these matrices to study the invariant coupling of four octets. This is useful in the analysis of meson–baryon scattering, and also in the analysis of non-leptonic hyperon decays, using octet spurions. Let us represent the four octets by four 3×3 traceless matrices, $A_i \ (i = 1 \ldots 4)$.

The possible invariants fall into two classes. There are those of the form

$$\text{Tr } A_1 A_2 A_3 A_4.$$

There are twenty-four permutations of the four As, but due to the invariance of the trace under cyclic permutations, only six of these lead to distinct invariants. Secondly, there are also invariants of the form

$$\text{Tr } A_1 A_2 \text{ Tr } A_3 A_4.$$

Here, only three permutations lead to distinct invariants.

Thus we have nine invariants in all. Unfortunately, we know that there are only eight independent couplings of four octets. (This was one of the consequences of the first example given of the Clebsch–Gordan algorithm.) This is an apparent contradiction; we can only escape if there is a linear relation among our nine matrix invariants.

Indeed there is such a relation; it is

$$\sum_{\substack{\text{six distinct perms}}} \text{Tr } A_1 A_2 A_3 A_4 = \sum_{\substack{\text{three distinct perms}}} \text{Tr } A_1 A_2 \text{ Tr } A_3 A_4,$$

for any four 3×3 traceless matrices. This identity was no doubt known to Cayley; however, it was first shown to me by Burgoyne, and I will embarrass him by calling it Burgoyne's identity.

Proof. Let A be any 3×3 matrix, with eigenvalues a_1, a_2, a_3; A satisfies its own characteristic equation

$$(A - a_1)(A - a_2)(A - a_3) = 0.$$

Let us write the coefficients of the powers of A in terms of invariants. We then obtain

$$A^3 - (\text{Tr } A)A^2 - \tfrac{1}{2}[\text{Tr } A^2 - (\text{Tr } A)^2]A - (\det A) = 0.$$

Multiplying this equation by A and taking the trace, we find

$$\text{Tr } A^4 = \tfrac{1}{2}(\text{Tr } A^2)^2,$$

if A is traceless. Now let

$$A = \sum_i \lambda_i A_i,$$

and let us extract the coefficient of $\lambda_1 \lambda_2 \lambda_3 \lambda_4$ from the above equation. We then obtain Burgoyne's identity.

2.14 *A mixed notation*

Actually, for many purposes, it is unnecessary to reduce an SU(3)-invariant interaction to an expression which explicitly involves isospin–hypercharge eigenstates; it suffices to reduce the coupling to a sum of SU(2)-invariant interactions. This can easily be done within our framework. For example, instead of explicitly writing down the matrix for the baryon fields, we can display the decomposition in the following way:

$$\psi_1^1 = \sqrt{(\tfrac{2}{3})}\,\Lambda,$$

$$\psi_a^1 = \Xi_a,$$

$$\psi_1^a = N^a,$$

and

$$\psi_b^a = \Sigma_b^a - \sqrt{(\tfrac{1}{6})}\,\delta_b^a\Lambda,$$

where a, $b = 2$, 3, and the terms on the right are the appropriate SU(2) tensors. If we insert this (and the corresponding expression for the mesons) in the invariant coupling, we will obtain the desired result.

This procedure is evidently insensitive to the convention one adapts for phases within an SU(2) multiplet. This is an advantage because there are two such conventions which are widely used in the literature. The convention we have used here is the one that is common in elementary particle physics and the one which is used in most of the earlier SU(3) literature. There is another convention, which is designed to agree with the standard angular-momentum convention, and which is used in de Swart's tables and in some of the later literature. In this convention, the field conjugate to π^+ is minus π^-.

Using the methods we have explained here, we can construct similar decompositions for any representation. (But the process is tedious.) For example, the decomposition for the (3, 0) representation is

$$\psi^{111} = \Omega,$$

$$\psi^{11a} = \sqrt{(\tfrac{1}{3})}\,\Xi^a,$$

$$\psi^{1ab} = \sqrt{(\tfrac{1}{3})}\,\Sigma^{ab},$$

and

$$\psi^{abc} = \Delta^{abc},$$

where Ξ and Σ label the states with the same isospin–hypercharge assignments as the corresponding baryons, Ω is a $(0)^{-2}$ state, and Δ a $(\tfrac{3}{2})^1$ multiplet, and where we have omitted components obtainable from the ones we have listed by trivial permutations of indices.

A problem. Before we leave these matters and go on to some applications, I would like to give you a problem to think about. You may find it amusing.

Consider a field theory in which there are eight pseudoscalar fields, of the same mass, forming an SU(3) octet, coupled together by an SU(3)-invariant interaction Lagrangian, involving no derivatives, and of fourth order in the fields. There are no other fields.

    (i) Show that the interaction Lagrangian involves only one arbitrary coupling constant;

(ii) Assume that for sufficiently strong coupling two pseudoscalar mesons bind to make a scalar bound state. These states must, of course, fall into SU(3) representations. Show that the only representations that can occur are **1**, **8** and **27**.

(iii) Show that **8** and **27** are necessarily degenerate in mass.

3 Applications

The applications of SU(3) fall into two classes: those that deal with the limit of exact symmetry, and those which attempt to treat the three kinds of symmetry-breaking interactions. All the applications I will consider here will be of the second class.

There are three types of symmetry-breaking interactions: the mysterious medium-strong interactions, the electromagnetic interactions, and the weak interactions*. The general procedure one adopts for these problems is as follows: one treats the symmetry-breaking interactions as perturbations to an exactly symmetric world. One calculates the effect of interest to lowest non-vanishing order in a perturbative expansion, and then uses the known (or conjectured) SU(3) transformation properties of the symmetry-breaking interaction to connect the coefficients in the expansion. In this way one obtains sum rules for mass splittings, magnetic moments, leptonic-decay amplitudes, etc.

The computational reason for this reliance on lowest-order perturbation theory is clear: as one goes to higher orders in perturbation theory, more free parameters appear, until, eventually, one has no predictions left. The physical justification is somewhat more obscure. Lowest-order perturbation theory is certainly very plausible for the weak and electromagnetic interactions, but it is much less so for the medium-strong interactions. We will use it, nevertheless, and keep our fingers crossed.

3.1 *Electromagnetism*

The electromagnetic interaction Lagrange density is

$$ej_\mu A^\mu,$$

where j_μ is the electric current and A_μ the electromagnetic field. A_μ does not involve hadronic fields; therefore it commutes with the generators of SU(3), and the transformation properties of the interaction are given only in terms of those of the current.

---

* Only medium-strong and electromagnetic interactions are discussed here. The lectures as given contained a brief discussion of the Cabibbo theory for semi-leptonic decay, but since this was discussed in more detail by other lecturers, I have not included it in the notes. See *Strong and Weak Interactions – Present Problems* (Academic Press, New York and London, 1966).

Now, we do not know the transformation properties of j_μ; however, we do know those of the electric charge

$$Q = \int j_0(\mathbf{x}, t) d^3 \mathbf{x},$$

because this is a sum of SU(3) generators,

$$Q = \frac{Y}{2} + I_z.$$

Thus, Q transforms like a group generator, that is to say, like a member of a (1, 1) representation. Which member? This can be most easily determined by inspecting the (1, 0) representation. The three members of this representation have hypercharge $(-\frac{2}{3}, \frac{1}{3}, \frac{1}{3})$, and I_z $(0, \frac{1}{2}, -\frac{1}{2})$. Thus, they have charge $(-\frac{1}{3}, \frac{2}{3}, -\frac{1}{3})$, and Q therefore transforms like the diagonal matrix

$$E = \frac{1}{3} \begin{pmatrix} -1 & 0 & 0 \\ 0 & 2 & 0 \\ 0 & 0 & -1 \end{pmatrix}.$$

We now assume that j_μ has the same transformation properties as Q. This is a very plausible assumption; it would be true, for example, if we began with an SU(3)-invariant field theory of baryons and mesons and constructed the minimal electromagnetic coupling.

3.2   Magnetic moments: baryons

The electromagnetic form factors for the baryon octet are defined by

$$\langle B|j_\mu(x)|B'\rangle = e^{ik \cdot x} \bar{u}(F_1^{BB'}(k^2)\gamma_\mu + F_2^{BB'}\sigma_{\mu\nu}k^\nu)u',$$

where B and B' are one-baryon states with corresponding spinors u and u', and k is the four-momentum transfer. We wish to find the constraints placed on these quantities, calculated to first order in electromagnetism and to zeroth order in all other symmetry-breaking interactions.

The baryon state on the right transforms like (1, 1), as does that on the left. The electric current transforms like (1, 1). $(1, 1) \otimes (1, 1)$ contains (1, 1) twice. Thus there are two invariant couplings. Thus, we can write the electric (magnetic) form factors of all eight baryons – as well as the electric (magnetic) form factor for the Σ^0–Λ transition – in terms of two unknown functions of k^2. It is easy to write down the explicit expression in terms of the matrix methods of Part 2:

$$F_{1,2}^{BB'}(k^2) = F_{1,2}^{(1)}(k^2) \operatorname{Tr} \bar{B}EB' + F_{1,2}^{(2)}(k^2) \operatorname{Tr} \bar{B}B'E,$$

where B and B' are the 3×3 matrices associated with the baryon states and $F_{1,2}^{(1)}$ and $F_{1,2}^{(2)}$ are unknown functions. In particular, $F_2^{BB'}(0)$, which tells us the eight baryon magnetic moments and the Σ^0–Λ transition moment, is given in terms of two unknown constants. It is trivial to do the traces and eliminate the two constants; we then find equations for seven moments in terms of $\mu(\text{p})$ and $\mu(\text{n})$. These are:

$$\mu(\Lambda) = \tfrac{1}{2}\mu(\text{n}),$$
$$\mu(\Sigma^+) = \mu(\text{p}),$$
$$\mu(\Xi^0) = \mu(\text{n}),$$
$$\mu(\Xi^-) = \mu(\Sigma^-) = -[\mu(\text{p}) + \mu(\text{n})],$$
$$\mu(\Sigma^0) = -\tfrac{1}{2}\mu(\text{n}),$$

and

$$\mu(\Sigma^0 \to \Lambda) = \tfrac{1}{2}\sqrt{3}\,\mu(\text{n}).$$

Only the first of these formulae has been checked experimentally. The formula predicts $\mu(\Lambda) = -0.95$ nuclear magnetons; the experimental value is -0.73 ± 0.17.

3.3 *Electromagnetic mass splittings*

Electromagnetic mass splittings are second-order effects; the electromagnetic Hamiltonian must act twice. At first glance this would seem to give nine independent parameters:

$$(0, 0) \oplus (1, 1) \oplus (1, 1) \oplus (3, 0) \oplus (0, 3) \oplus (2, 2)$$
$$\downarrow \quad \downarrow \times \downarrow \quad \downarrow \quad \downarrow \quad \downarrow$$
$$(0, 0) \oplus (1, 1) \oplus (1, 1) \oplus (3, 0) \oplus (0, 3) \oplus (2, 2),$$

where the first line is the decomposition of the two Hamiltonians, the second that of the two baryon states, and the arrows the independent couplings. However, this neglects the fact that the mass splitting is symmetric under interchange of the two Hamiltonians. If we analyze the symmetry of the terms in $(1, 1) \otimes (1, 1)$, we find that $(0, 0)$, $(2, 2)$ and one of the $(1, 1)$s are symmetric, while $(0, 3)$, $(3, 0)$, and the other $(1, 1)$ are antisymmetric. So the figure above should read:

$$(1, 1) \otimes (1, 1) = (0, 0) \oplus (1, 1) \oplus (2, 2) \oplus \text{antisymmetric terms}$$
$$\downarrow \quad \downarrow \quad \searrow \quad \searrow$$
$$(1, 1) \otimes (1, 1) = (0, 0) \oplus (1, 1) \oplus (1, 1) \oplus (2, 2) \oplus (0, 3) \oplus (3, 0).$$

Furthermore, the first of these (the singlet term) contributes equally to all baryon masses. Thus, there are only three significant terms for the mass splittings. Since there are four observable splittings, this means we can find one relation among them.

The three invariants are

$$\mathrm{Tr}\ \bar{B}E^2 B,\ \mathrm{Tr}\ \bar{B}BE^2,\quad \text{and}\quad \mathrm{Tr}\ \bar{B}EBE.$$

Again, it is trivial to evaluate the traces. The resulting formula is

$$\Xi^- - \Xi^0 = \Sigma^- - \Sigma^+ + \mathrm{p} - \mathrm{n},$$

where the names of the particles stand for their masses. Experimentally, the right-hand side of this equation is 6.6 ± 0.1 MeV, while the left-hand side is 6.5 ± 1.0 MeV.

3.4   *Electromagnetic properties of the decuplet*

The analysis of the form factors of the $\frac{3}{2}^+$ decuplet proceeds along exactly the same lines as that for the $\frac{1}{2}^+$ octet (except that there are four form factors instead of two). The decuplet state on the right transforms like $(0, 3)$, that on the left like $\overline{(0, 3)} = (3, 0)$:

$$(3, 0) \otimes (0, 3) = (3, 3) \oplus (2, 2) \oplus (1, 1) \oplus (0, 0).$$

Thus, we have to find only one SU(3) invariant. But we already know one – the electric charge! Thus, for the decuplet, *all moments are proportional to the charge.*

Electromagnetic mass splittings are just as simple. Here there are only two possible invariants. Once again, we know of two objects with the right transformation properties, Q and Q^2. Thus

$$\delta M_{\mathrm{em}} = \alpha Q + \beta Q^2,$$

with α and β unknown constants.

3.5   *The medium-strong interactions*

In his classic first paper on SU(3), Gell-Mann proposed that the medium-strong interaction Hamiltonian should transform like a component of an octet. This assumption is, of course, completely independent from the assumption of SU(3) invariance of the very strong interactions. Nevertheless, the remarkable success of the mass formula that can be derived from the two assumptions offers evidence for both of them.

If the medium-strong interaction transforms like a component of an octet, it must be the Λ-like component, for no other carries zero hypercharge and isospin. It follows from our Clebsch–Gordan algorithm that there are at most two ways of coupling an octet to the product of an IR and its conjugate; therefore, for any hadron multiplet there are, at most, three independent coefficients in the mass formula – one for the common mass, and two for the splitting.

It turns out to be possible to explicitly calculate the two splitting terms, once and for all, for every unitary multiplet. (This is a curious fact; the electromagnetic mass formulae do not have this property.) This was first done by Okubo, using a lengthy method. I shall give here a short proof due to Smorodinski.

For any representation of SU(3), the eight matrices which form the generators of the representation transform, under the action of the group, like an octet. Therefore, we can arrange them in a 3×3 matrix (a matrix of matrices!):

$$G = \begin{pmatrix} -2\alpha Y & — & — \\ — & \alpha Y + I_z & \sqrt{2}\,I_+ \\ — & \sqrt{2}\,I_- & \alpha Y - I_z \end{pmatrix},$$

where the Is are the usual isospin generators, Y is the hypercharge, and α is an unknown real constant. The blank spaces are occupied by strangeness-changing generators in which we are not interested. To determine α, we observe that those transformations which mix the first and third axes, and leave the second invariant, do not change the electric charge. Electric charge is a generator of the group, and it is evident that the only member of the matrix of generators which is left invariant by these transformations is the 22 entry. Therefore, this must be proportional to the electric charge, which means

$$\alpha = \tfrac{1}{2}.$$

We want to find objects which transform like the 11 components of octets. One such object is clearly

$$G_1^1 = -Y.$$

Since matrix operations are invariant under SU(3), any matrix function of G will also transform like a matrix, that is to say, like a mixture of octet and singlet. In particular, let us consider the co-factor matrix. (The co-factor matrix is the matrix composed of the determinants of the minors of a matrix; it occurs in the standard expression for the inverse of a matrix):

$$(\text{cof } G)_1^1 = G_2^2 G_3^3 - G_2^3 G_3^2$$

$$= \frac{Y^2}{4} - \mathbf{I}^2.$$

This is a second object with the desired transformation properties. (It is not a pure octet, but the singlet part will not affect a formula for mass differences.) Thus, for any unitary multiplet the three terms in the mass formula are

$$m = a + bY + c\left(\frac{Y^2}{4} - I(I+1)\right).$$

This is the famous Gell-Mann–Okubo formula.

I have written this formula in terms of mass; however, it turns out that one obtains a better fit with experiment if one uses masses for fermions and squares of masses for bosons. Nobody knows why this is so. (Covariant perturbation theory certainly suggests that masses are the 'natural' parameters for fermions, and squares of masses for bosons, but this is not much of an argument.)

The experimental tests of the formula are discussed in Professor Barbaro-Galtieri's lectures.

4   Ideas of octet enhancement

Up to now, we have been considering the three kinds of symmetry-breaking interactions as independent effects. I would now like to turn to the attempts that have been made to connect them and, in particular, to the remarkable relations discovered a few years ago between the medium-strong and electromagnetic mass splittings.

'One measures a circle beginning anywhere.' Let us begin with the mass spectrum of the Σ hyperons. These particles have isospin one; therefore the mass operator is the sum of three parts, with isospins zero, one, and two, respectively. The last two are electromagnetic in origin. Let us attempt to estimate their relative magnitude, in the most naive way. Naively, we would expect the major portion of the electromagnetic self-mass of a particle to be given by diagrams of the form shown in Fig. 2 where the blobs represent electromagnetic form factors. If we estimate the magnitude of these diagrams for the nucleon system, using experimental form factors, we find that the magnetic form factors fall off too rapidly to make any significant contribution to the integral giving the self-mass. Therefore the self-mass is, to a good approximation, an integral over a quadratic form in the electric form factors only. Now, for the nucleons. The electric form factors are proportional to the charge, to a very good approximation; therefore, by unitary symmetry, the same will be true for the Σs. Thus, we would expect the two charged Σs to have the same self-mass, that is to say, we would expect the mass splitting to be mostly isospin two.

However, when we examine the experimental situation, we find that

$$\Delta m^{(1)} = (\Sigma^- - \Sigma^+)/\sqrt{2} \approx 5.6 \text{ MeV},$$

Fig. 2

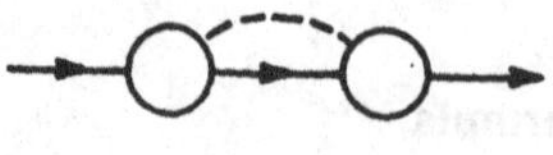

and

$$\Delta m^{(2)} = (\Sigma^+ + \Sigma^- - 2\Sigma^0)/\sqrt{6} \approx 0.7 \text{ MeV}.$$

Precisely the reverse of our naive prediction!

Clearly, then, there is something in the structure of the strong interactions that tends to enhance the isospin one part of the mass splitting (and perhaps, also, to suppress the isospin two part). The strong interactions are the sum of the very strong interactions and the medium-strong interactions. In the last section we derived sum rules for electromagnetic mass splittings by neglecting the effects of the medium-strong interactions, and obtained good agreement with experiment. So, we would be surprised if the medium-strong interactions had an important effect on the mass splittings. Thus, whatever it is that enhances the isospin-one mass splitting would be present even if we turned off the medium-strong interactions.

However, the very strong interactions do not know isotopic spin; they only know SU(3). They do not have the option of enhancing $I = 1$ perturbations. They have only the option of enhancing the octet or the 27-plet part of the perturbation. If they enhanced the 27-plet, we would see enhancement of both isospins, for the 27-plet contains both neutral isotriplets and neutral isoquintuplets. The octet, however, contains only isotriplets.

Thus, we are forced to conclude that there is something in the structure of the very strong interactions that causes them to enhance the octet part of the electromagnetic mass splittings. This is called octet enhancement.

Notice that octet enhancement is a property of the very strong interactions, not of electromagnetism. The observed octet pattern of electromagnetic mass splittings has nothing at all to do with the transformation properties of the electromagnetic interaction. Once we have come to this discovery, it is very tempting to speculate that the same thing is true for the medium-strong interactions. Perhaps the observed octet pattern of medium-strong mass splittings is due, not to the transformation properties of these interactions, but to an octet enhancement mechanism. In fact, perhaps it is the same mechanism for both interactions – perhaps there is a mechanism for *universal octet enhancement*.

Is this a plausible idea or just a wild speculation? Well, one way of checking its plausibility is to construct dynamical models that embody universal octet enhancement. I will discuss three such models here.

(i) First, there is the tadpole model, proposed by Glashow and myself. This model assumes the existence of an octet of scalar mesons. (These need not be stable particles, of course; they would be resonances or anti-bound states. The last is a more likely situation for s-wave channels.) These are labeled with the names of the corresponding pseudoscalar mesons, with a prime. Thus, K', $\bar{K}'$, π', η'. In a field theory involving scalar mesons there arises the possibility of a class of Feynman diagrams that do not otherwise occur. These are diagrams with only one external line. Fig. 3 shows such a diagram. For obvious reasons, these are called tadpole diagrams. In the absence of the symmetry-breaking interactions these diagrams vanish, for they represent a transition between a unitary octet (the scalar mesons) and a unitary singlet (the vacuum). However, if we turn on the medium-strong interactions we can make an η' tadpole, and if we turn on electromagnetism we can make a $\pi^{0'}$ tadpole. These diagrams, by themselves, are of very little interest, but they are important as internal parts of other diagrams. Fig. 4 shows two tadpoles occurring as internal parts of baryon self-energy diagrams. The tadpole on the left makes a contribution to the medium-strong self-mass of the Λ; the one on the right makes a contribution to the electromagnetic self-energy of the proton. Our fundamental (and completely unjustified) assumption is that these diagrams dominate symmetry breaking. Because the scalar mesons form an octet, we get universal octet dominance. If they formed a 27-plet, we would get 27-plet dominance.

(ii) Next, there is the mixing model, proposed by Sakurai, Radicati, Zanello and Picasso. These authors posit that the medium-strong mass splittings are dominated by diagrams that involve the mixing of a singlet vector meson and an octet vector meson. The electromagnetic splittings are, correspondingly, dominated by ρ^0–ω_1 mixing. These mixing matrix elements transform like the components of an octet; thus we obtain universal octet dominance.

Fig. 3

Fig. 4

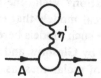

(iii) Finally, there is the bootstrap model, introduced by Cutkosky and Tarjanne, and investigated in more detail by Dashen and Frautschi. Let us suppose we have a system of particles, interacting through some interaction with a high degree of symmetry, under which the particles form degenerate bound states. Let us introduce a new interaction which breaks the symmetry. Then the energies of the bound states will be changed by two mechanisms: the masses of the components will be changed, and the forces will also be changed. In lowest-order perturbation theory this can be written as

$$\delta m_B = A \, \delta m_c + f,$$

where m_B is a vector composed of the masses of the bound states, m_c is a vector composed of the masses of the components, f is a vector composed of the contribution of the change in the forces, and A is a matrix. Now let us suppose the problem is, in fact, a bootstrap. (The example studied by Dashen and Frautschi is the static baryon-octet resonance-decuplet bootstrap.) Then m_B and m_c are identical, and the equation becomes

$$(1 - A)\delta m = f.$$

If A has an eigenvalue close to one, then there will be enhancement, independent of f. If this eigenvalue corresponds to a set of eigenvectors that transform like an octet, then it will be octet enhancement. Dashen and Frautschi have found, in the static bootstrap, that there *is* an octet of eigenvectors with eigenvalue close to one. This is a strong point in favour of the bootstrap model. In the other theories, octet enhancement is fed in at the beginning; in the bootstrap model it is calculated. The calculation could lead to 27-plet enhancement; it does not.

If universal octet enhancement was the only consequence of these models, they would not really be worth much; there would be hardly more conclusions than assumptions. However, all the models I have discussed possess a second feature which is most easily demonstrated for the tadpole model. Let us return to Fig. 4. The diagram on the left gives the principal part of the medium-strong mass splittings of the hadrons in terms of the strength of their coupling to η' and the magnitude of the η' tadpole. Conversely, we can deduce from the observed splittings the η' hadron–hadron coupling constants, to within an unknown multiplicative factor. Since η' and $\pi^{0\prime}$ are part of the same octet, we can then calculate the $\pi^{0\prime}$ hadron–hadron coupling constants, just by doing a unitary rotation. But if we know these, we know the values of the diagrams on the right, to within an unknown multiplicative constant.

Thus, we can calculate the principal part of the electromagnetic mass splittings in terms of the principal part of the medium-strong mass splittings, and one free parameter. (I emphasize, *not* one free parameter for every unitary multiplet, but one free parameter for all the hadrons.) This is a new phenomenon, not a consequence of universal octet enhancement. For example, if we had two octets of scalar mesons, we would have two sets of unknown coupling constants, which would contribute in different proportions to the medium-strong and electromagnetic splittings. Thus, we would not be able to deduce the electromagnetic splittings from the medium-strong ones.

We call this second desirable property of a dynamical model *non-degeneracy*. Non-degeneracy occurs in mixing models as long as there is a unique mixing mechanism. Thus, as long as we have either dominant vector mixing or dominant pseudoscalar mixing, for example, we can play the same game as above. However, if both play an important role, we are out of luck. In the bootstrap approach the symmetry-violating solutions are found by solving an eigenvalue equation. If one of the eigenvalues is close to one, near spontaneous symmetry breakdown occurs. If *only* one is close to one, we have non-degeneracy. (This is the origin of the term.)

Let us try to get a rough estimate of the accuracy of the algorithm sketched out above. We know that for the baryons the medium-strong mass splitting is mainly F type, i.e., proportional to the hypercharge. Non-degeneracy tells us that the electromagnetic splitting must also be mainly F type, i.e., proportional to the electric charge. We know

$$n > p,$$

so we expect

$$\Sigma^- > \Sigma^0 > \Sigma^+,$$

and

$$\Xi^- > \Xi^0,$$

both of which are in agreement with experiment. If we do the detailed calculation, we find that the general features of this estimate are preserved; we get the right signs, but the magnitudes are not too reliable. This is encouraging; it makes us believe that we indeed have a reliable way of estimating the principal part of the electromagnetic splittings. However, in order to go further to get a real comparison between theory and experiment, we need a theory of the terms beyond the principal term. We need *a theory of the next corrections.*

This is the third desirable feature of a dynamical model. Such a theory has been constructed for the tadpole model. It has not yet been constructed for the other models. This is not a matter of principle: it is certainly possible; it just has not been done yet. Let me explain how things work with tadpoles.

It is well known that the electromagnetic mass of any particle can be written in terms of the forward scattering of unphysical photons off that particle, summed over all polarizations and integrated over all photon four-momenta. Any approximation to this unphysical amplitude generates an approximation to the self-mass. One reasonable approximation to the amplitude is to include all poles in all three Mandelstam variables. Fig. 5 shows all the Feynman diagrams that have poles. All the particles are on the mass-shell except the photons. The first diagram shows the pole in s, the second the pole in u. When we sum and integrate these terms, we just obtain the conventional expression for the self-mass in terms of electromagnetic form factors, first suggested by Feynman and Speisman, and first derived from dispersion relations by Cini, Ferrari, and Gatto. The third diagram shows the pole in t. There are many such diagrams; however, the only ones which will survive the summation and integration to make a contribution to the self-mass are those in which the particle exchanged is a scalar meson. These yield the tadpoles.

Thus, at least in the tadpole model, the prescription for finding the next corrections is simple. We just calculate the conventional contributions to the self-mass, which are given unambiguously in terms of experimentally measurable form factors, and add them to the tadpole terms. Of course, to actually perform the calculation we need some estimates for the form factors. The nucleon form factors are known from experiment; from these we can find those for the hyperons, using unitary symmetry. For the pion, there are reasons to believe that the ρ pole dominates the form factor; the kaon form factor is then determined by unitary symmetry.

Fig. 5

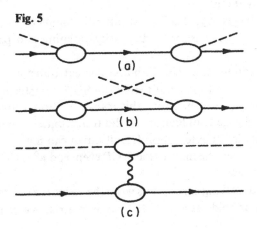

(a)

(b)

(c)

Table 3 shows the results of such a calculation, done by Schnitzer, Socolow and myself.

As you can see, except for the kaon mass splitting, which is badly off, the agreement with experiment is excellent.

Table 3 *Contributions to electromagnetic mass splittings*

	Non-tadpole	Tadpole	Total	Experiment
n–p	−1.1	2.4	1.3	1.3
Σ^0–Σ^+	−0.7	3.6	2.9	3.1±0.1
Σ^-–Σ^0	1.4	3.6	5.0	4.8±0.1
Ξ^-–Ξ^0	1.2	4.8	6.0	6.2±0.7
π^+–π^0	4.9	0.0	4.9	4.6
K^+–K^0	2.8	−4.2	−1.4	−4.0±0.2

Finally, I would like to make a few remarks about non-leptonic weak interactions. The most natural interaction for these processes is $J_\mu J_\mu^+$, where J_μ is the Cabibbo current. However, this interaction does not guarantee the most striking experimental regularity of non-leptonic decays, the $\Delta I = \frac{1}{2}$ rule. (As you probably know, to get the $\Delta I = \frac{1}{2}$ rule, one needs at least two pairs of conjugate currents.) This smells very much like another case of octet dominance, and it is very natural to ask whether the mechanisms we have been discussing can be extended to non-leptonic decays. At first glance this seems very easy to do; in the tadpole model, for example, dominance of diagrams involving $K^{0\prime}$ tadpoles would give the $\Delta I = \frac{1}{2}$ rule for parity-conserving decays. To explain parity-violating decays, we would need to hypothesize an octet of pseudoscalar mesons to make pseudoscalar tadpoles, that is to say, we would need to hypothesize them if they were not already there.

This is a very nice suggestion; it is a great pity it is completely wrong. It breaks down for separate reasons in the parity-violating and parity-conserving cases.

First, the parity-violating case. The Cabibbo current transforms like $\cos\theta\,\pi^+ + \sin\theta\,K^+$. It is easy to see that there is an SU(3) transformation that turns this into pure π^+. Furthermore, this transformation commutes with charge and with charge conjugation. In the transformed frame, the only pseudoscalar tadpole the interaction Hamiltonian can make is that associated with π^0. However, the interaction is CP even, and π^0 is CP odd. Therefore, there is no tadpole.

Second, the parity-conserving case. Let us write the vacuum expectation values of the eight scalar fields as a 3×3 Hermitian matrix. We can find an SU(3) transformation that diagonalizes this matrix. Let us change our definition of hypercharge to mean hypercharge rotations in this new coordinate system. Then, there are no hypercharge-changing tadpoles. That is to say, the only effect of the $K^{0'}$ tadpoles is to mix states in such a way as to change the natural definition of strangeness. They do not cause decays.

I have given these arguments in terms of the tadpole model. Corresponding optimistic assumptions can be framed and destroyed within the bootstrap model. I leave this task as an exercise for you.

Bibliography

Part 1

The arguments referred to in Part 1 can be found in my thesis (Cal. Inst. of Tech., 1962). They can also be constructed, with a little labour, from the results of D. R. Speiser and J. Tarski, *J. Math. Phys.* **4**, 588 (1963).

Part 2

This part is based on my contribution to *High-energy Physics and Elementary Particles* (I.A.E.A., Vienna, 1965), p. 331.

A good reference for group theoretical background is E. P. Wigner, *Group Theory* (Academic Press, New York).

The results attributed to Burgoyne and Smorodinski are private communications.

Part 3

Many applications of SU(3) (and many of the classic papers in the field) are found in the invaluable anthology of Gell-Mann and Ne'eman: M. Gell-Mann and Y. Ne'eman, *The Eightfold Way* (Benjamin, New York, 1964).

Part 4

The tadpole theory is explained in S. Coleman and S. L. Glashow, *Phys. Rev.* **134**, B671 (1964).

The mixing theory is explained in: L. E. Picasso, L. Radicati, D. Zanello and J. J. Sakurai, *Nuovo Cimento* **37**, 187 (1965).

An excellent review of the bootstrap theory is the contribution of D. H. Sharp to *Recent Developments in Particle Symmetries, Proc. of the 1965 Int. School of Physics "Ettore Majorana"* (ed. A. Zichichi), (Academic Press, N.Y., 1966).

Table 3 is taken from R. Socolow, *Phys. Rev.* **137**, B1221 (1965).

2
Soft pions
(1967)

The purpose of these lectures is to explain certain techniques, developed in the last few years by Adler, Weisberger, Weinberg, and others, for the analysis of processes involving low-energy pions. I have tried, as far as possible, to make the lectures self-contained; the only background required of the reader is an understanding of field theory on the Feynman diagram level. In particular, no previous knowledge of current commutators or low-energy theorems is assumed.

In Sec. 1 the reduction formula is developed and some of its consequences discussed. Sec. 2 is a brief summary of the relevant parts of weak-interaction theory. Soft pions first appear in Sec. 3, a discussion of the Goldberger–Treiman relation. Sec. 4 is an analysis of the various definitions of PCAC. Sec. 5 is a discussion of Lagrangian models in general, and the gradient-coupling model in particular. In Sec. 6, Adler's rule for the emission of one soft pion is derived. The current commutation relations are introduced in Sec. 7. In Sec. 8 the formula for the s-wave pion–hadron scattering length is derived. In Sec. 9, the special case of pion–pion scattering is treated. In Sec. 10, a few remarks are made about leptonic decays of kaons.

I have done no original work in this field; most of what I know I have learned in conversations with S. Adler, H. Schnitzer, and S. Weinberg. Those who were lucky enough to attend Weinberg's Loeb lectures at Harvard will know how much these lectures owe to him. I am also indebted to J. Bernstein, who allowed me to read the manuscript of his forthcoming book on currents.

1   The reduction formula

Let us suppose we have a Lagrangian field theory, in which the Lagrange density, $\mathscr{L}$, depends only on a single scalar field, ϕ, and its derivatives. (We assume a single scalar field only for simplicity; everything we say will be readily generalizable to a theory of many fields of arbitrary spin.) Given $\mathscr{L}$, we know, in principle, how to calculate S-matrix elements: we sum all Feynman diagrams contributing to the process of interest.* (Fig. 1 represents such a sum for a two-particle scattering process; the

Fig. 1

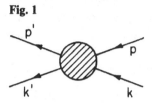

arrows distinguish incoming and outgoing particles.) To obtain the S-matrix element, we multiply this sum by appropriate kinematic factors. (For the Feynman conventions used here, see Appendix 1.)

However, we can also calculate the sum of diagrams with the external lines not on the mass shell. The expression thus obtained clearly has nothing to do with any S-matrix element. Does it have any meaning at all?

One way to give it a meaning is to introduce a linear coupling of ϕ to an external c-number source ρ. That is to say, we change the Lagrange density in the following way:

$$\mathscr{L} \to \mathscr{L} + \rho(x)\phi(x). \tag{1.1}$$

Now, let us consider the matrix element $\langle 0|S|0\rangle$ to fourth order in ρ. This expression is given graphically by Fig. 2, where the crosses represent the

Fig. 2

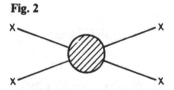

interaction with the source. But this is the same sum of diagrams that occurs in Fig. 1, except that every external line in Fig. 1 is now multiplied by a Feynman propagator, a factor of $i(p^2 - m^2)^{-1}$, because it has now become an internal line.

---

* This is not strictly true. There are always terms in the S-matrix where one or more particles do not interact at all; these must be added to the diagrams. Thus in elastic scattering, the diagrams give not S, but $S - 1$. We will avoid the complications introduced by this phenomenon by always assuming that no two momenta are equal.

However, we can calculate Fig. 2 in another way. Remember Dyson's formula for the S-matrix

$$S = T \exp i \int d^4x \, \mathcal{H}_1(x), \tag{1.2}$$

where $\mathcal{H}_1$ is the interaction Hamiltonian density in the interaction picture, and T is the time-ordering operator. Conventionally, this formula is applied to a field theory by calling all the quadratic terms in the Hamiltonian the free Hamiltonian and calling the remainder the interaction. However, we can just as well call the entire old Hamiltonian the free Hamiltonian; then only the extra term in Eq. (1.1) is the interaction.

Thus we obtain:

$$\langle 0|S|0\rangle = T \left\langle 0 \Big| \exp - i \int \rho(x)\phi(x)\, d^4x \Big| 0 \right\rangle, \tag{1.3}$$

where ϕ is, as always, the field operator in the 'interaction picture', which is, in our case, the Heisenberg picture when ρ vanishes. To find the object of interest (Fig. 2), we must differentiate four times with respect to ρ and set ρ equal to zero. This gives

$$T \langle 0|\phi(x_1)\phi(x_2)\phi(x_3)\phi(x_4)|0\rangle \tag{1.4}$$

We can now forget about diagrams and external sources, for we have established our principal result: S-matrix elements, extrapolated off the mass shell, with every external line multiplied by a free propagator, are simply the Fourier transforms of vacuum expectation values of time-ordered products of Heisenberg fields.

This lengthy sentence may also be expressed by an equation, giving the S-matrix element in terms of the Fourier transform. In our case, this equation is

$$\langle k', p'|S|k, p\rangle = (\text{K.F.}) \int d^4x_1 \ldots d^4x_n (i)^4 (\Box_1^2 + m^2) \ldots (\Box_4^2 + m^2)$$
$$e^{+ik'\cdot x_1} e^{+ip'\cdot x_2} e^{-ik\cdot x_3} e^{-ip\cdot x_4}$$
$$\times T \langle 0|\phi(x_1)\phi(x_2)\phi(x_3)\phi(x_4)|0\rangle, \tag{1.5}$$

where (K.F.) stands for kinematic factors (defined in Appendix 1). The i's and Klein–Gordon operators just serve to cancel the extra propagators. Eq. (1.5) is the famous reduction formula. As we have seen, it is just the Feynman prescription for calculating an S-matrix element, written in coordinate space. It is useful because it enables us to use coordinate-space information (e.g. commutation relations) to place restrictions on the S-matrix.

We need not take all the particles off the mass-shell; for example, by applying reasoning similar to that above to $\langle p'|S|p\rangle$, we can establish that

$$\langle k', p'|S|k, p\rangle = (\text{K.F.}) \int d^4x_1 d^4x_2 (i)^2 (\Box_1^2 + m^2)(\Box_2^2 + m^2)$$
$$e^{+ik'\cdot x_1} e^{-ik\cdot x_2} T \langle p'|\phi(x_1)\phi(x_2)|p\rangle, \qquad (1.6)$$

as well as thirteen other formulae, in which other combinations of particles are taken off the mass shell.

Although we have proven the reduction formula for the case in which ϕ is the canonical field, the field that appears in the Lagrangian, the formula is still true if ϕ is any local scalar field, provided ϕ is properly normalized; that is to say, provided

$$\langle p|\phi(x)|0\rangle = (2\pi)^{-3/2}(2E)^{-1/2}e^{ip\cdot x}. \qquad (1.7)$$

(Notice that everything in this equation, except the scale, is determined by Lorentz invariance.)

We will show this for the simple case where only one particle is taken off the mass shell; the generalization is straightforward. Let us return to momentum space. The reduction formula tells us to calculate the matrix element given by Fig. 3(a), multiply by $(p^2 - m^2)$, and go on to the mass

Fig. 3

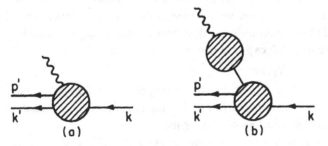

(a)   (b)

shell. (In the figure, the wiggly line represents the new field.) Thus, only those diagrams which have a pole at $p^2 = m^2$ can contribute to the final result. The only diagrams with poles are those with one-particle inter-mediate states (Fig. 3(b)). The residue at the pole is the product of the two shaded blobs. The two-pronged blob is one, by Eq. (1.7). The four-pronged blob is the S-matrix element. Q.E.D.

Thus, if all we know are S-matrix elements, we cannot uniquely assign a field to a particle. If ϕ is a good field for a given particle, so are such peculiar objects as

$$m^{-16}\Box^{16}\phi, \quad \phi + (\Box^2 + m^2)^3\phi^5,$$

etc. These define different off-the-mass-shell extrapolations, but they all lead to the same S-matrix on the mass shell.

2   The weak interactions: first principles

The weak interaction Hamiltonian is the space integral of the Hamiltonian density, $\mathcal{H}_W$. We assume that $\mathcal{H}_W$ has the usual current–current form

$$\mathcal{H}_W = \frac{G}{2}\,\mathcal{J}_\mu\,\mathcal{J}^{\mu+}. \tag{2.1}$$

$\mathcal{J}_\mu$ is the sum of a leptonic part and a hadronic part,

$$\mathcal{J}_\mu = \mathcal{J}_\mu^H + \mathcal{J}_\mu^L. \tag{2.2}$$

Although we know very little about the form of $\mathcal{J}_\mu^H$, that of $\mathcal{J}_\mu^L$ has been well-established:

$$\mathcal{J}_\mu^L = \bar{v}_e \gamma_\mu (1 + i\gamma_5)e + (e \to \mu). \tag{2.3}$$

(We use a notation in which $\gamma_5^2 = -1$.) G can then be determined from muon decay experiments. It is approximately $10^{-5} M_p^{-2}$. $\mathcal{J}_\mu^H$ can be written as the sum of a vector current and an axial-vector current:

$$\mathcal{J}_\mu^H = V_\mu + A_\mu. \tag{2.4}$$

Each of these, in turn, can be written as the sum of a set of currents with definite strangeness-changing properties. For a major part of these lectures we will concentrate on strangeness-conserving processes, and thus we will need to consider only the strangeness-conserving parts of the current.

Semi-leptonic decays are processes of the form

$$i \to f + \text{leptons},$$

where i and f are hadrons. If we neglect electromagnetic effects and higher-order weak corrections, the matrix elements for such a decay factors into the product of two terms:

$$\langle f, l | \mathcal{J}_\mu^H \mathcal{J}_\mu^{L\,+} | i \rangle = \langle f | \mathcal{J}_\mu^H | i \rangle \langle l | \mathcal{J}_\mu^{L\,+} | 0 \rangle. \tag{2.5}$$

The second term is known exactly by virtue of Eq. (2.3). Unfortunately, the first term is not so easy to calculate. We can use symmetry principles – Lorentz invariance, CP conservation, assumed isospin or SU(3) transformation properties of the current – to express this matrix element in terms of a few unknown functions of four-momentum-transfer; however, to calculate these functions we need strong-interaction dynamics.

Let me give two examples of semi-leptonic processes. The first is neutron β-decay. The relevant matrix element is

$$\langle p | \mathcal{J}_\mu^H(x) | n \rangle = (\text{K.F.})e^{-ik \cdot x}\bar{u}_p[\gamma_\mu g_V(k^2) + i\gamma_\mu \gamma_5 g_A(k^2) + \cdots]u_n, \tag{2.6}$$

where k is the four-momentum transfer, the us are the appropriate Dirac bispinors, and the three dots represents other terms that are of order k^2, and hence do not make a significant contribution to this process. (Although they are important in muon capture by nuclei, and also in high-energy neutrino reactions.) The A and V terms clearly come from the axial-vector and vector currents, respectively. Experimentally,

$$g_V(0) \approx 1, \, g_A(0)/g_V(0) \approx 1.25.$$

We shall often denote these quantities simply by g_V and g_A.

The second example is pion decay. Here, by parity, only the axial current can contribute. The relevant matrix element is

$$\langle 0|A_\mu(x)|\pi^- \rangle = (\text{K.F.})e^{-ip\cdot x}(ip_\mu F_\pi/\sqrt{2}). \tag{2.7}$$

where p is the pion four-momentum. Unlike the previous example, here the quantity F_π has no p^2 dependence, since p^2 is fixed at m_π^2. From the observed pion lifetime F_π is readily calculated to be $0.19 \, M_p$.

If we take the divergence of Eq. (2.7), we find that

$$\langle 0|\partial^\mu A_\mu(x)|\pi^- \rangle = (\text{K.F.})m_\pi^2 F_\pi e^{-ip\cdot x}/\sqrt{2}. \tag{2.8}$$

The most important property of this equation is that the right-hand side is not zero! For, in view of the arguments of the proceeding section, this means that

$$\phi_{\pi^-} = \sqrt{2} \, \partial_\mu A^\mu / F_\pi m_\pi^2, \tag{2.9}$$

is a perfectly good pion field, suitable for use in the reduction formula. We shall begin to exploit this fact in the next section.

3 The Goldberger–Treiman relation and a first glance at PCAC

Let us define $g(k^2)$ by

$$\langle p|\phi_{\pi^-}(x)|n \rangle = (\text{K.F.})e^{-ik\cdot x}(k^2 - m_\pi^2)^{-1}\bar{u}_p\gamma_5 u_n g(k^2)\sqrt{2}. \tag{3.1}$$

The only thing we know about $g(k^2)$ experimentally is its value at m_π^2. For, by the arguments of Sec. 1, this must be the renormalized pion–nucleon coupling constant, the quantity we measure by extrapolating to the pion pole in nucleon–nucleon scattering, or by extrapolating to the nucleon pole in pion photoproduction. (The $\sqrt{2}$ in the formula is just an isospin factor.) Thus,

$$g(m_\pi^2) = g = 13.5. \tag{3.2}$$

Now let us return to the axial-vector contribution to neutron beta-decay. This time I will write out all the invariants.

$$\langle p|A_\mu(x)|n \rangle = i(\text{K.F.})e^{-ik\cdot x}\bar{u}_p$$
$$\times [\gamma_\mu\gamma_5 g_A(k^2) + k_\mu\gamma_5 g_p(k^2) + \sigma_{\mu\nu}\gamma_5 k^\nu g_M(k^2)]u_n. \tag{3.3}$$

Taking the divergence, we find

$$\langle p|\partial^\mu A_\mu(x)|n\rangle = (\text{K.F.})e^{-ik\cdot x}\bar{u}_p\gamma_5 u_n[-2Mg_A(k^2)+k^2g_p(k^2)], \quad (3.4)$$

where M is the nucleon mass. Comparing this with Eqs. (2.9) and (3.1), we obtain

$$\sqrt{2}g(k^2)/(k^2-m_\pi^2) = [-2Mg_A(k^2)+k^2g_p(k^2)]\sqrt{2}/F_\pi m_\pi^2. \quad (3.5)$$

This equation is simply a consequence of our definition of ϕ_π, and is without predictive power unless supplemented by further assumptions. We now make such an assumption. We assume that $g(k^2)$ is 'slowly varying' over a distance of the order of m_π^2, that is to say, that

$$g(m_\pi^2) \approx g(0). \quad (3.6)$$

This is a special case of the PCAC hypothesis which we will discuss in more detail shortly. (PCAC is an acronym for 'partially conserved axial current'. We will explain this peculiar phrase in the next section.)

I stress that Eq. (3.6) is pure assumption. It would certainly be false, for example, if ϕ_π were like the peculiar fields involving Klein–Gordon operators which we constructed at the end of Sec. 1.

However, if we accept Eq. (3.6), then, evaluating Eq. (3.5) at $k^2=0$, we instantly obtain

$$F_\pi g \approx 2Mg_A. \quad (3.7)$$

This is the famous Goldberger–Treiman relation. It is in excellent agreement with experiment:

$$2.56M \approx 2.50M.$$

4   A hard look at PCAC

Many statements about the meaning of PCAC and the explanation of the Goldberger–Treiman relation exist in the literature and in the folklore of physics. Here are some of them:

1. 'The Goldberger–Treiman relation is just pology. It is simply the statement that neutron beta-decay is dominated by the one-pion pole.' This is *wrong*. If the one-pion pole diagram (Fig. 4) made the only contribu-

Fig. 4

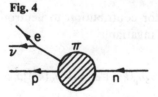

tion to neutron beta-decay, we would predict

$$g_p = gF_\pi(k^2 - m_\pi^2)^{-1}, \tag{4.1}$$

and

$$g_A = 0. \tag{4.2}$$

This is (1) not the Goldberger–Treiman relation and (2) in flat contradiction with experiment.

2. 'PCAC is the statement that the matrix elements of $\partial_\mu A^\mu$ are slowly varying.' This is almost right. (That is to say, it is very close to the viewpoint I will espouse in these lectures.) However, I would prefer to say rather that the matrix elements are *normally varying*: that, just as for S-matrix elements and electromagnetic form factors, about which we do have empirical information, the rate of variation is determined by the distance to the nearest singularity. Thus, in the particular case of $g(k^2)$, the nearest singularity is the beginning of the three-pion cut. This should induce a derivative on the order of $(9m_\pi^2)^{-1}$ at the origin, just as the ρ pole induces a derivative in the electromagnetic form factor of the order of m_ρ^{-2}. Thus the error made in extrapolating from zero to m_π^2 should be of the order of $(m_\pi^2/9m_\rho^2)$ or approximately 10%. The advantage of stating things this way is twofold. (1) It gives a clear idea of how to use PCAC in processes where the kinematics are more complicated than in neutron decay, and where the invariant functions may depend on several variables. (2) It emphasizes that $\partial^\mu A_\mu$ is in no way especially 'smooth'; the derivatives of its matrix elements are no smaller than those of many familiar operators. What is special is that the pion mass is small, compared to the characteristic masses of strong interaction physics; thus extrapolation over a distance of m_π^2 introduces only small errors.

3. 'PCAC is the statement that if the pion mass were zero the axial vector current would be exactly conserved.' This does not look at all like the preceding version of PCAC; nevertheless, it also leads to the Goldberger–Treiman relation. Let us see how this works.

Let us imagine ourselves in a world in which the pion mass is zero and the axial current is conserved. We can still define F_π, G_A, etc. by Eqs. (2.8) and (3.3). (Although their physical meanings are, of course, quite different – F_π is now not the pion decay constant, but the electron decay constant.) Because the axial current is conserved, Eq. (3.4) becomes

$$-2Mg_A(k^2) + k^2 g_p(k^2) = 0. \tag{4.3}$$

However, one-pion exchange (Fig. 4 again!) now produces a pole in g_p:

$$g_p(k^2) = gF_\pi/k^2 + \text{non-singular terms}. \tag{4.4}$$

Thus, when we evaluate (4.3) at $k^2 = 0$, we obtain

$$-2Mg_A + gF_\pi = 0, \tag{4.5}$$

which is the Goldberger–Treiman relation!

Of course, in the real world the pion mass is not zero and the axial current is not conserved, so Eq. (4.5) is not exact. However, the pion mass is small; thus if it is the only thing that keeps the axial current from being conserved, we might expect Eq. (4.5) to be accurate up to terms of the order of m_π^2. One might express this viewpoint by saying that the axial current is 'almost conserved'. Actually, the phrase used is 'partially conserved' – hence the term PCAC.

Note that although this formulation of PCAC is quite different from the preceding one, it leads to the same result and the same rough estimate of the error. Although in these lectures I will try and stick with the preceding formulation (principally because I believe it offers a clearer idea of the sources of possible errors in the more complicated applications which we shall do shortly), this way of looking at things is also a good one. Sometimes we will refer back to it, and it will give us new insights.

4. 'PCAC is the statement that the pion field is the divergence of the axial-vector current.' This statement occurs frequently in the literature. Unfortunately, it is completely free of content, since, as we have seen, it is true by definition. Of course, if we are working within the framework of a Lagrangian field theory with a specific strong-interaction Lagrangian, it is a well-defined statement to say that $\partial^\mu A_\mu$ is proportional to the canonical pion field that occurs in the Lagrangian. However, it is a statement without predictive power, since we know no more about the variation of the matrix elements of canonical fields than we do about those of any other local operator.

Nevertheless, we will sometimes have occasion to write down Lagrangian field theories which possess PCAC in this sense. Our motivation is this: when we make assumptions about the properties of the weak interaction currents, we do not have absolute freedom; we must always be sure that our assumptions do not contradict each other or the general principles of relativistic local field theory. The easiest way to check consistency is to construct a Lagrangian field theory that embodies all of our assumptions. However, since we are *only* using this theory to check consistency, there is no reason why the strong interaction coupling constants must be large. Indeed, we will usually take them to be small. In this case, we can calculate the matrix elements of the canonical pion field in lowest-order perturbation theory, and verify that the definition of PCAC given here implies the correct version (statement 2 above).

In the next section we will construct such a Lagrangian model to verify the consistency of the assumptions that lie behind the Goldberger–Treiman relation.

5   The gradient-coupling model

In the early days of pion physics, a popular model of pion decay was the one expressed in Fig. 5. In this model, F_π is clearly proportional to

Fig. 5

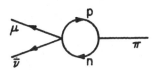

g. The Goldberger–Treiman relation, however, makes precisely the opposite assertion:

$$F_\pi = 2Mg_A/g. \tag{5.1}$$

This strongly suggests that the Goldberger–Treiman relation is a characteristically strong-interaction result – that it would not be true if the strong interactions were weak. We will show that this suggestion is misleading by displaying a Lagrangian model in which the Goldberger–Treiman relation holds even if the strong interactions are weak. (Indeed, because of the remarks at the end of the last section, this is the only domain in which, for a Lagrangian field theory, we can be sure it holds.)

However, first I will review the familiar method for obtaining currents from a Lagrangian. A local Lagrangian field theory is determined by a Lagrange density, $\mathscr{L}$, which is a function of a set of fields ϕ^α and their first derivatives $\partial_\mu \phi^\alpha$. The dynamics of the theory is given by Hamilton's principle,

$$\delta \int \mathscr{L} \, d^4x = 0, \tag{5.2}$$

for variations of the fields which vanish at infinity.

Let us consider an infinitesimal transformation of the fields of the form

$$\delta\phi^\alpha = F^\alpha \delta\lambda, \tag{5.3}$$

where the Fs are arbitrary functions of the ϕs at the point x, and $\delta\lambda$ is an infinitesimal constant. Let us assume that under this transformation

$\mathscr{L}$ is invariant. That is to say,

$$\delta\mathscr{L} = \left(\frac{\partial\mathscr{L}}{\partial\phi^\alpha}F^\alpha + \frac{\partial\mathscr{L}}{\partial(\partial_\mu\phi^\alpha)}\partial_\mu F^\alpha\right)d\lambda = 0, \qquad (5.4)$$

where the sum on repeated indices is implied. Now let us consider a transformation of the same form as (5.3), but with $\delta\lambda$ an arbitrary function of space and time. Then

$$\delta\mathscr{L} = \frac{\partial\mathscr{L}}{\partial(\partial_\mu\phi^\alpha)}F^\alpha\partial_\mu\delta\lambda, \qquad (5.5)$$

no longer vanishes. However, by (5.2), its integral must still vanish;

$$\int d^4x\left[\frac{\partial\mathscr{L}}{\partial(\partial_\mu\phi^\alpha)}F^\alpha\right]\partial_\mu\delta\lambda = 0, \qquad (5.6)$$

which implies, since $\delta\lambda$ is arbitrary,

$$\partial_\mu\left[\frac{\partial\mathscr{L}}{\partial(\partial_\mu\phi^\alpha)}F^\alpha\right] = 0. \qquad (5.7)$$

Thus, for every invariance of $\mathscr{L}$ we have a conserved current

$$\mathscr{J}^\mu = \frac{\partial\mathscr{L}}{\partial(\partial_\mu\phi^\alpha)}F^\alpha. \qquad (5.8)$$

If $\mathscr{L}$ is the sum of two terms,

$$\mathscr{L} = \mathscr{L}_s + \mathscr{L}', \qquad (5.9)$$

where the first term is invariant under (5.3) and the second term does not depend on the $\partial_\mu\phi$s, then, by the same reasoning, Eq. (5.7) is replaced by

$$\partial_\mu\mathscr{J}^\mu = \frac{\partial\mathscr{L}'}{\partial\phi^\alpha}F^\alpha, \qquad (5.10)$$

with $\mathscr{J}^\mu$ defined as above.

We are now in a position to analyze our model. It involves an isodoublet of nucleon fields, denoted by ψ, and an isotriplet of pion fields, denoted by ϕ. The Lagrange density is

$$\mathscr{L} = \mathscr{L}_0 - i\frac{g}{2M}\bar{\psi}\gamma_\mu\gamma_5\boldsymbol{\tau}\psi\cdot\partial^\mu\boldsymbol{\phi}, \qquad (5.11)$$

where $\mathscr{L}_0$ is the usual free Lagrange density and $\boldsymbol{\tau}$ are the standard 2×2 isospin matrices. This is called the gradient-coupling model. The coupling constant has been chosen such that $g(k^2)$, defined by Eq. (3.1), is (in lowest order) a constant equal to g. Now let us consider the infinitesimal transformations

$$\delta\boldsymbol{\phi} = \delta\boldsymbol{\lambda}, \quad \delta\psi = 0. \qquad (5.12)$$

Associated with these transformations is an isotriplet of axial-vector currents:

$$\mathscr{A}_\mu = \partial_\mu \phi - i \frac{g}{2M} \bar{\psi} \gamma_\mu \gamma_5 \tau \psi. \qquad (5.13)$$

The only part of $\mathscr{L}$ not invariant under (5.12) is the pion mass term; therefore,

$$\partial^\mu \mathscr{A}_\mu = -m_\pi^2 \phi. \qquad (5.14)$$

This current, as it stands, is not a suitable candidate for A_μ; its one-nucleon matrix elements do not have the right value. Therefore, we define

$$A_\mu = \frac{2M g_A}{g} \mathscr{A}_\mu. \qquad (5.15)$$

and take A_μ to be the positively-charged component of this triplet.

I leave it to you to verify that in this model (1) all three statements of PCAC given in the preceding section are true, and (2) the Goldberger–Treiman relation is valid. (Always, of course, working only to lowest order in g.)

Problem. Suppose, instead of choosing (5.15) for the axial current, we choose

$$A_\mu + \alpha \partial_\mu \phi,$$

with α some undetermined constant. Is the Goldberger–Treiman relation still valid? If so, why? If not, why not? Answer the question using all three versions of PCAC, if possible.

6   **Adler's rule for the emission of one soft pion**
       We will now develop a formalism, due to S. Adler, for calculating the matrix element of any hadronic process of the form

$$i \to f + \pi, \qquad (6.1)$$

where i and f are any hadronic states, in terms of that for the process

$$i \to f. \qquad (6.2)$$

For example, we will be able to relate pion production,

$$N + N \to N + N + \pi, \qquad (6.3)$$

to nucleon–nucleon scattering

$$N + N \to N + N. \qquad (6.4)$$

Our method will be to obtain an *exact* formula for the process (6.1), with the pion off the mass shell and with the pion four-momentum, k, close to zero. We will then extrapolate this expression to the mass shell. It is clear that this extrapolation can be trusted only if all invariants of the form $p \cdot k$ are small, where p is any momentum in the initial or final state. Thus, in the example, the formalism will give an expression for (6.4) valid in the neighborhood of threshold. This kinematic situation is sometimes described by saying that the pion is 'soft'. The terminology comes from the concept of a soft photon – defined in precisely the same way – which arises in the theory of infrared corrections.

Before proceeding to the detailed analysis, it will be convenient to rewrite some of our fundamental formulae in an isospin-symmetric form. We will assume that the axial current is part of an isotriplet of currents A_μ^a ($a = 1, 2, 3$), normalized such that A_μ is $(A_\mu^1 + i A_\mu^2)/2$. Thus for, example, Eq. (2.7) becomes

$$\langle \pi^b | A_\mu^a | 0 \rangle = (\text{K.F.}) e^{ip \cdot x} i \delta_{ab} p_\mu F_\pi, \tag{6.5}$$

and our other formulae are altered correspondingly.

Now for the analysis. The S-matrix element for (6.1) is clearly related by the reduction formula to

$$\langle f | \partial^\mu A_\mu^a | i \rangle = i k^\mu \langle f | A_\mu^a | i \rangle. \tag{6.6}$$

We want to investigate this object near $k = 0$. At first glance, it might seem that (6.6) vanishes at this point, because of the multiplicative factor of k. However, this is not necessarily the case. Because of energy–momentum conservation, we must alter the momenta of the initial and final states as we send k to zero. Thus there is a possibility that the matrix element of A_μ^a blows up – develops a pole – as k goes to zero, and the product has a finite limit. Fig. 6 shows this happening for a typical Feynman diagram

Fig. 6

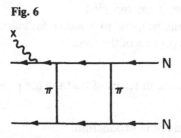

contributing to the process (6.3). (The wiggly line terminating in a cross represents the current.) As k goes to zero, the internal nucleon line is forced on to the mass shell, and the diagram becomes infinite. It requires only an elementary Feynman calculation to show that (6.6) then has a non-zero limit.

This example can readily be generalized. Let us divide all Feynman diagrams that contribute to the matrix element of A_μ^a into two classes: (1) 'pole diagrams' in which the current is attached to an external line, as in Fig. 6; (2) 'guts diagrams' in which the current is attached to an internal line, as in Fig. 7. The guts diagrams cannot develop poles as k goes to zero. (Proof: the locations of the singularities of a diagram are determined by the Landau rules. The Landau rules are purely kinematic; they involve only the possible values of the internal momenta. For a guts diagram, when k is zero, these are the same as for the diagram without the current. The diagram without the current is a diagram for a scattering process in the physical region. Scattering processes do not have poles in the physical region.*) Thus, their contribution to (6.6) vanishes at zero. By PCAC, they may therefore be neglected on the mass shell. Therefore, to calculate the scattering amplitude, we need only compute the pole diagrams.

(A side remark: Suppose, in Fig. 6, the pion is charged, and further suppose that we give the proton and neutron slightly different masses, as is indeed the case in nature. Then the pole diagram would vanish at zero. However, then its contribution to (6.6) would be rapidly varying near zero – since it would have a very-near-by pole – and we would not be able to extrapolate it. Thus we would again be led to the same conclusion.)

But it is trivial to calculate the pole diagrams. For these simply consist of the diagrams which contribute to (6.2), with a current hooked on to an external line. Using the Goldberger–Treiman relation, it is easy to see that, when we take the divergence of this current, we simply reproduce the pion–nucleon coupling of the gradient-coupling theory. Thus we obtain

Fig. 7

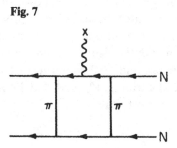

---

* This is not strictly true. For processes with three or more particles in both the initial and final states, trivial macrocausality poles may enter the physical region. In the following, we assume we are working in a range of energies and momenta such that these poles may be neglected.

Adler's rule. In any strong-interaction process, to calculate the matrix element for the emission of one soft pion, take the matrix element for the process without the pion, and sum all the terms obtained by attaching the pion to each of the external lines, using gradient coupling.

In other words, lowest-order perturbation in the gradient-coupling theory is exact for the emission of one soft pion.

7   Current commutators

Suppose we attempt to extend the formalism of the previous section to processes involving two soft pions. Clearly, in this case we would have to analyze objects like

$$T\langle f|\partial^\mu A_\mu^a(x)\partial^\nu A_\nu^b(y)|i\rangle.$$

But now when we attempt to pull out the differential operators we obtain not only a double divergence of a matrix element of two currents, but also an equal time commutator from the derivative of the time-ordering operator. Thus, to go further, we need some information about the equal-time commutator.

7.1   *Vector–vector commutators*

Let me review the CVC hypothesis of Feynman and Gell-Mann. Recall our observation that, experimentally, g_V is very close to one. Why should this be so? Feynman and Gell-Mann offered the following explanation; suppose that the weak current for nucleons is of exactly the same structure as that for leptons. Then, the bare value of g_V, the number that occurs in the interaction Lagrangian, would be one, as would the bare value of g_A. Further suppose that the vector current is conserved. (CVC = conserved vector current). Then in this case it can be shown that the g_V is not renormalized by the strong interactions. (For a proof, see Appendix 2). Now, currents conserved by the strong interactions are not that easy to find; the only one known with the right quantum numbers is the isospin current. Therefore, Feynman and Gell-Mann further postulated that V_μ is part of an isotriplet of currents, V_μ^a, and that V_μ^a is proportional to the isospin current I_μ^a; that is to say

$$V_\mu^a = \alpha I_\mu^a, \tag{7.1}$$

with α a constant. This is a stronger hypothesis than conservation, but it has been checked in the famous weak-magnetism experiment, and is now generally accepted.

Originally, α was taken to be two; that is to say, the coupling of the vector current to nucleons was assumed to be exactly equal to the coupling to leptons. The observed small deviation of g_V from one was then ascribed to electromagnetic corrections. However, ever since the famous SU(3) analysis of the weak currents by Cabibbo, it has been realized that this assumption is based on a too-naive formulation of universality; Cabibbo theory predicts

$$\alpha = 2g_V, \tag{7.2}$$

where g_V is identified with the cosine of the Cabibbo angle.

Now, in most simple models of the strong interactions (e.g. Yukawa-type Lagrangians), the equal-time commutators of the fourth components of the isospin currents have the same algebraic structure as those of the associated charges; i.e.

$$[I_0^a(\mathbf{x}, 0), I_0^b(\mathbf{y}, 0)] = i\varepsilon_{abc}I_0^c(\mathbf{x}, 0)\delta^3(\mathbf{x} - \mathbf{y}). \tag{7.3}$$

In more complicated models, there may be additional terms on the right, proportional to gradients of delta-functions, which vanish when we integrate the currents to make the charges. For simplicity, we will ignore the possible occurrence of these terms. However, as you may readily check, none of the calculations we will do will depend on the assumption of their absence.

From the preceding equations, we find

$$[V_0^a(\mathbf{x}, 0), V_0^b(\mathbf{y}, 0)] = 2ig_V\varepsilon_{abc}V_0^c(\mathbf{x}, 0)\delta^3(\mathbf{x} - \mathbf{y}). \tag{7.4}$$

7.2 Vector–axial commutators

These follow directly from the identification of V_μ^a with the isospin current, and the statement that A_μ^a is an isotriplet:

$$[V_0^a(\mathbf{x}, 0), A_0^b(\mathbf{y}, 0)] = 2ig_V\varepsilon_{abc}A_0^c(\mathbf{x}, 0)\delta^3(\mathbf{x} - \mathbf{y}). \tag{7.5}$$

7.3 Axial–axial commutators

There is no direct experimental check on these objects (other than the applications we are going to discuss). However, there are some general theoretical arguments that make the commutators we are going to assume particularly attractive. I will sketch two lines of argument.

(1) *Chirality principle.* Suppose we follow the original suggestion of Feynman and Gell-Mann and introduce parity-violation into the weak interactions only through the use of the projection matrix $1 + i\gamma_5$ in the definition of the currents. This means that the parity-transformed currents, $V_\mu - A_\mu$, only involve the matrix $1 - i\gamma_5$. These are orthogonal projection matrices; therefore

$$[(V_0^a(\mathbf{x}, 0) + A_0^a(\mathbf{x}, 0)), (V_0^b(\mathbf{y}, 0) - A_0^b(\mathbf{y}, 0))] = 0. \tag{7.6}$$

Now let us abstract Eq. (7.6) from the Feynman–Gell-Mann theory and adopt it as a general rule. (I will call this 'the chirality principle'.) Then we deduce that

$$[A_0^a(\mathbf{x}, 0), A_0^b(\mathbf{y}, 0)] = 2ig_V\varepsilon_{abc} V_0^c(\mathbf{x}, 0)\delta^3(\mathbf{x} - \mathbf{y}). \qquad (7.7)$$

(2) Universality principle. Roughly, universality is the statement that all the weak interactions have the same strength. Originally, when only nuclear beta-decay was known, this was formulated as the requirement that the baryon–lepton coupling have the same strength as the lepton–lepton coupling. However, with the current plethora of hadrons, it is difficult to unambiguously generalize this statement. Should one require that all hadronic coupling constants be equal? Or should the sum of the squares be equal to the lepton constant squared? Anyway, how does one compare coupling constant for particles of different spin? To avoid these difficulties, Gell-Mann suggested the following definition of universality: 'The algebra generated by repeated equal-time commutation of the fourth components of the total weak-current with its adjoint must be the same as the corresponding algebra generated from the lepton currents alone.'

This definition has several obvious advantages: (1) if the only hadrons were nucleons, it would lead to the old definition of universality; (2) it is independent of the details of the hadron spectrum and the structure of the strong interactions; (3) within any particular model of the strong interactions, it fixes the relative scale of the hadron and lepton currents, as well as the relative scale of the baryon and boson parts of the hadron current.

It can be shown that this definition of universality, together with the Cabibbo theory, leads to the commutators we have written down. However, since the argument involves considerable use of SU(3), I will not give it here.

I should complete this section by giving some examples of Lagrangian models in which these commutators hold and which also obey PCAC. (Note that the gradient-coupling model will not do; in it, the fourth components of axial currents commute.)* Fortunately, I am spared this labor; any of the Lagrangians discussed by Professor Zumino in his lectures will do the job.†

8   The Weinberg–Tomozawa formula and the Adler–Weisberger relation

In this section we will derive a formula for the s-wave threshold scattering length in the elastic scattering of a pion off any hadron target.

---

* They are canonical momentum densities.
† See *Hadrons and Their Interactions* (Academic Press, New York and London, 1968).

Our general method will be the same as in the derivation of Adler's rule; we will expand the matrix element in a power series in the pion momenta (plus pole terms) near zero momentum, and then extrapolate to the mass shell. In this case, however, we will take two pions off the mass shell; therefore, the current commutators will play an important role.

We consider a process of the type

$$\pi^a + i \rightarrow \pi^b + f, \tag{8.1}$$

where the superscripts indicate the isospin of the pions, and the momenta are as indicated in Fig. 8. The reduction formula tells us that the object

Fig. 8

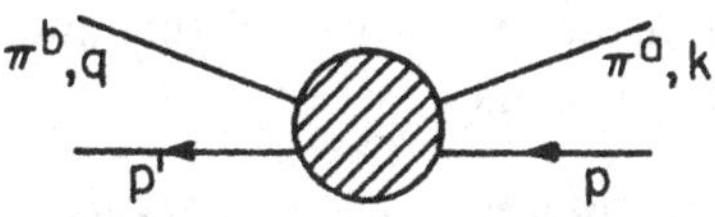

we need to study is

$$
\begin{aligned}
I &= \int \mathrm{d}^4x\, \mathrm{d}^4y\, \mathrm{e}^{\mathrm{i}q\cdot x} \mathrm{e}^{-\mathrm{i}k\cdot y} T \langle f | \partial^\mu A^a_\mu(x) \partial^\nu A^b_\nu(y) | i \rangle \\
&= -\frac{(2\pi)^4 \delta^4(p + k - p' - q) F^2_\pi m^4_\pi \mathcal{M}}{(q^2 - m^2_\pi)(k^2 - m^2_\pi)(2\pi)^3 (4EE')^{1/2}},
\end{aligned} \tag{8.2}
$$

where $\mathcal{M}$ is the invariant matrix element for the scattering process, the object to which we wish to apply PCAC. It is related to the S-matrix by

$$\langle fb | S - 1 | ia \rangle = (\text{K.F.}) \mathcal{M} (2\pi)^4 \delta^4(p + k - p' - q'). \tag{8.3}$$

We now wish to expand I, and hence $\mathcal{M}$, near $q = k = 0$. We will lump together, and eventually neglect, all terms of second order and higher. Note that the invariants k^2, q^2, and $k \cdot q$ are of second order, while the invariants $p \cdot k$, $p' \cdot q$, $p \cdot q$, and $p' \cdot k$ are all of first order, and, as a consequence of energy–momentum conservation, all equal, aside from terms of second order.

Pulling the differential operators through the time ordering symbol, we may write I as the sum of three terms,

$$I = I_1 + I_2 + I_3, \tag{8.4}$$

where

$$I_1 = -\int \mathrm{d}^4x\, \mathrm{d}^4y\, \mathrm{e}^{\mathrm{i}q\cdot x} \mathrm{e}^{-\mathrm{i}k\cdot y} \delta(x_0 - y_0) \langle f | [A^b_0(x), \partial^\nu A^a_\nu(y)] | i \rangle, \tag{8.5}$$

$$I_2 = \int \mathrm{d}^4x\, \mathrm{d}^4y\, \mathrm{e}^{\mathrm{i}q\cdot x} \mathrm{e}^{-\mathrm{i}k\cdot y} \partial^\mu_x \partial^\nu_y T \langle f | A^a_\mu(x) A^b_\nu(y) | i \rangle, \tag{8.6}$$

and

$$I_3 = - \int d^4x\, d^4y\, e^{iq\cdot x} e^{-ik\cdot y} \partial^\mu_x \delta(x_0-y_0) \langle f|[A^a_0(x), A^b_\nu(y)]|i\rangle$$

$$= +i \int d^4x\, d^4y\, e^{iq\cdot x} e^{-ik\cdot y} q^\mu \delta(x_0-y_0) \langle f|[A^a_0(x), A^b_\nu(y)]|i\rangle, \quad (8.7)$$

where we have integrated by parts in the last equation. Each of these terms leads to a corresponding term in $\mathcal{M}$; we will call these terms $\mathcal{M}_1$, $\mathcal{M}_2$, and $\mathcal{M}_3$.

We will begin by analyzing I_3; this will turn out to be the most important term. Let us choose q and k such that their space components are zero. (Note that we lose no information by this choice, since, as we have argued above, there is only one first-order invariant.) Then the commutator becomes one we know (Eq. (7.7)) and I_3 becomes

$$I_3 = -2g_\text{v}\varepsilon_{abc} \int d^4x\, d^4y\, e^{-ik_0 y_0} e^{-iq_0 y_0} \delta^4(x-y) q^0 \langle f|V^c_0(x)|i\rangle. \quad (8.8)$$

The δ-function may be trivially integrated away. Likewise the space integration is trivial since

$$\int d^3x \langle f|V^c_0(x)|i\rangle = 2g_v \langle f|I^c_t|i\rangle = 2g_\text{v} I^c_t \delta^3(\mathbf{p}-\mathbf{p}'), \quad (8.9)$$

where $\mathbf{I}$ is the total isospin and I^c_t the isospin of the target, is the pure iso-spin part of the matrix element, with the space part of the wave-function neglected. The time integration simply gives a δ-function. Thus we obtain

$$I_3 = -4g_\text{v}^2\varepsilon_{abc} I^c_t 2\pi\delta^4(p+k-p'-q')q_0, \quad (8.10)$$

which leads to

$$\mathcal{M}_3 = +\frac{8g_\text{v}^2}{F_\pi^2}\, \varepsilon_{abc} I^c_t p\cdot q. \quad (8.11)$$

This may be further simplified by noting that, for isotriplet states,

$$\langle b|I^c|a\rangle = i\varepsilon_{abc}. \quad (8.12)$$

This transforms $\mathcal{M}_3$ into

$$\mathcal{M}_3 = -i\frac{8g_\text{v}^2}{F_\pi^2}\, (\mathbf{I}_\pi \cdot \mathbf{I}_t) p\cdot q. \quad (8.13)$$

Next we turn to $\mathcal{M}_2$. In I_2, all the derivatives are outside the time-ordering symbol. Thus, $\mathcal{M}_2$ is of the same form as the matrix elements we discussed in Sec. 6; just as we did there, we may divide $\mathcal{M}_2$ into pole terms (Fig. 9) and a remainder. The remainder is of second order, and we will neglect it. The pole terms, however must be calculated explicitly.

Fig. 9

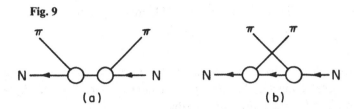

(a)   (b)

When this is done, though, a surprising thing happens: *at threshold*, they
are of the order of m_π^2, and therefore can be neglected.

I will show this explicitly for the case of a $\frac{1}{2}^+$ target (e.g. the nucleon).
The first diagram in Fig. 9 gives a matrix element proportional to

$$\bar{u}'\gamma_5 \not{k} \frac{1}{\not{p}+\not{q}-m_t} \not{q}\gamma_5 u. \tag{8.14}$$

(I have suppressed isospin factors.) This is equal to

$$\bar{u}' \not{k} \frac{1}{\not{p}+\not{q}+m_t} \not{q} u. \tag{8.15}$$

Now, *at threshold*,

$$k_\mu = q_\mu = \frac{m_\pi}{m_t} p_\mu. \tag{8.16}$$

Thus, $\not{p}$, $\not{q}$, and $\not{k}$ all commute, and, using the Dirac equation, we find that
(8.16) is equal to

$$\bar{u}' m_\pi \frac{1}{2m_t + m_\pi} m_\pi u, \tag{8.17}$$

which is indeed of order m_π^2. The second diagram may be treated in the
same way.

The only term left to study is $\mathcal{M}_1$. I will give two arguments to show that
it is of order m_π^2, and can be neglected. (1) I_1 explicitly involves $\partial^\mu A_\mu$.
According to one version of PCAC, this is of order m_π^2. (2) By straightforward manipulation of I_1, $\mathcal{M}_1$ can be shown to be equal to a constant
term plus terms of second order. Only the constant term need concern
us. By Adler's Rule, if we send the incoming pion to zero, keeping the
outgoing pion on the mass shell, the matrix element should be given
exclusively by the pole terms. On the other hand, our preceding analysis
shows that, in this limit, the matrix element consists of the sum of the pole
terms, the constant term in $\mathcal{M}_1$, and terms manifestly of order m_π^2. Therefore the constant term is of order m_π^2. (The constant term is sometimes
called the σ term, because in the σ model it is related to the matrix element
of the σ field.)

The upshot of all this is that, at threshold, only $\mathcal{M}_3$ is important:

$$\mathcal{M} = -i \frac{8g_V^2}{F_\pi^2} (\mathbf{I}_\pi \cdot \mathbf{I}_t) p \cdot q + O\left(\frac{m_\pi^2}{m_t^2}\right). \qquad (8.18)$$

Near threshold, the elastic scattering matrix element is simply related to the s-wave scattering lengths. Doing the kinematics (for details see Appendix 3), we find for the scattering lengths

$$a = -L\left(1 + \frac{m_\pi}{m_t}\right)^{-1} 2\mathbf{I}_\pi \cdot \mathbf{I}_t = -L\left(1 + \frac{m_\pi}{m_t}\right)^{-1}$$
$$\times [I(I+1) - I_t(I_t+1) - 2], \qquad (8.19)$$

where I is the total isotopic spin, and L is a constant, called 'Weinberg's universal length', defined by

$$L = \frac{g_V^2 m_\pi}{2\pi F_\pi^2}. \qquad (8.20)$$

This can be rewritten using the Goldberger–Treiman relation;

$$L = \frac{g^2 m_\pi}{8\pi M^2}\left(\frac{g_V}{g_A}\right)^2 = 0.11 m_\pi^{-1}. \qquad (8.21)$$

Eq. (8.19) is called the Weinberg–Tomozawa formula. It was derived in its full generality, using the method I have explained here, by Weinberg, and was found for certain special cases, using a different method, by Tomozawa. It is valid for the scattering of pions off any target, subject to the following two restrictions. (1) The pole terms must be negligible at threshold. If they are not, they must be added by hand, altering the formula. Notice that this makes our approach impracticable for, say, the scattering of pions off carbon, since each of the excited states of carbon up to a few hundred MeV contributes a rapidly-varying pole term. (2) The mass of the target must be much heavier than the mass of the pion, in order that the terms we have neglected will be small compared to the term we have calculated. The only case this excludes is that of pion–pion scattering, which we will discuss in some detail in the next section.

Now let us apply (8.19) to pion–nucleon scattering. We find

$$a_{1/2} = 0.20 m_\pi^{-1},$$
$$a_{3/2} = -0.10 m_\pi^{-1}. \qquad (8.22)$$

where the subscripts indicate total isospin. This is in excellent agreement with the experimental numbers, 0.17 and -0.09.

It is instructive to rewrite Eq. (8.22) in the following way:

$$a_{1/2} + 2a_{3/2} = 0,$$
$$a_{1/2} - a_{3/2} = 0.30 m_\pi^{-1}. \qquad (8.23)$$

The first of these equations depends only on the $\mathbf{I} \cdot \mathbf{I}$ form of the matrix element, not upon the coefficient in front. It was obtained long ago by Sakurai, from a model in which pion–nucleon scattering was dominated by ρ exchange. The derivation we have given is much more general; it is independent of the very existence of the ρ, let alone the role it plays in pion–nucleon scattering.

The second equation can be rewritten, with the aid of dispersion relations, as an integral over total cross sections. Furthermore, we can use Eq. (8.21) to turn this into a formula that expresses g_V/g_A in terms of this integral and hadronic masses and coupling constants. This formula is the famous Adler–Weisberger relation. It is curious that this relation was first derived directly as an integral formula, and only later seen to be a disguised low-energy theorem.

9 Pion–pion scattering *à la* Weinberg

In the last section we derived a formula for the scattering of a pion off an arbitrary target. We remarked, however, that if the target were itself a pion, the formula would break down, because the terms we neglected would then be of the same magnitude as the terms we retained. Thus, pion–pion scattering would appear to be a hopelessly intractable problem. However, Weinberg showed that, by exploiting Bose statistics and crossing symmetry, one could obtain a definite prediction for the threshold scattering lengths, provided only that one made an assumption about the isospin transformation properties of the so-called σ-term. $\pi\pi$ scattering is unique in that all known symmetries (including crossing) connect it to itself. (Compare, e.g., πN scattering.) The calculation is both very simple and very sophisticated: simple because it involves only elementary algebra, sophisticated because it used the very 'modern' concepts of analyticity, crossing, and current commutation relations.

The calculation yields scattering lengths much smaller than those predicted by other techniques. This is important for, if the scattering lengths were large, it would be very difficult to justify our extrapolation procedures for processes involving two or more soft pions. Of course, since we will use these very extrapolation techniques in the calculation, it by no means proves that the scattering lengths are small. It is, however, an important consistency check.

We now turn to the actual calculation. We label the pion momenta and isospin as in Fig. 10. If all four pions are off the mass shell, there are six independent scalar invariants. For convenience, we will use the over-complete set of seven invariants formed by

$$s = (k+p)^2, \, t = (k-q)^2, \, u = (p-q)^2,$$

Fig. 10

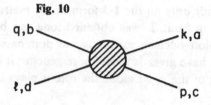

and the four masses k^2, p^2, q^2 and l^2. Energy–momentum conservation relates these:

$$s+t+u=p^2+k^2+q^2+l^2. \tag{9.1}$$

We will now write down the expansion of the invariant amplitude $\mathcal{M}_{abcd}$, in a power series in the momenta, discarding terms of the fourth order or higher.

There are three independent isospin invariants, which we will choose to be

$$\delta_{ac}\delta_{bd}, \quad \delta_{ab}\delta_{cd}, \quad \text{and } \delta_{ad}\delta_{bc}.$$

Let us consider the coefficient of the first of these. Suppose it contains a term linear in k^2. Then, by Bose statistics it must contain a term linear in p^2 with the same coefficient. Time reversal says the same must be true of q^2 and l^2. Thus, the masses can enter only in the combination

$$k^2+p^2+q^2+l^2.$$

But, by (9.1), this is $s+t+u$. Thus, this term (and, *mutatis mutandis*, the other two also) can be written in terms of s, t, and u alone.

It is now easy to write down the most general expression for $\mathcal{M}$ allowed by Bose statistics and crossing:

$$i\mathcal{M} = \delta_{ac}\delta_{bd}[Am_\pi^2 + B(u+t)+Cs] + \delta_{ab}\delta_{cd}[Am_\pi^2+B(u+s)+Ct]$$
$$+ \delta_{ad}\delta_{bc}[Am_\pi^2+B(s+t)+Cu], \tag{9.2}$$

where A, B, and C are unknown constants. (The i is inserted in the definition to keep the constants real.)

By Adler's rule, if we send k to zero, and keep all the other pions on the mass shell, $\mathcal{M}$ must vanish. (There are no pole terms; parity forbids a three-pion vertex.) At this point,

$$s=t=u=m_\pi^2. \tag{9.3}$$

Therefore,

$$A+2B+C=0. \tag{9.4}$$

By the analysis of the preceding section, if we send both k and q to zero, the only term which survives is the σ term. We will assume this is pure $I=0$, i.e. proportional to δ_{ab}. This is the extra assumption mentioned earlier; without it, the σ term could be any combination of $I=0$ and $I=2$. The only motivation for this assumption is that this is indeed what happens in the σ-model.

At this point $(k=q=0)$,

$$s=u=m_\pi^2, \quad t=0. \tag{9.5}$$

Therefore,

$$A+B+C=0. \tag{9.6}$$

From Eq. (9.4) and (9.6) we can already deduce that

$$B=0$$

and

$$A=-C. \tag{9.7}$$

Finally, we observe that if both q and k are close to zero, Eq. (8.11) tells us that the linear part of the matrix element is given by

$$i\mathcal{M}=\frac{8g_V^2}{F_\pi^2}\,i\varepsilon_{abe}(I^e)_{dc}p\cdot q=\frac{8g_V^2}{F_\pi^2}\,(i\varepsilon_{abe})(i\varepsilon_{dce})p\cdot q$$

$$=\frac{8g_V^2}{F_\pi^2}\,(\delta_{ac}\delta_{bd}-\delta_{ad}\delta_{bc})p\cdot q. \tag{9.8}$$

In this region

$$s=m_\pi^2+2p\cdot q, \quad u=m_\pi^2-2p\cdot q, \quad t=0. \tag{9.9}$$

Therefore

$$C=\frac{4g_V^2}{F_\pi^2}=8\pi L/m_\pi. \tag{9.10}$$

We have now determined all the unknown constants; thus we can evaluate $\mathcal{M}$ at threshold. Here,

$$s=4m_\pi^2, \quad t=u=0. \tag{9.11}$$

Thus, at threshold

$$i\mathcal{M}=m_\pi^2C[3\delta_{ac}\delta_{bd}-\delta_{ab}\delta_{cd}-\delta_{ad}\delta_{bc}]. \tag{9.12}$$

We wish to express this result in terms of isospin states. We can think of $\mathcal{M}_{bdac}$ as a 9×9 matrix acting on two-index isospin wave functions, associated with two-particle isospin states by the rule

$$|\psi\rangle=\psi_{ac}|a\rangle|c\rangle$$

(The sum on repeated indices is, as always, implied.) $\mathcal{M}$ has one eigenvalue, $\mathcal{M}^{(0)}$, for isospin zero states, five equal eigenvalues $\mathcal{M}^{(2)}$, for isospin two states, and three eigenvalues, evidently zero, for isospin one states. We wish to find $\mathcal{M}^{(0)}$ and $\mathcal{M}^{(2)}$.

An evident isospin zero state is

$$\psi_{ac} = \delta_{ac}. \tag{9.13}$$

Applying $\mathcal{M}$ to this, we find

$$i\mathcal{M}_{bdac}\delta_{ac} = m_\pi^2 C(9-1-1)\delta_{bd}. \tag{9.14}$$

Thus,

$$i\mathcal{M}^{(0)} = 7m_\pi^2 C. \tag{9.15}$$

To find $\mathcal{M}^{(2)}$, we take the trace:

$$i(5\mathcal{M}^{(2)} + \mathcal{M}^{(0)}) = \mathcal{M}_{acac} = m_\pi^2 C(9-9-3), \tag{9.16}$$

whence,

$$i\mathcal{M}^{(2)} = -2m_\pi^2 C. \tag{9.17}$$

Doing the kinematics (see Appendix 3 for the relevant formulae) we find

$$a_0 = \tfrac{7}{4}L = 0.20 m_\pi^{-1},$$

and

$$a_2 = \tfrac{1}{2}L = -0.06 m_\pi^{-1}. \tag{9.18}$$

These scattering lengths are quite small – much smaller than those predicted by any previous calculation. This should not be surprising; previous calculations treated pion dynamics as essentially an autonomous system; the characteristic range of variation; therefore, was on the order of m_π, and the scattering lengths of order m_π^{-1}. (They had to be – there was no other length in the problem.) Our extrapolation procedure, on the other hand is valid only if the characteristic range of variation is several times greater than m_π (for example, on the order of m_ρ). Eventually, the scattering lengths will be measured (from the phases in K_{l4} decay, if in no other way.) It will be interesting to see which approach is correct.

10   Kaon decays

We now turn to a class of processes which, although they only involve one soft pion, nevertheless require the evaluation of commutators. These are weak decays with a soft pion in the final state. These decays may be either leptonic

$$i \to f + \pi^a + \text{leptons}, \tag{10.1}$$

or non-leptonic

$$i \to f + \pi^a, \tag{10.2}$$

where i and f are hadronic states. In either case, we have to evaluate a matrix element of the form

$$\langle f | \mathcal{J} | i \rangle, \tag{10.3}$$

where $\mathcal{J}$ is either a weak current or the weak non-leptonic Hamiltonian density.

To analyse this matrix element, we use the same methods as before. We write it in terms of

$$T\langle f|\partial^\mu A_\mu^a(x)\,\mathcal{J}(y)|i\rangle. \tag{10.4}$$

and pull the divergence through the time-ordering operator. This gives us, as usual, pole terms, plus the equal-time commutator

$$\langle f|[A_0^a(\mathbf{x}, 0),\,\mathcal{J}(\mathbf{y}, 0)]|i\rangle. \tag{10.5}$$

Unfortunately, in all the cases of interest (kaon decays, hyperon decays, etc.), $\mathcal{J}$ is a strangeness-changing operator, and the commutators we have used so far are of no help in evaluating this expression. However, we can appeal to the chirality principle of Sec. 7 and assume, whatever the detailed structure of $\mathcal{J}$, that it is made up only of $V + A$ currents. We may then write

$$[A_0^a(\mathbf{x}, 0),\,\mathcal{J}(\mathbf{y}, 0)] = [V_0^a(\mathbf{x}, 0),\,\mathcal{J}(\mathbf{y}, 0)]. \tag{10.6}$$

This is an object about which we have some empirical information, for we usually know the isospin transformation properties of $\mathcal{J}$. I will briefly sketch the application of these methods to leptonic kaon decays. Let us begin with K_{l2} and K_{l3} decays – that is to say, with the processes

$$K^+ \to \text{leptons}$$

and

$$K^+ \to \pi^0 + \text{leptons}.$$

If we define $\mathcal{J}_\mu$ to be the strangeness-changing weak current, the relevant matrix elements are

$$\langle 0|\,\mathcal{J}_\mu(0)|K^+\rangle = -ip_\mu(\text{K.F.})F_K, \tag{10.7}$$

and

$$\langle\pi^0|\,\mathcal{J}_\mu(0)|K^+\rangle = i(\text{K.F.})[f_+(p_K + p_\pi)_\mu + f_-(p_K - p_K - p_\pi)_\mu]. \tag{10.8}$$

We know the commutator (10.6) from the $\Delta I = 1/2$ rule, the statement $\mathcal{J}_\mu$ is the charged component of an isospinor. There are no pole terms, since parity forbids a three-pseudo-scalar vertex. Thus we deduce

$$f_+ + f_- = g_V F_K/F_\pi. \tag{10.9}$$

This relation was first found by Callan and Treiman. It is in good agreement with experiment.

Callan and Treiman attempted to apply the same techniques to K_{l4} decay

$$K^+ \to \pi^+ + \pi^- + \text{leptons}.$$

The relevant form factors are

$$\langle \pi^+\pi^- | \mathscr{J}_\mu(0)|K^+\rangle = (K.F.) \times [f_1(q_+ +q_-)_\mu + f_2(q_+ -q_-)_\mu$$
$$+ f_3(k-q_+ -q_-) + f_4\varepsilon_{\mu\nu\lambda\sigma}k^\nu q_+^\lambda q_-^\sigma], \quad (10.10)$$

where $q_\pm$ and k are the momenta of the pions and the kaon, respectively.

Here we have the option of reducing either the π^+ or the π^-. If we reduce the π^-, the commutator is zero, and we obtain, in the limit $p_- \to 0$,

$$f_3 = 0, \qquad f_1 = f_2. \qquad\qquad\qquad ((10.11)$$

If we reduce the π^+, the commutator is not zero, and we obtain in the limit $p_+ \to 0$,

$$f_3 = \sqrt{2}g_V(f_+ + f_-)/F_\pi,$$

and

$$f_1 + f_2 = \sqrt{2}\,2g_V f_+/F_\pi. \qquad\qquad (10.12)$$

Callan and Treiman were puzzled by the discrepancy between the two predictions for f_3. They ascribed the large variation of f_3 to a strong $\pi\pi$ s-wave interaction. However, this is unacceptable for two reasons. (1) If there is such an interaction, all soft pion calculations involving two or more pions are invalid. (2) All the assumptions of the calculation would be true if the strong interactions were weak, as we pointed out in Sec. 4. In such a case, Eqs. (10.11) and (10.12) would still hold, but there could be no strong final-state interaction.

This apparent paradox was resolved by Weinberg, who observed that there was a neglected pole diagram (Fig. 11). This makes a contribution

Fig. 11

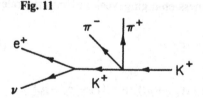

only to f_3, proportional to

$$\frac{(p_+ -p_-)\cdot(2k-p_+ -p_-)}{(p_+ +p_-)^2 - 2k\cdot(p_+ +p_-)}. \qquad (10.13)$$

(The form of the four-boson interaction is taken from Sec. 8). Note that (10.13) is one when p_+ is zero, and minus one when p_- is zero. This diagram thus supplies the desired rapid variation of f_3; there is no need to introduce a strong final-state interaction. (For the details of the analysis, see Weinberg's paper, cited in the bibliography.)

Appendix 1. Notational conventions

1 We use the metric $g_{\mu\nu}$ for which

$$g_{00} = -g_{11} = -g_{22} = -g_{33} = 1.$$

2 The differential operator ∂_μ is defined as $\partial/\partial x^\mu$; likewise

$$\Box^2 = \partial^\mu \partial_\mu = \partial_0^2 - \mathbf{V}^2,$$

and the Klein–Gordon equation is

$$(\Box^2 + m^2)\phi = 0.$$

2 The γ-matrices are defined by

$$\gamma_\mu \gamma_\nu + \gamma_\nu \gamma_\mu = 2g_{\mu\nu}.$$

Also, for any vector a,

$$\rlap{/}{a} = a_\mu \gamma^\mu.$$

The Dirac equation is

$$(i\rlap{/}{\partial} - m)\psi = 0$$

4 The propagators for spin 0 and spin 1/2 fields are

$$i(p^2 - m^2)^{-1} \quad \text{and} \quad i(\rlap{/}{p} - m)^{-1},$$

respectively.

5 In a matrix element, every particle carries with it a kinematic factor of $(2\pi)^{-3/2}(2E)^{-1/2}$. To shorten equations, the product of these factors is frequently denoted simply by (K.F.).

Appendix 2. No-renormalization theorem.

In this appendix we wish to show that if the vector current is proportional to the isospin current, the value of g_V does not depend on the strong interactions. In other words, g_V is not renormalized by the strong interactions.

Let us define the total isospin, I, by

$$I^a = \int d^3x I_0^a(\mathbf{x}, 0).$$

If we neglect electromagnetism, I is conserved; it commutes with the Hamiltonian and turns states of a given energy into states of the same energy. Likewise, from its definition, it commutes with the momentum, and turns states of a given momentum into states of the same momentum. Therefore, I must turn one-nucleon states at rest into one-nucleon states at rest. Now the components of I are a set of operators closed under commutation; in mathematical language they form a Lie algebra, one that is, in fact, the same as the angular-momentum Lie algebra. Therefore, the one-nucleon states at rest must form the basis for a two-dimensional representation of this algebra.

However, we know from ordinary angular-momentum theory that there is only one such representation, the spin-$\frac{1}{2}$ representation, unique up to a similarity transformation. We fix the similarity transformation by defining the neutron and proton states to be eigenstates of I_z, and choosing the relative phase such that the matrix elements of the raising and lowering operators are real. Thus, the matrix elements of $\mathbf{I}$ between one-nucleon states at rest are uniquely determined by the isospin algebra; they can not depend on any strong-interaction parameters.

But now we are almost home. For, by Eq. (7.1)

$$I^a = \alpha \int d^3x \, V_0^a(\mathbf{x}, 0),$$

and the right-hand side of this equation, evaluated between one-nucleon states at rest, is clearly linearly related to g_V, since the momentum transfer, k, is zero. Thus, g_V can not depend on the strong interactions, and we have proved the theorem.

Appendix 3. Threshold S-matrix and threshold scattering lengths

In this appendix, we will consider a two-particle scattering process with initial momenta (p, q) and final momenta (p', q'), and relate the value of the S-matrix near threshold to the s-wave scattering length.

Near threshold

$$\langle p', q'|(S-1)|p, q\rangle = A\delta^4(p' + q' - p - q).$$

with A some constant. Let us introduce a new basis, labeled by the total and relative momenta, P and k.

$$\langle P', k'|(S-1)|P, k\rangle = A\delta(E - E')\delta^3(\mathbf{P} - \mathbf{P}').$$

If we go to the center-of-mass frame (i.e. restrict ourselves to states with $P = 0$) we may write the center of mass S-matrix as

$$\langle \mathbf{k}'|(S-1)|\mathbf{k}\rangle = A\delta(E - E').$$

Note that for small $\mathbf{k}$,

$$E = \frac{k^2}{2m_1} + \frac{k^2}{2m_2} = \frac{k^2}{2\mu},$$

where m_1 and m_2 are the masses of the two particles, and μ is the reduced mass. The s-wave state, $|k\rangle$, is defined by

$$|k\rangle = \frac{1}{4\pi k} \int d^3k \, \delta(k - |\mathbf{k}|)|\mathbf{k}\rangle.$$

We have normalized $|k\rangle$ such that

$$\langle k'|k\rangle = \delta(k - k').$$

It follows that, near $k = 0$,

$$\langle k'|S - 1|k\rangle = 4\pi k^2 A\delta(E' - E)$$
$$= 4\pi k_\mu A\delta(k' - k).$$

The s-wave phase shift is defined by

$$S|k\rangle = e^{2i\delta}|k\rangle.$$

Near threshold

$$\delta = ak,$$

where a is the scattering length. Therefore

$$(S - 1)|k\rangle = 2iak|k\rangle.$$

Comparison with the earlier formulae shows that

$$a = -2\pi i A\mu,$$

the desired result.

If we use invariant Feynman amplitudes, defined as in the lectures, then

$$A = (2\pi)^4 (\text{K.F.}) \mathcal{M}.$$

At threshold in the center of mass frame,

$$(\text{K.F.}) = \frac{1}{(2\pi)^6} \frac{1}{4m_1 m_2}.$$

Thus

$$a = -i\frac{\mu \mathcal{M}}{8\pi m_1 m_2} = -i\frac{\mathcal{M}}{8\pi(m_1 + m_2)}.$$

If the two particles are identical, there is an extra factor of $1/\sqrt{2}$ in the definition of $|k\rangle$, because $|\mathbf{k}\rangle$ and $|-\mathbf{k}\rangle$ are identical states. This introduces an extra factor of $1/2$ into our final formulae.

Bibliography

Introduction:

The background material on Feynman graphs and field theory can be found in many texts. One good one is: Bjorken and Drell, *Relativistic Quantum Fields* (McGraw-Hill).

Sec. 1:

The reduction formula was first found in the context of LSZ field theory: H. Lehmann, K. Symanzik and W. Zimmerman, *Nuovo Cimento* 1, 205 (1955). The insensitivity of the formula to the choice of fields was found by H. Borchers, *Nuovo Cimento* 15, 784 (1960).

Sec. 3:

M. Goldberger and S. Treiman, *Phys. Rev.* **110**, 996 (1958).

Sec. 4:

The following papers (a few among many) analyze different versions of PCAC and give several Lagrangian models: M. Gell-Mann and M. Levy, *Nuovo Cimento* **16**, 705 (1960). J. Bernstein, S. Fubini, M. Gell-Mann and W. Thirring, *Nuovo Cimento* **17**, 757 (1960).

Sec. 6:

S. L. Adler, *Phys. Rev.* **139B**, 1638 (1965). Y. Nambu and D. Lurié, *Phys. Rev.* **125**, 1429 (1962).

Sec. 7:

M. Gell-Mann, *Phys. Rev.* **125**, 1064 (1962). M. Gell-Mann, *Physics* **1**, 63 (1964).

Secs. 8 and 9:

S. L. Adler, *Phys. Rev.* **143**; 1144 (1966). W. I. Weisberger, *Phys. Rev.* **143**, 1302 (1966). S. Weinberg, *Phys. Rev. Letters* **17**, 616 (1966). Y. Tomozawa, *Nuovo Cimento* **46A**, 707 (1966).

Sec. 10:

C. G. Callan and S. B. Treiman, *Phys. Rev. Letters* **16**, 153 (1966). S. Weinberg, *Phys. Rev. Letters* **17**, 336 (1966). Many further applications are discussed, and references to the literature given, in N. Cabibbo, *"Weak Interactions"*, *Proceedings of the 13th International Conference on High-Energy Physics* (U. of Calif. Press).

3

Dilatations
(1971)

1   Introduction
It is an old idea in particle physics that, in some sense, at sufficiently high energies the masses of the elementary particles should become unimportant. In recent years this somewhat vague hope has acquired a more definite form in the theory of scale transformations, or dilatations. These are transformations that would be exact invariances of the world if all elementary particle masses (more generally, all dimensionful couplings) vanished. The hope is that, by studying these approximate symmetries in Lagrangian field theories, we can gain knowledge about how scale transformations behave in the real world, and learn something about those kinematic realms where the effects of masses are indeed unimportant, or at least simply calculable. This is a hope that has been fulfilled for broken chiral symmetries, by the study of models such as the sigma-model.

The purpose of these lectures is to report on the progress of such investigations. As we shall see, there is still much that is obscure. In particular, the connection between the sorts of things I will be talking about and the famous experimental scaling of deep inelastic electroproduction remains to be unravelled. Nevertheless, what has been done so far has already yielded some remarkable theoretical surprises.

Section 2 is a description of the formal Lagrangian theory of scale transformations. Some of the Ward identities that express broken scale invariance are derived, models of Nambu–Goldstone scale-symmetry breaking are discussed, and the connection between scale invariance and conformal invariance is explained. These last two topics will not be returned to in the remainder of these lectures, and a reader who is only interested in the main line of the argument may skip the Sections (2.3 and 2.4) that discuss them.

Section 3 is an investigation of broken scale invariance in renormalized perturbation theory. We find that the Ward identities derived in the preceding section are lies (in the polite current parlance, they contain anomalies), and we find the true equations that replace them, the Callan–Symanzik equations.

Section 4 explains how, despite the Callan–Symanzik anomalies (in fact, because of them) scale invariance may still be regained in appropriate kinematic regions. This astonishing magic trick is due to Ken Wilson.

My own contributions to this field have been minor. I have learned most of what I know from conversations with Curtis Callan, John Ellis, Roman Jackiw, Kurt Symanzik, Kenneth Wilson, and Bruno Zumino. I would like to express both my gratitude to them and my hope that I have not distorted their ideas too badly.

2   The formal theory of broken scale invariance
2.1   Symmetries, currents, and Ward identities

We will deal in these lectures with Lagrangian field theories:[1] that is to say, theories involving a set of fields, which we will assemble into a big vector, ϕ, and whose dynamics are determined by a function of the fields and their first derivatives, $\mathscr{L}(\phi, \partial_\mu\phi)$, called the Lagrangian, via Hamilton's principle:

$$\delta I \equiv \delta \int d^4x \, \mathscr{L} = 0, \tag{2.1}$$

for solutions of the equations of motion. I is called the action integral.

Let us consider some infinitesimal transformation of the fields:

$$\phi \to \phi + \delta\phi. \tag{2.2}$$

(We will always suppress the infinitesimal parameter that should properly multiply $\delta\phi$.) Let us suppose that under this transformation

$$\delta I = \int d^4x \, \Delta, \tag{2.3}$$

where Δ is some function of the fields and their derivatives. (Eq. (2.3) is supposed to be true for general fields, not just for solutions of the equations of motion.) If Δ vanishes, the transformation has no effect on the dynamics and is called a symmetry.

Then one can show from the general formalism of Lagrangian field theory,[2] without any assumptions other than those stated that:

(a) It is always possible to define an object, j^μ, called 'the current associated with the transformation' such that

$$\partial_\mu j^\mu = \Delta. \tag{2.4}$$

Do not be misled by the notation into assuming that the current is always a four-vector. Its Lorentz-transformation properties depend upon those of the transformation (2.2). If the transformation is an internal-symmetry transformation (Lorentz scalar), then the current will indeed be a four-vector, but for more complicated cases, it will have more complicated transformation properties. For example, the four-parameter group of space-time translations leads to a set of four currents, which, together, form the components of the energy–momentum tensor.

(b) It is always possible to define a generalization of the ordinary time-ordered product of a string of fields, called the T^*-ordered product. It has the interesting property that

$$\frac{\partial}{\partial y^\mu} T^* \langle 0| j^\mu(y)\phi(x_1) \dots \phi(x_n)|0\rangle$$

$$= T^* \langle 0|\Delta(y)\phi(x_1) \dots \phi(x_n)|0\rangle$$
$$- i\delta^{(4)}T^*(x_1 - y)\langle 0|\delta\phi(x_1)\phi(x_2) \dots \phi(x_n)|0\rangle$$
$$- i\delta^{(4)}T^*(x_2 - y)\langle 0|\phi(x_1)\delta\phi(x_2) \dots \phi(x_n)|0\rangle$$
$$+ \cdots. \tag{2.5}$$

These equations are called Ward identities.

In the case of internal symmetries (including chiral symmetries) these equations follow directly from the definition of the time-ordered product and the equal-time commutation relations. In more complicated cases, the direct definition of the time-ordered product is ambiguous when two space-time arguments coincide, and the commutators have peculiar terms (Schwinger terms) in them. Nevertheless, there is always a way of removing the ambiguity such that the Ward identities remain the same as in the simplest case.[3]

A particularly useful consequence of the Ward identities can be obtained by integrating them over all space with respect to the variable y. The integral of the left hand side vanishes by integration by parts. (If there are no massless particles in the theory – these can give surface terms in the parts integral.) Thus we obtain

$$\int d^4y \, T^* \langle 0|\Delta(y)\phi(x_1) \dots \phi(x_n)|0\rangle$$

$$= iT^* \langle 0|\delta\phi(x_1) \dots \phi(x_n)|0\rangle$$
$$+ iT^* \langle 0|\phi(x_1)\delta\phi(x_2) \dots \phi(x_n)|0\rangle$$
$$+ \cdots. \tag{2.6}$$

These equations are called zero-energy theorems. The reason for the terminology is clear in Fourier space, where the divergence Δ carries zero energy (and zero momentum). The one-soft-pion theorems of current algebra are special cases of (2.6).

2.2   *Scale transformations and scale dimensions*

Scale transformations, or dilatations, are transformations on space-time of the form

$$\alpha: x \rightarrow e^{\alpha} x, \tag{2.7}$$

where x is a space-time point and α is a real number.[4] Like all other transformations of physical interest (e.g. chiral transformations) scale transformations can be implemented in a wide variety of ways in field theories. We will be most interested in theories in which these transformations act linearly on the fields:

$$\alpha: \phi(x) \rightarrow e^{\alpha d} \phi(e^{\alpha x}), \tag{2.8}$$

with d some matrix (In Sect. 2.4, we will discuss briefly some theories in which scale transformations act non-linearly.) The infinitesimal transformation is

$$\delta \phi = (d + x^{\lambda} \partial_{\lambda}) \phi. \tag{2.9}$$

For a large class of theories (including all renormalizable field theories) these transformations are symmetries, if all non-dimensionless coupling constants (including the masses) are set equal to zero, and if d is chosen to be a matrix that multiplies all Bose fields by one and all Fermi fields by $\frac{3}{2}$. In realistic models, of course, the masses are not zero, so the symmetry is broken.

As an example of how this works, let us consider the ever-popular model of a pseudoscalar meson interacting with a spin-one-half nucleon through Yukawa coupling. We will divide the Lagrangian into a scale-symmetric part and a scale-breaking part:

$$\mathscr{L} = \mathscr{L}_{s} + \mathscr{L}_{B}, \tag{2.10}$$

where

$$\mathscr{L}_{s} = i \bar{\psi} \gamma_{\mu} \partial^{\mu} \psi + \tfrac{1}{2} \partial_{\mu} \phi \partial^{\mu} \phi + g_{0} \bar{\psi} \gamma_{5} \psi \phi - \frac{\lambda_{0}}{4!} \phi^{4}, \tag{2.11}$$

and

$$\mathscr{L}_{B} = -m_{0} \bar{\psi} \psi - \tfrac{1}{2} \mu_{0}^{2} \phi^{2}. \tag{2.12}$$

(The subscripts are to remind you that these are bare masses and coupling constants.) As stated, we choose the transformation laws

$$\delta \psi = (\tfrac{3}{2} + x_{\lambda} \partial^{\lambda}) \psi \tag{2.13}$$

and

$$\delta \phi = (1 + x_{\lambda} \partial^{\lambda}) \phi. \tag{2.14}$$

From these equations, it follows directly that

$$\delta \mathscr{L}_{s} = (4 + x_{\lambda} \partial^{\lambda}) \mathscr{L}_{s}$$

this vanishes in the integral (2.1) upon integration by parts. On the other hand,

$$\delta \mathscr{L}_B = -(3 + x_\lambda \partial^\lambda) m_0 \bar{\psi}\psi - \tfrac{1}{2}(2 + x_\lambda \partial^\lambda)\mu_0^2 \phi^2. \tag{2.15}$$

Upon integration by parts, this does not vanish, but becomes

$$\Delta = m_0 \bar{\psi}\psi + \mu_0^2 \phi^2. \tag{2.16}$$

If you have accepted the dogma of Sect. 2.1, you realize that this means that we can define a scale current, s_μ, such that

$$\partial_\mu s^\mu = \Delta. \tag{2.17}$$

This is the local version of the statement that only the masses break scale invariance. Also, without even bothering to explicitly construct s_μ, we can immediately write down the low-energy theorems that follow from broken scale invariance, the Eq. (2.6). I will not bother to write them out, though, until we will need them, at the beginning of Sect. 3.

You have probably noticed that the numbers $\tfrac{3}{2}$ and 1, which occur in the transformation laws for the fields, are just the dimensions of the fields, in the sense of ordinary dimensional analysis. You may be wondering, therefore, whether what we have been doing is just dimensional analysis, disguised by a fancy formalism. If this is the case, you are wrong: the transformations of dimensional analysis not only scale the dynamical variables of a physical theory (in our case, the fields), they also scale all non-dimensionless numerical parameters (in our case, the masses). Phrased somewhat more abstractly, the transformations of dimensional analysis turn one physical theory into another, different theory (e.g. with different masses), and are always exact symmetries – given the exact solutions to the first theory, they yield the exact solutions to the second. Scale transformations, as we have defined them, are very different animals: they do not change numerical parameters – that is to say, they stay within a given physical theory – and they are not exact symmetries, except in special cases (vanishing masses).

To emphasize this difference, we will call the numbers that occur in field transformation laws like (2.13) and (2.14), 'scale dimensions'. Since, at the current stage of our investigation, they appear to be identical with the dimensions of the fields in the sense of dimensional analysis, this may seem mere nitpicking. However, I assure you that the distinction will be important in the future.

2.3 *More about the scale current and a quick look at the conformal group*

In the last Section we introduced the scale current, s^μ, but said nothing about its explicit form. It turns out that for a large class of theories,[5] it is possible to define an energy–momentum tensor, $\theta^{\mu\nu}$, such that

$$s^\mu = x_\nu \theta^{\mu\nu}. \qquad (2.18)$$

This should not be too much of a surprise. After all, the conserved currents associated with other geometrical transformations, such as translations and Lorentz transformations, are also written in terms of the energy-momentum tensor. Also, Eq. (2.18) implies that exact scale invariance is equivalent to the vanishing of the trace of the energy-momentum tensor:

$$\partial^\mu s_\mu = \theta^\mu_\mu = 0. \qquad (2.19)$$

This should strike a familiar chord, if you remember that for free electromagnetism (the prototype of a scale-invariant field theory), the energy-momentum tensor is traceless. (This is why, in scalar theories of gravitation, there is no bending of light by the sun; there is nothing for the scalar graviton to couple to.)

On the other hand, Eq. (2.19) is far from self-evident. The energy–momentum tensor $\theta_{\mu\nu}$ is not, in general, the conventional symmetric energy–momentum tensor of Belinfante; it is another conserved symmetric tensor which differs from the Belinfante tensor by extra terms which do not affect the construction of the total four-momentum nor the Lorentz generators, but which are important in the scale current. Also, (2.19) is not valid for a general field theory; a theory allows the construction of an energy–momentum tensor obeying (2.19) only if a certain condition is met. I will first state this (obscure) condition and then attempt to clarify its meaning.

Let $\Sigma^{\mu\nu}$ be the spin matrix, that matrix that occurs in the transformation law of the fields under infinitesimal Lorentz transformations,

$$\delta^{\mu\nu}\phi = [x^\mu\partial^\nu - x^\nu\partial^\mu + \Sigma^{\mu\nu}]\phi.$$

Then, we can find an energy–momentum tensor such that (2.19) is valid if and only if

$$\frac{\partial\mathscr{L}}{\partial(\partial_\nu\phi)}\cdot[g^{\mu\nu}d + \Sigma^{\mu\nu}]\phi = \partial_\nu\sigma^{\mu\nu}, \qquad (2.20)$$

where $\sigma^{\mu\nu}$ is some tensor function of the fields and their derivatives. It is easy to check that all renormalizable field theories, indeed all theories of the interactions of fields of spin $\leqslant 1$ where the derivative interactions are of the same form as in renormalizable theories, satisfy (2.21).

This is, as promised, obscure. To understand what is going on, let us consider the case of exact scale invariance, Eq. (2.10). In this case, a little computation reveals a surprise: we can construct from the energy-momentum tensor, not only the conserved scale current, but four other conserved currents:

$$K^{\lambda\mu} = x^2\theta^{\lambda\mu} - 2x^{\lambda}x_{\rho}\theta^{\rho\mu}. \tag{2.21}$$

$$\partial_{\mu}K^{\lambda\mu} = 2x_{\mu}\theta^{\lambda\mu} - 2x_{\rho}\theta^{\rho\lambda} - 2x^{\lambda}\theta^{\rho}_{\rho} = 0. \tag{2.22}$$

Thus it appears that, for theories for which (2.19) holds, exact scale invariance implies invariance under four other (at the moment mysterious) infinitesimal transformations. This leads to a conjecture: the left hand side of Eq. (2.21) is the change in the Lagrangian under these mysterious infinitesimal transformations (with the possible additions of some terms which vanish as a consequence of exact scale invariance); the condition that it be a divergence is just the condition that the action integral be unchanged.

What can these mysterious transformations be? To gain some insight, let us study the simplest scale-invariant theory, the theory of a free massless scalar field,

$$\mathcal{L} = \tfrac{1}{2}\partial_{\mu}\phi\partial^{\mu}\phi. \tag{2.23}$$

The equation of motion is the wave equation

$$\partial_{\mu}\partial^{\mu}\phi = 0. \tag{2.24}$$

Since this is a free field theory, its complete quantum dynamics is determined by the two-point function,

$$\langle 0|\phi(y)\phi(x)|0\rangle = \frac{1}{2\pi^2}\frac{1}{(x-y)^2}. \tag{2.25}$$

(The ambiguity at the pole is removed by giving x a small imaginary part, lying in the forward light cone).

Now, the wave equation, (2.25), is very similar to the three-dimensional Laplace equation. This equation, as you know, possesses geometrical symmetries beyind the usual Euclidean transformations; to be precise, it is invariant under the three-dimensional inversion of coordinates:

$$I: \mathbf{x} \to \frac{\mathbf{x}}{\mathbf{x}^2}. \tag{2.26}$$

This suggests that the theory of the free massless scalar field may be invariant under the Minkowski-space inversion:

$$I: x \to -x/x^2. \tag{2.27}$$

(The reason for the minus sign will become clear shortly.)

Let us check this idea. Under the inversion,

$$I: (x-y)^2 \to \frac{(x-y)^2}{x^2y^2}. \tag{2.28}$$

Thus, the two-point function is invariant if we define the field to transform in the following way:

$$I: \phi(x) \to \frac{1}{x^2} \phi\left(\frac{-x}{x^2}\right). \tag{2.29}$$

We still have to check that the statement that the imaginary part of x is in the forward light cone is invariant under the inversion. With no loss of generality, we can take both the real and the imaginary part of x to be in the 0–1 plane. Let us define

$$x_{\pm} = x^0 \pm x^1. \tag{2.30}$$

Then in these coordinates, the statement that the imaginary part is inside the forward cone becomes

$$\text{Im } x_{\pm} > 0. \tag{2.31}$$

This is obviously invariant under the inversion

$$I: x_{\pm} \to -1/x_{\mp}. \tag{2.32}$$

The inversion (2.28) is a discrete transformation, and is thus of no use in obtaining conserved currents; for these, continuous transformations are needed. However, we can use the inversion to find new continuous invariances, by applying it to the right and left of old ones. For example, if we take the four-parameter group of space-time translations,

$$a: x \to x + a, \tag{2.33}$$

and apply (2.28) to the right and left of these transformations, we obtain

$$a: x \to \frac{x - a \cdot x^2}{1 - 2a \cdot x + a^2 x^2}. \tag{2.34}$$

These are called conformal transformations. The associated four infinitesimal transformations are

$$\delta^\mu x^\nu = -g^{\mu\nu} x^2 + 2x^\mu x^\nu. \tag{2.35}$$

However, no new transformations are obtained by applying the inversion to the right and left of an infinitesimal Lorentz transformation or dilatation; one just obtains Lorentz transformations and dilatations again.

It is straightforward to verify that this set of infinitesimal transformations (six Lorentz transformations, four translations, four conformal transformations, and one dilatation) is closed under commutation; that is to say, it forms the basis of a fifteen-dimensional Lie algebra. (The only commutator that is not immediately evident is that of a conformal transformation and a translation; this turns out to be the sum of an infinitesimal Lorentz transformation and an infinitesimal dilatation.) The fifteen-parameter group obtained by exponentiating this algebra is called the conformal group. As we have shown, it can be defined as the connected part of the smallest group of transformations on space-time containing both the Poincaré group and the inversion. As we have also shown, the whole conformal group can be realized as a group of symmetries of the free massless scalar field.

A Lorentz-invariant field theory that is invariant under the infinitesimal conformal transformations (2.36) is automatically also scale-invariant, since the commutator of an infinitesimal conformal transformation and an infinitesimal translation contains a scale transformation. However, there is no corresponding group theoretical reason for the converse to be true; indeed, it is fairly easy to find examples of field theories that are scale-invariant but not conformally invariant.

It is straightforward (but lengthy and tedious) to work out the transformation properties of a field of arbitrary spin under infinitesimal conformal transformations, and to find the condition that a scale-invariant theory be also conformally invariant. This turns out to be Eq. (2.21). This verifies the conjecture we made at the beginning of this section: the four extra conserved currents that appear in the limit of exact scale invariance for a theory for which (2.19) holds are simply the four currents associated with infinitesimal conformal transformations.[5]

For theories of this type (which, I remind you, include all renormalizable field theories) this has an interesting consequence even when scale invariance is broken. In this case, Eq. (2.23) becomes

$$\partial_\mu K^{\lambda\mu} = -2x^\lambda \theta_\mu^\mu. \tag{2.36}$$

Thus we get two sets of low-energy theorems from the general formula (2.6). From broken scale invariance

$$\partial_\mu s^\mu = \theta_\mu^\mu, \tag{2.37}$$

we obtain formulae for the matrix elements of θ_μ^μ at zero momentum transfer, while from Eq. (2.37) we obtain formulae for the first derivative of this object with respect to momentum transfer at zero momentum transfer. This is, in principle, much more information than we have from the corresponding statements in the case of chiral symmetry, the one-soft-pion theorems. These only give us formulae for the matrix elements of the divergences of the axial currents at zero momentum transfer, and tell us nothing about their derivatives.

However, despite this bright promise, we will have, in the remainder of these lectures, so much trouble keeping straight what really happens with the scale-invariance low-energy theorems that we will have very little to say about the conformal ones.

2.4   *Hidden scale invariance*

It has long been known that there are two ways a symmetry that commutes with the Lorentz group (an internal symmetry or a chiral symmetry) can be realized in a quantum field theory:

(1) The symmetry may be manifest. The ground state of the theory (the vacuum) is invariant under the symmetry group, particles arrange themselves into degenerate multiplets corresponding to irreducible representations of the group, and S-matrix elements are invariant. If a small term is added to the Lagrangian to break the symmetry, the masses of the particles in the multiplets split, and the symmetry of S-matrix elements becomes only approximate. This seems to be the case in nature for isospin.

(2) The symmetry may be hidden. (This case, first investigated by Nambu and Goldstone, is usually called 'spontaneously broken symmetry', but this terminology is deceptive; the symmetry is not really broken – the currents associated with the symmetry are still conserved and the Ward identities are still valid – it is just that the consequences of the symmetry are less obvious than in the other case.) The ground state of the theory (the vacuum) is not invariant and particles do not arrange themselves into multiplets: however, there do appear a set of special particles, the 'Goldstone bosons'. These are massless spinless mesons, scalar or pseudoscalar depending on whether the symmetry current is vector or axial. Instead of a symmetric S-matrix, we obtain low-energy theorems, statements relating an S-matrix element for a process involving one zero-energy, zero-momentum Goldstone boson to an S-matrix element with the boson absent. If a small term is added to the Lagrangian to break the symmetry, the Goldstone bosons acquire a small mass and the low-energy theorems require an off-mass-shell extrapolation, become only approximate on the mass shell. This seems to be the case in nature for the chiral transformations of $SU(2) \times SU(2)$; the would-be Goldstone bosons are the pions.

A cheap way to get some feeling for symmetries realized in the Nambu–Goldstone manner is by the study of 'phenomenological Lagrangians'. These are Lagrangian field theories that embody the Goldstone phenomenon, at least in lowest order of perturbation theory. Frequently these models are hard to take seriously as real physical theories; for example, the non-linear chiral Lagrangians, which have played such a prominent role in investigations of current algebra, are non-renormalizable, so that it is impossible to proceed beyond lowest-order perturbation theory in an unambiguous way. Nevertheless, they have at least two uses: they can offer clues to general theorems that one might try to establish by more careful methods, and they can offer counter-examples to wrong arguments, by fulfilling all the premises but not obeying the supposed consequence. I would like to briefly discuss here a few simple phenomenological Lagrangians for which scale invariance (and conformal invariance) is realized as a Nambu–Goldstone symmetry, rather than as a manifest symmetry.

The basic trick is very simple. We begin with a scalar field χ that transforms in the conventional linear way under scale transformations,

$$\alpha: \chi(x) \rightarrow e^\alpha \chi(e^\alpha x), \tag{2.38}$$

and define a new field, σ, by

$$f^{-1} e^{f\sigma} = \chi, \tag{2.39}$$

where f is a constant with the dimensions of length. The new field now transforms in a non-linear way

$$\alpha: \sigma(x) \rightarrow \sigma(e^\alpha x) + \alpha/f. \tag{2.40}$$

The corresponding infinitesimal transformation is[6]

$$\delta\sigma = x^\lambda \partial_\lambda \sigma + f^{-1}. \tag{2.41}$$

(If you are familiar with non-linear chiral Lagrangians, this transformation law may remind you of the chiral transformation law for the pions, the Goldstone bosons of chiral symmetry, which also has an inhomogeneous term in it.) A scale (and conformally) invariant Lagrangian for the σ is

$$\mathscr{L} = \frac{1}{2} \partial_\mu \chi \, \partial^\mu \chi = \frac{1}{2f^2} (\partial_\mu e^{f\sigma})(\partial^\mu e^{f\sigma}). \tag{2.42}$$

We can now use the σ field to make any Lagrangian into a scale-invariant Lagrangian: we simply multiply the scale-breaking parts by appropriate powers of $\exp[f\sigma]$ to make them scale invariant, and add the free σ Lagrangian (2.43). For example, for the meson–nucleon theory discussed in Sect. 2.2, we write

$$\mathscr{L} = \mathscr{L}_s - m_0 \bar\psi \psi e^{f\sigma} - \frac{\mu_0^2}{2} \phi^2 e^{2f\sigma} + \frac{1}{2f^2} \partial_\mu e^{f\sigma} \partial^\mu e^{f\sigma}, \tag{2.43}$$

where $\mathscr{L}_s$ is the unchanged scale-invariant part of the original Lagrangian Eq. (2.11). This Lagrangian is manifestly scale-invariant. (It is easy to check that it is also conformally invariant.) Despite this, the meson and the nucleon have non-zero masses in lowest-order perturbation theory. Only the σ is massless; it is the Goldstone boson of scale invariance. (Somewhat surprisingly, it is also the Goldstone boson for the four conformal transformations. A naive generalization of the situation for internal symmetries would lead us to believe that five Goldstone bosons are necessary if we are to have five hidden symmetries, but as this example shows, this is not so.)

In the case of chiral symmetries, phenomenological Lagrangians are also used to study symmetry breaking. Usually the symmetry-breaking term is chosen such that the divergence of the axial-vector current is proportional to the pion field. This is called naive PCAC. (The acronym stands for 'partially conserved axial current'.) The advantage of doing this is that it guarantees that the model will possess, in lowest-order perturbation theory, pion pole dominance of the matrix elements of the divergence of the axial vector currents, real PCAC.

Likewise, in the sort of model we have been considering, it is obviously useful to choose the scale-symmetry-breaking term such that the divergence of the scale current is proportional to the σ field. This might be called PCDC – partially conserved dilatation current. This is easily arranged. For example, we can choose the symmetry-breaking term

$$\mathscr{L}_B = -\frac{m_\sigma^2}{16f^2}\left[e^{4f\sigma} - 4f\sigma - 1\right].\tag{2.44}$$

If we expand this in powers of the σ field, we find that

$$\mathscr{L}_B = -\frac{m_\sigma^2}{2}\sigma^2 + O(\sigma^3).\tag{2.45}$$

This is in proper form (there are no terms linear in the σ field), and we see that m_σ is indeed the σ mass. (In lowest order perturbation theory only, as always.) Just as in Sect. 2.2, we can compute the contribution of this term to the divergence of the scale current:

$$\delta\mathscr{L}_B = x_\lambda\partial^\lambda\mathscr{L}_B - \frac{m_\sigma^2}{16f^2}\left[4e^{4f\sigma} - 4\right].\tag{2.46}$$

Integrating by parts, we find

$$\partial_\mu s^\mu = \theta_\mu^\mu = -\frac{m_\sigma^2}{f^2}\sigma,\tag{2.47}$$

the desired result.

There is obviously much more that can be done along these lines. For example, I have said nothing about the interplay between scale symmetry and chiral symmetry, nor have I derived any experimental consequences of these ideas. (These are most readily obtained by identifying the σ with some observed particle. The most popular candidate is the evanescent ε, the scalar isoscalar dipion resonance, last seen somewhere between 700 and 1000 MeV.) For these matters I refer you to the literature.[7]

3 The death of scale invariance

3.1 *Some definitions and technical details*

Let us return to the low-energy theorems discussed in Sect. 2.1. For simplicity, I will restrict myself to the theory of a single self-interacting scalar meson,

$$\mathcal{L} = \frac{1}{2} \partial_\mu \phi \partial^\mu \phi - \frac{\lambda_0}{4!} \phi^4 - \frac{1}{2} \mu_0^2 \phi^2, \tag{3.1}$$

but I will try to conduct the discussion in such a way that the generalization to more complicated theories, such as the meson–nucleon theory of Sect. 2.2, is obvious.

Let us denote the one-particle-irreducible[8] renormalized Green's function with n external lines by

$$\Gamma^{(n)}(p_1 \ldots p_n)$$

where the ps are the momenta carried by the external lines, oriented so they all go inward. (The ps are, of course, not all independent; their sum must be zero.) Likewise, let us denote the one-particle-irreducible Green's function with n external meson lines and one insertion of Δ, the divergence of the scale current (in this case, $\mu_0^2 \phi^2$) by

$$\Gamma_\Delta^{(n)}(k; p_1 \ldots p_n),$$

where the ps are defined as before, and k is the momentum carried by the insertion. Then it is a trivial exercise in Fourier transforms to see that in this case the low-energy theorem (2.6) becomes

$$\left(\sum_{r=1}^{n-1} p_r \cdot \frac{\partial}{\partial p_r} + nd - 4 \right) \Gamma^{(n)}(p_1 \ldots p_n) = -i\Gamma_\Delta^{(n)}(0; p_1 \ldots p_n). \tag{3.2}$$

where d is the scale dimension of the scalar field. The analysis of the preceding section has told us that d is one, of course, but I would like to suppress this information momentarily, both to make it easy for you to see the generalization to Fermi fields, for which d is $\frac{3}{2}$, and for another reason, which will become clear shortly.

Equation (3.2) is fairly easy to understand. The first two terms on the left are just the transformation law of the fields, Eq. (2.9) written in momentum space; the sum only runs over $n-1$ momenta because only $n-1$ of the momenta are independent. The factor of four appears because in passing from the Fourier transform of a T^*-product to the conventionally defined $\Gamma^{(n)}$, we must factor out a four-dimensional δ-function, which leaves the four behind as it passes through the differential operator. The right-hand side of the equation is just the left-hand side of (2.6).

Fig. 1

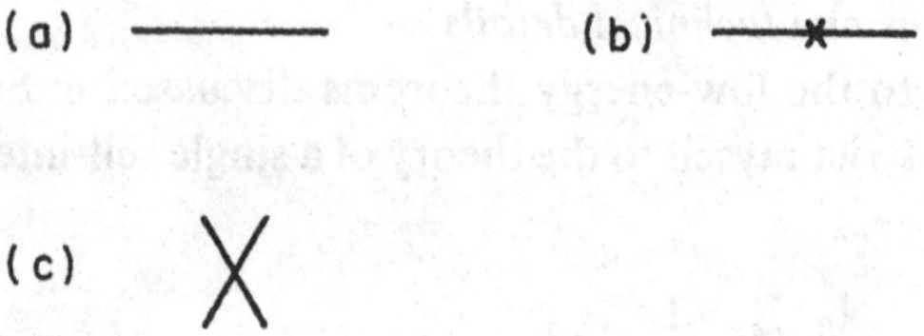

(a) ——————   (b) ——✗——

(c)   X

Just to check that we have not made a sign error, let us check (3.2) in lowest-order perturbation theory. To zeroth order, $\Gamma^{(2)}$ is given by Fig. 1(a), and $\Gamma_\Delta^{(2)}$ is given by Fig. 1(b), where the cross denotes the mass insertion. Thus Eq. (3.2) becomes

$$\left(p \cdot \frac{\partial}{\partial p} - 2\right)(-i)(p^2 - \mu^2) = -i(2\mu^2), \tag{3.3}$$

which is correct. Likewise, to first order, $\Gamma^{(4)}$ is given by Fig. 1(c), and there is no contribution to $\Gamma_\Delta^{(4)}$. Thus Eq. (3.2) becomes

$$\sum_{i=1}^{3}\left(p_i \cdot \frac{\partial}{\partial p_i}\right)(-i\lambda) = 0 \tag{3.4}$$

which is again correct.

We can write (3.2) in another form, which will be useful in the sequel. Let us trade the ps for a set of variables consisting of

$$s = \sum_{r=1}^{n} p_r^2, \tag{3.5}$$

and the dimensionless kinematic variables $p_i \cdot p_j / s$. Then ordinary dimensional analysis tells us that $\Gamma^{(n)}$ is of the form

$$\Gamma^{(n)} = s^{(4-n)/2} F^{(n)}\left(\frac{s}{\mu^2}, \lambda, \frac{p_i \cdot p_j}{s}\right), \tag{3.6}$$

where μ is the renormalized mass, and λ is the renormalized coupling constant. (Remember, λ is dimensionless.) From this it is easy to see that (3.2) can equivalently be written as

$$\left[\mu \frac{\partial}{\partial \mu} + n(1-d)\right]\Gamma^{(n)}(p_1 \ldots p_n) = i\Gamma_\Delta^{(n)}(0; p_1 \ldots p_n). \tag{3.7}$$

3.2   *A disaster in the deep Euclidean region*

The Euclidean region is that region in multi-particle momentum space in which all four-momenta are Euclidean; that is to say, they all have real space parts and imaginary time parts. The deep Euclidean region is that part of the Euclidean region in which the magnitude of s

(defined by Eq. (3.5)) gets very large, while the dimensionless variables $p_i \cdot p_j/s$ stay fixed; furthermore, no partial sum of the ps is zero. The deep Euclidean region is the maximally unphysical limit in which to study Green's functions: all external lines are far off the mass shell, and, furthermore, no matter how the diagram is cut in two, the momentum transferred between the two halves is far off the mass shell.

There are famous bounds on the behaviour of Feynman amplitudes in the deep Euclidean region, first established by Weinberg.[9] For the functions that appear in Eq. (3.7), these Weinberg bounds say that $\Gamma^{(n)}$ grows no faster than $s^{(4-n)/2}$, times a polynomial in $\ln(s/\mu^2)$, to any finite order in renormalized perturbation theory. (The coefficients in the polynomial are, in general, functions of the coupling constant and of the dimensionless variables $p_i \cdot p_j/s$. Also, the order of the polynomial grows with the order of perturbation theory.) Likewise, $\Gamma_\Delta^{(n)}$ grows no faster than $s^{(2-n)/2}$, again times a polynomial in $\ln(s/\mu^2)$. Crudely, the reason why $\Gamma_\Delta^{(n)}$ grows less rapidly than $\Gamma^{(n)}$ is that, in the deep Euclidean region, all internal momenta are getting large; adding a mass insertion adds an internal propagator, which knocks out one power of s.

We can now combine this with the broken-scale-invariance low-energy theorems, Eq. (3.7), to get a much more powerful statement about the asymptotic behaviour of the Green's functions than is given by the Weinberg bounds alone. For the Weinberg bounds tell us that in the deep Euclidean region, the right-hand side of (3.7) is negligible compared to the individual terms on the left-hand side. Thus, we can neglect it, and obtain an equation for the asymptotic form of $\Gamma^{(n)}$,

$$\left[\mu \frac{\partial}{\partial \mu} + n(1-d) \right] \Gamma_{as}^{(n)} = 0, \tag{3.8}$$

where the subscript indicates the asymptotic form in the deep Euclidean region. Or, since d is one,

$$\mu \frac{\partial}{\partial \mu} \Gamma_{as}^{(n)} = 0. \tag{3.9}$$

That is to say, there are no logarithmic factors in every order of renormalized perturbation theory. This is indeed a powerful statement; unfortunately it is also a false one; anyone who has ever done any Feynman calculation involving a closed loop knows that the logarithms are in fact present. Therefore, Eq. (3.7), from which we deduced the false statement (3.9), must itself be false. In the current technical language the Ward identities of broken scale invariance must contain anomalies. Phrased more straightforwardly, *the entire theoretical structure of Sect. 2 is a lie!*

I want to make the logic of this argument clear. I am not saying that the formal theory of broken scale invariance is wrong because it makes asymptotic predictions that differ from the asymptotic behaviour found in perturbation theory; only a madman would take the asymptotic behaviour of perturbation theory so seriously.[10] I need only assume that perturbation theory properly gives the successive derivatives of the Green's functions with respect to the coupling constant at zero coupling constant. This seems to me to be a very weak and extremely reasonable assumption, at least for renormalizable field theories. Under this assumption, if Eq. (3.7) is generally valid, it must be true order by order in perturbation theory. If the right- and left-hand sides of this equation are equal in a fixed order in perturbation theory, they must be equal in the deep Euclidean limit. They are not.

The technical reason for the occurrence of the anomalies is not difficult to understand: the formal canonical manipulation required to prove Ward identities is justified only if we introduce a cutoff to remove the divergences from the theory. However, this does us no good unless the cut-off is chosen in such a way that the cutoff theory still obeys the Ward identities. For such familiar cases as quantum electrodynamics or the sigma model, for example, this condition presents no difficulties; it is easy to introduce a cutoff in such a way that the relevant equations (gauge invariance in one case and PCAC and current algebra in the other) remain true. For scale invariance, though, the situation is hopeless; any cutoff procedure necessarily involves a large mass, and a large mass necessarily breaks scale invariance in a large way. This argument does not show that the occurrence of anomalies is inevitable, but it does show that there is no reason to believe it is impossible.

3.3   *Anomalous dimensions and other anomalies*

Last year, Roman Jackiw and I got interested in these anomalies and decided to get some information about them by the most simple-minded method imaginable. In meson–nucleon theory (the Lagrangian (2.10)), we simply computed separately the right- and left-hand sides of the low-energy theorem (3.7), to lowest non-trivial order in perturbation theory, order g^2, to see how they differed. We found[2] that, to this order, all anomalies could be absorbed in a change in the scale dimension of the fields. In particular, for the meson field, Eq. (3.7) is changed from a falsehood to a truth if we replace the naive value of d (one) by

$$d = 1 + \frac{g^2}{8\pi^2}. \tag{3.10}$$

For the self-energy operator, for example, the solution to Eq. (3.8) becomes

$$\Gamma^{(2)}_{as} \sim s(s/\mu^2)^{1-d}. \tag{3.11}$$

Expanding the exponent and discarding terms of order g^4 and higher, we find

$$\Gamma^{(2)}_{as} \sim s - \frac{g^2}{8\pi^2} s \ln(s/\mu^2). \tag{3.12}$$

This is the correct asymptotic behaviour, to this order in perturbation theory. Similar remarks apply to the nucleon field; here d must be changed from its naive value ($\frac{3}{2}$) to

$$d = 3/2 + \frac{g^2}{32\pi^2}. \tag{3.13}$$

(As it happens, these changes in d not only fix up (3.7), but also take care of all the anomalies in the Ward identities for arbitrary momentum transfers.)

This phenomenon is frequently described by saying that the fields acquire anomalous dimensions. This is a slightly misleading way of putting things, for it tempts us to confuse scale dimensions and dimensions in the sense of dimensional analysis, and to think that something counter to common sense has occurred. It is only the scale dimensions of the fields that have changed as a result of the interactions; the dimensions in the sense of dimensional analysis remain firmly fixed at one and $\frac{3}{2}$. I emphasize again that there is no logical connection between these two sets of numbers; the (discredited) analysis of Sect. 2 is the only thing that ever led us to believe they were equal.

It would be very pleasant if this phenomenon persisted in higher orders of perturbation theory, if all anomalies could be absorbed into a redefinition of the scale dimension. If this were the case, then we would have a correct theory of broken scale invariance with almost as much predictive power as the false theory of Sect. 2; the only price would be the introduction of a new, dynamically determined parameter for each field in the theory, its anomalous dimension.

Unfortunately, this is not what happens. To see this, let us return to the theory of a self-interacting meson field, Eq. (3.1), and assume that Eq. (3.8) is valid in all orders of perturbation theory, except that d is not one. Then it is easy to see that

$$\mu \frac{\partial}{\partial \mu} \left[\Gamma^{(4)}_{as} / (\Gamma^{(2)}_{as})^2 \right] = 0. \tag{3.14}$$

That is to say, although both the numerator and denominator of this expression have factors of $\ln(s/\mu^2)$ in their asymptotic expansions, the

Fig. 2

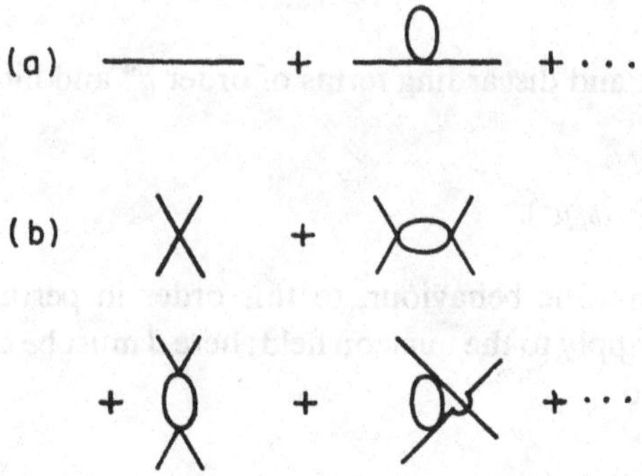

(a) ——— + —◯— + · · ·

(b)

logarithms cancel in the ratio, order by order in perturbation theory. Let us check this prediction by calculating the ratio to order λ^2. The relevant diagrams for $\Gamma^{(2)}$ are shown in Fig. 2(a); the second diagram is just a constant, cancelled by mass renormalization. Thus, we obtain, in the deep Euclidean region,

$$\Gamma_{as}^{(2)} \sim s + O(\lambda^2). \qquad (3.15)$$

Figure 2(b) shows the relevant diagrams for $\Gamma^{(4)}$. Every child knows that the three second-order diagrams grow logarithmically for high s; thus we obtain

$$\Gamma_{as}^{(4)} \sim \lambda + a\lambda^2 \ln(s/\mu^2) + b\lambda^2. \qquad (3.16)$$

where a and b are constants. (Actually, b is a function of the dimensionless kinematic variables, but, for our purposes, that is as good as being a constant.) Putting this together we find

$$\Gamma_{as}^{(4)}/(\Gamma_{as}^{(2)})^2 \sim \lambda s^{-2} + a\lambda^2 s^{-2} \ln(s/\mu^2) + b\lambda^2 s^{-2} + O(\lambda^3). \qquad (3.17)$$

This is in contradiction to our prediction; thus there must be further anomalies in addition to anomalous dimensions.

3.4   The last anomalies: the Callan–Symanzik equations

We are lost in a dark wood. Whenever we look at a more complicated situation, investigate higher orders of perturbation theory, new anomalies appear. Happily, there exists a way to order this chaos, discovered by Curtis Callan and Kurt Symanzik,[11] who independently found the true equations that replace the false low-energy theorems (3.7). These equations are surprisingly compact. For the theory of a self-interacting meson field, for example, they involve only two dynamically determined parameters. This is only one more than there would be if the only anomalies were anomalous dimensions.[12] I will now derive the Callan–Symanzik equations for this theory.

The simplest place to begin is in unrenormalized perturbation theory: here one computes the unrenormalized Green's functions (which I will distinguish by a subscript u) in a power series in the bare charge, with the bare mass held fixed. I remind you of the relation between the renormalized Green's functions, which we have been using until now, and the unrenormalized ones

$$\Gamma^{(n)}(p_1 \ldots p_n) = (Z_3)^{n/2} \Gamma_u^{(n)}(p_1 \ldots p_n). \tag{3.18}$$

where Z_3 is the wave-function renormalization constant. Of course, unlike renormalized perturbation theory, unrenormalized perturbation theory is full of divergences. To control these, we will imagine that we have put a cut-off in the theory in some way, say by modifying the Feynman propagator in the standard manner. Once we re-express things in terms of the physical mass and coupling constant, and perform the multiplicative renormalization (3.18), everything is cutoff-independent in the limit of high cutoff; this is just the statement that (3.1) is a renormalizable Lagrangian. I will denote the unrenormalized Green's function with one mass ($\mu_0^2 \phi^2$) insertion by $\Gamma_{u\Delta}^{(n)}(k; p_1 \ldots p_n)$. Then it is easy to see, just by looking at diagrams, that

$$i\Gamma_{u\Delta}^{(n)}(0; p_1 \ldots p_n) = \mu_0 \frac{\partial}{\partial \mu_0} \Gamma_u^{(n)}(p_1 \ldots p_n). \tag{3.19}$$

This is the unrenormalized version of the low-energy theorem (3.7). As it stands, it is useless, since the differential operator cannot be turned into a scaling operator; the unrenormalized Green's functions depend on another parameter besides μ_0 with the dimensions of a mass, the cutoff.

Simple power counting[13] shows that if we define

$$\Gamma_\Delta^{(n)}(k; p_1 \ldots p_n) = Z(Z_3)^{n/2} \Gamma_{u\Delta}^{(n)}(k; p_1 \ldots p_n), \tag{3.20}$$

then, for an appropriate (cutoff-dependent) choice of the constant Z, we can make the left-hand side of (3.20) cutoff-independent in the limit of high cutoff. This condition leaves Z undetermined up to a finite (cutoff-independent) multiplicative factor. We will choose this factor later, to make our final equations look as simple as possible.

Putting all this together, we find the consequence of (3.19) for the renormalized Green's functions,

$$i\Gamma_\Delta^{(n)}(0, \ldots) = Z\mu_0 \frac{\partial}{\partial \mu_0} \Gamma^{(n)}(\ldots) - \frac{n}{2} Z\mu_0 \frac{\partial \ln Z_3}{\partial \mu_0} \Gamma^{(n)}(\ldots). \tag{3.21}$$

We have not yet finished, since we have the derivative of a renormalized Green's function with respect to a bare mass, and a renormalized Green's function is cutoff-independent only when expressed in terms of renormalized masses and coupling constants. This is easily taken care of with the aid of the chain rule of differentiation:

$$\left[\left(Z\mu_0\frac{\partial\mu}{\partial\mu_0}\right)\frac{\partial}{\partial\mu}+\left(Z\mu_0\frac{\partial\lambda}{\partial\mu_0}\right)\frac{\partial}{\partial\lambda}-\frac{n}{2}\left(Z\mu_0\frac{\partial\ln Z_3}{\partial\mu_0}\right)\right]\cdot\Gamma^{(n)}(\dots)$$
$$=i\Gamma^{(n)}_\Delta(0,\dots).\tag{3.22}$$

where all derivatives are taken with fixed cutoff. The only terms in these equations which retain a possible trace of cutoff-dependence are the three terms in parentheses. These, however, are constants, independent of n and the momenta. Therefore, it is easy to show, by evaluating the equations at three independent points and solving for these terms as functions of cutoff-independent quantities, that they are also cutoff-independent, in the limit of high cutoff. (An especially simple choice is the three points at which the renormalized mass, renormalized charge, and scale of the renormalized field are defined.)

We now choose Z (until now undetermined up to a finite multiplicative constant) such that

$$Z\mu_0\frac{\partial\mu}{\partial\mu_0}=\mu,\tag{3.23}$$

and define

$$\beta=Z\mu_0\frac{\partial\lambda}{\partial\mu_0},\tag{3.24}$$

and

$$\gamma=\frac{1}{2}Z\mu_0\frac{\partial\ln Z_3}{\partial\mu_0}.\tag{3.25}$$

(Dimensional analysis shows that these functions can only depend on λ.) We thus obtain

$$\left[\mu\frac{\partial}{\partial\mu}+\beta(\lambda)\frac{\partial}{\partial\lambda}-n\gamma(\lambda)\right]\Gamma^{(n)}(\dots)=i\Gamma^{(n)}_\Delta(0,\dots).\tag{3.26}$$

These are the Callan–Symanzik equations.

Some remarks should be made about these equations:

(1) As promised, these are the equations that replace the naive low-energy theorem (3.7). If β were zero, (3.25) would be identical to (3.7), except that the scale dimension, d, would be anomalous:

$$d=1+\gamma.\tag{3.27}$$

Unfortunately, as we have already seen, β is not zero; there are further anomalies beyond anomalous dimensions. A direct computation[10] shows that

$$\beta = \frac{3\lambda^2}{16\pi^2} + O(\lambda^3). \tag{3.28}$$

(It will be important in the sequel that this first non-zero term in the perturbation expansion of β is positive.)

(2) The generalization to more complicated field theories is obvious: For every dimensionless coupling constant, there is a β-like term, and, for every field, there is a γ-like term. Thus, for example, in meson–nucleon theory, if we denote a Green's function with n nucleon fields, n antinucleon fields, and m meson fields, by $\Gamma^{(n,n,m)}$, the Callan–Symanzik equations are

$$\left[\mu \frac{\partial}{\partial \mu} + m \frac{\partial}{\partial m} + \beta_1 \frac{\partial}{\partial \lambda} + \beta_2 \frac{\partial}{\partial g} - 2n\gamma_1 - m\gamma_2 \right] \Gamma^{(n,n,m)}(\ldots)$$
$$= i\Gamma_\Delta^{(n,n,m)}(0; \ldots)$$

where the βs and the γs are functions of λ, g and μ/m.

(3) Likewise, if we attempt to study Green's functions for objects other than canonical fields (for example, conserved or partially-conserved currents), this will, in general, change the structure of the γ-terms, which refer to a particular field. The β-terms, though, which do not make reference to a choice of fields, but only to the underlying dynamics, will remain unchanged.

(4) Unlike the low-energy theorems of current algebra, which they so closely resemble, the Callan–Symanzik equations are practically useless for low-energy phenomenology. It is the β-terms that make the difference. A current-algebra low-energy theorem is useful because it expresses one Green's function in terms of another in a way that does not depend on strong-interaction dynamics. The Callan–Symanzik equations, on the other hand, express one Green's function in terms of another *and* its derivatives with respect to coupling constants. If you know how to compute *these*, you have already solved the strong-interaction dynamics, and there is no reason for you to be piddling around with low-energy theorems.

(5) Finally, there remains the possibility that β has a zero, and that, for some reason (the bootstrap?) the real value of λ is at this zero. (This is a speculation of Kenneth Wilson, among others). If this is the case, we would regain the naive theory of scale invariance (with anomalous dimensions). There is very little to be said for or against this possibility from a study of perturbation theory alone.

4   The resurrection of scale invariance

4.1   *The renormalization group equations and their solution*

By the arguments we used in Sect. 3.2, in the deep Euclidean region, we can neglect the right-hand side of the Callan–Symanzik equations. Thus, for the case of a self-interacting meson field, we obtain the following differential equations for the asymptotic forms of the Green's functions:

$$\left[\mu \frac{\partial}{\partial \mu} + \beta(\lambda) \frac{\partial}{\partial \lambda} - n\gamma(\lambda) \right] \Gamma_{as}^{(n)} = 0. \tag{4.1}$$

These equations are old friends; they are the equations associated with the so-called renormalization group, first devised by Gell-Mann and Low to study quantum electrodynamics, and later extended to general renormalizable field theories.[14]

The renormalization group Eq. (4.1) can be derived much more simply than the Callan–Symanzik equations. As I stated earlier, the deep Euclidean region is the maximally unphysical region; all external lines and all momentum transfers are far from the mass shell. One's first thought would be that in this regime all memory of the actual meson mass would be lost, and all Green's functions would be independent of the mass. However, this is too naive: the objects we are studying are Green's functions for renormalized fields, expressed as functions of the renormalized coupling constant; both the normalization of the field and the value of the renormalized coupling constant are defined on the mass shell; these quantities remember the mass shell no matter how far we flee into the deep Euclidean region. Therefore, the correct statement is that all memory of the actual value of the mass is lost, except for that which is contained in the scale of the fields and the value of λ. In other words, in the deep Euclidean region, a small change in the mass can always be compensated for by an appropriate small change in λ and an appropriate small rescaling of the fields. The equations (4.1) are just the mathematical expression of this statement.

This argument seems to me to owe very little to asymptotic estimates derived from perturbation theory; therefore I will assume from now on that the renormalization group equations are valid independent of perturbation theory, and try to use them to get information about the asymptotic forms of Green's functions in a non-perturbative way.

The renormalization group equations are fairly easy to solve. To motivate the solution, let us consider a similar equation with a direct dynamical interpretation:

$$\frac{\partial \rho}{\partial t} + v(x) \frac{\partial \rho}{\partial x} = L(x)\rho, \tag{4.2}$$

this is an equation of hydrodynamic type. We can think of ρ as the density of a population of bacteria moving with a fluid along a pipe. The fluid has a velocity $v(x)$, a known function of position along the pipe. As the bacteria move along, they are subjected to a changing illumination, $L(x)$, which determines their rate of reproduction.

To solve this equation is now trivial. The solution proceeds in two steps. First we find the position, x', at time t, of an element of the fluid that is at position x at time zero. That is to say, we solve the ordinary differential equation

$$\frac{\mathrm{d}x'(x, t)}{\mathrm{d}t} = v(x'),\tag{4.3}$$

with the boundary condition

$$x'(x, 0) = x.\tag{4.4}$$

Because of the time-translation invariance of the equations, this function also tells us the position at time t_1 of an element of the fluid that was at position x at time t_2, to wit, $x'(x, t_1 - t_2)$. We now find the bacterial density by integrating along a fluid element as it travels down the pipe. Thus we obtain

$$\rho(x, t) = f(x'(x, -t)) \exp\left[\int_{-t}^{0} \mathrm{d}t' L(x'(x, t')) \right],$$

where f is an arbitrary function, the density of bacteria in the fluid element at the starting time, $t = 0$, as a function of its position at that time, $x'(x, -t)$.

We can now leave our hydrodynamic–bacteriological analogy and return to the renormalization group Eq. (4.2). Making the obvious substitutions, we see that the general solution of (4.2) is obtained by first solving the ordinary differential equation

$$\frac{\mathrm{d}\lambda'(\lambda, t)}{\mathrm{d}t} = \beta(\lambda'),\tag{4.5}$$

with the boundary condition

$$\lambda'(\lambda, 0) = \lambda.\tag{4.6}$$

The general solution of (4.1) is then

$$\Gamma_{as}^{(n)} = s^{(4-n)/2} f^{(n)} \left(\lambda'\left(\lambda, \frac{1}{2} \ln \frac{s}{\mu^2} \right), \frac{p_i \cdot p_j}{s} \right)$$

$$\times \exp\left[-n \int_{0}^{\frac{1}{2} \ln(s/\mu^2)} \mathrm{d}t\, \gamma(\lambda'(\lambda, t)) \right].\tag{4.7}$$

where $f^{(n)}$ is an arbitrary function. Here we have used dimensional analysis, Eq. (3.6), to explicitly restore the dependence on the variables

Fig. 3

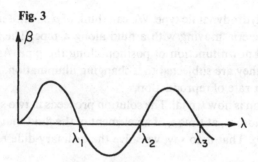

the figure so I can talk about them when I have finished the main argument.

Now, let us assume that the physical value of λ lies between zero and λ_1. Then, when we integrate the differential equation

$$\frac{d\lambda'(\lambda, t)}{dt} = \beta(\lambda'),$$ (4.5)

starting from the boundary condition

$$\lambda'(\lambda, 0) = \lambda,$$ (4.6)

we will find that λ' is a monotonically increasing function of t, for β is positive in this region. Indeed, λ' stays monotonically increasing for all t, for its derivative can change sign only when λ' exceeds λ_1, which it cannot do because $\beta(\lambda_1)$ is zero. In other words, λ' asymptotically approaches λ_1 from below as t goes to infinity,

$$\lim_{t \to \infty} \lambda'(\lambda, t) = \lambda_1.$$ (4.9)

This is the essential point: under the stated assumptions, for a range of values of λ, λ' approaches a fixed limit as t goes to infinity. Furthermore, the value of this limit is independent of the value of λ; as t goes to infinity, λ' loses all memory of where it started.

Likewise, if λ is between λ_1 and λ_2, the next zero of β (although it is not essential to the argument such a second zero exist), λ' monotonically approaches λ_1 from above. Continuing along the real axis, λ_2 is an exceptional point; if λ equals λ_2, λ' also equals λ_2 for all t. If λ is between λ_2 and λ_3, λ' approaches λ_3 from below as t goes to infinity, etc. In terms of the fluid analogy we discussed earlier, the odd zeros of β (λ_1 and λ_3) are sinks; everything is eventually drawn into them.

Now let us return to the case $(0 < \lambda < \lambda_2)$ in which the limiting value of λ is λ_1. Inserting this limit in (4.7), we obtain 'the asymptotic form of the asymptotic form',

$$\lim_{s \to \infty} \Gamma_{as}^{(n)} = s^{(4-n)/2} f^{(n)} \left(\lambda_1, \frac{p_i \cdot p_j}{s} \right) \times \left(\frac{s}{\mu^2} \right)^{-n\gamma(\lambda_1)/2} K^n,$$ (4.10)

where the constant K is given by

$$K = \exp \int_0^\infty dt \, [\gamma(\lambda_1) - \gamma(\lambda'(\lambda, t))]. \qquad (4.11)$$

(We will later argue that this integral is convergent.) But this is just simple scaling behavior again! Indeed, it is even simpler scaling behaviour than we thought we had back in Sect. 3.3, when we thought for a while that the only anomalies were anomalous dimensions. For if that were the case, the anomalous dimensions, the exponents that would appear in the asymptotic form (4.10), would depend continuously on λ. In fact, as (4.10) shows, the anomalous dimensions that appear in the asymptotic form are *independent* of λ, for a range of λ ($0 < \lambda < \lambda_2$), as is the whole asymptotic form, aside from the constant K. The β anomalies, which complicate things terribly at low energies, simplify things enormously in the deep Euclidean region. What a wonderful reversal! (This praise of the β-terms is justified, of course, only if our critical assumption – the existence of a zero in β – is true. If it is not, the β-terms remain troublemakers, at high energies as well as low.)

If we are willing to assume that the relevant functions are differentiable as well as continuous, then we can obtain an estimate of the error in (4.10), for then we can see how rapidly λ approaches λ_1. In the neighborhood of λ_1, eq. (4.5) can be approximated by

$$\frac{d\lambda'}{dt} = a(\lambda_1 - \lambda'), \qquad (4.12)$$

where a is a positive number,

$$a = -\left.\frac{d\beta}{d\lambda}\right|_{\lambda_1}. \qquad (4.13)$$

The solution to (4.12) is

$$\lambda' = \lambda_1 + c \exp[-at], \qquad (4.14)$$

where c is a constant. This immediately implies the convergence of the integral (4.11). Also, after some straightforward differentiation, it leads to the statement that the error in (4.10) is of the order $(s/\mu^2)^{-a/2}$, compared to the terms retained. Without a knowledge of a, it is impossible to say how small this is, but at least it is a good solid power, not just some logarithms. (All of this presumes, of course, that the zero at λ_1 is a single zero, as shown in Fig. 3. If it were a double or higher-order zero, the terms neglected in (4.10) would just be down by logarithms compared to the terms retained, and the limit would be much less interesting.)

4.3   Scaling and the operator product expansion

We have already heard much about the operator product expansion at this school;[16] therefore, rather than explain it in detail here, I will merely summarize its essential features. The operator product expansion was first introduced by Wilson[17] as a conjecture; it is an asymptotic expansion for the product of two local operators as their space-time arguments coincide, as a sum of local operators with c-number coefficients. In equations,

$$A(x)B(y) = \sum_n f_n(x-y)O_n\left(\frac{x+y}{2}\right),\qquad (4.15)$$

where A and B are two arbitrary local operators, x and y are the two arguments that are being brought close together, O_n is the infinite string of local operators, and f_n are the c-number coefficients. (We use the equality sign for notational simplicity, even though the series in supposed to be only asymptotic.) Wilson also conjectured that at small distances scale invariance became exact. That is to say that it was possible to assign a dimension $d(A)$, $d(B)$, etc. to every operator occurring in Eq. (4.15), such that for small values of $(x-y)$, f_n became a homogeneous function of $(x-y)$, of order $d(O_n) - d(A) - d(B)$. (We will refer to this statement as Wilson's rule.) In Wilson's original scheme, these dimensions could be anomalous (not equal to the dimensions of the operators in the sense of dimensional analysis), and were not defined in any way outside the operator product expansion; they were simply the numbers that made Wilson's rule work.

The obvious questions to ask are: (1) Is the operator product expansion true? (2) If so, is Wilson's rule true? (3) If so, are the anomalous dimensions that enter into Wilson's rule the same as the anomalous dimensions that govern the asymptotic forms of Green's functions in the deep Euclidean region? I will now give an ingenious argument due to Kurt Symanzik[18] that says the answer to all three questions is yes. The argument does require some speculative assumptions, but they are the same assumptions (the existence of a zero in β, etc.) that we used to get simple scaling behaviour in the deep Euclidean region; no additional assumptions are needed.

For simplicity, let us restrict ourselves to the easiest case, in which the two operators that approach each other are two scalar fields in the self-interacting meson theory we have already discussed so much. In this case, we might expect the first two terms in (4.15) to be

$$\phi(x)\phi(y) = f_0(x-y)I + f_1(x-y)\phi^2\left(\frac{x+y}{2}\right) + \cdots \qquad (4.16)$$

where I is the identity operator, and ϕ^2 is the renormalized square of the field. (Let me remind you that this is nothing so simple as the product of ϕ and ϕ; it is a complicated object defined as a limit in a cut-off theory.)[13]

In free-field theory, it is easy to establish this equation. In this case there are no anomalous dimensions, the renormalized ϕ^2 operator is just the normal-ordered product, and it is easy to check that for small arguments f_0 is an inverse square and f_1 is a constant. For the interacting theory, (4.16) has been established to any finite order in renormalized perturbation theory by Callan and Zimmerman.[19] For our purposes, it is most convenient to state their result in momentum space. (Bringing two space-time arguments together in position space is equivalent to letting a momentum transfer go to Euclidean infinity.) In our notation, the asymptotic expansion found by Callan and Zimmerman is

$$\Gamma^{(n+2)}\left(\frac{p+q}{2}, \frac{p-q}{2}, p_1 \ldots p_n\right) = f(q^2)\Gamma^{(n)}_{\phi^2}(p; p_1 \ldots p_n) + \ldots,$$

(4.17)

as q goes to Euclidean infinity, for n greater than zero. The terms neglected are smaller by one power of k^2 than the terms retained. The function $f(q^2)$ is just the Fourier transform of f_1. The identity term in (4.16) is missing from (4.17) because our Γs are connected Green's functions; thus it only contributes to the expansion when n is zero. In any finite order of perturbation theory, $f(q^2)$ is not a simple power; it is full of logarithms. This is reminiscent of the situation for scaling behaviour in the deep Euclidean region.

Please note that this limit, the limit that is relevant to the operator product expansion, does *not* take us into the deep Euclidean region. Only one momentum gets large and Euclidean; the others stay close to the mass shell. Therefore, this is not a limit in which the renormalization group equations are applicable. However, the Callan–Symanzik equations are still applicable; they hold at all momenta:

$$\left[\mu\frac{\partial}{\partial\mu} + \beta\frac{\partial}{\partial\lambda} - n\gamma\right]\Gamma^{(n)}(p_1 \ldots) = \Gamma^{(n)}_\Delta(0; p_1 \ldots).$$

(4.18)

In addition to these two equations, we will need two more. One is the generalization of (4.17) when there is a mass insertion, Δ,

$$\Gamma^{(n+2)}_\Delta\left(k; \frac{p+q}{2}, \frac{p-q}{2}, p_1 \ldots p_n\right)$$

$$= f(q^2)\Gamma^{(n)}_{\Delta,\phi^2}(k; p; p_1 \ldots p_n) + \ldots,$$

(4.19)

again, for n greater than zero, as q goes to Euclidean infinity. The function f is the same function that appears in (4.17). This is a reflection of the fact that (4.16) is an *operator* equation; the only purpose of the fixed momenta in (4.17) and (4.19) is to create the initial and final states between which the operators are to be evaluated. Since the equation is true whatever the states are, it does not matter if they are created by ϕs or Δs.

The other equation we will need is the Callan–Symanzik equation for a Green's function built of a string of ϕs and one ϕ^2. Simple power counting shows that this can be renormalized by multiplying by the usual factors of Z_3 for the ϕs, and by an independent renormalization constant for the ϕ^2. Thus, by the same reasoning as led to the original Callan–Symanzik equations,

$$\left[\mu \frac{\partial}{\partial \mu} + \beta \frac{\partial}{\partial \lambda} - n\gamma - \gamma_{\phi^2} \right] \Gamma^{(n)}_{\phi^2}(p; p_1 \ldots p_n) = \Gamma^{(n)}_{\Delta,\phi^2}(0; p; p_1 \ldots p_n),$$

(4.20)

where γ_{ϕ^2} is yet another function of λ. (You may think the notation is a bit over-complex, since it ignores the fact that ϕ^2 and Δ are proportional. However, I want to emphasize that this is in no way essential to the argument, which can be extended trivially to other cases than (4.17), for which this is not true.)

Finally, I remind you if we denote by $d(\phi)$ the number that gives the exponent in the deep Euclidean asymptotic form (4.10), then we showed in Sect. 4.2 that

$$d(\phi) = 1 + \gamma(\lambda_1),$$

(4.21)

where λ_1 is a zero of β. Likewise, in the same sense of deep Euclidean asymptotic behaviour,

$$d(\phi^2) = 2 + \gamma_{\phi^2}(\lambda_1).$$

(4.22)

We are now ready to go. The essential point is simple: $\Gamma^{(n)}_{\Delta,\phi^2}$ can be computed in two ways; either we first take the operator-product limit and then use the Callan–Symanzik equations, or we first use the Callan–Symanzik equations and then take the operator-product limit. Doing things in one order, we find

$$\left(\mu \frac{\partial}{\partial \mu} + \beta \frac{\partial}{\partial \lambda} - (n+2)\gamma \right) \Gamma^{(n+2)} \left(\frac{p+q}{2}, \frac{p-q}{2}, \ldots \right)$$
$$= \Gamma^{(n+2)}_{\Delta} \left(0; \frac{p+q}{2}, \frac{p-q}{2}, \ldots \right)$$
$$= f(q^2) \Gamma^{(n)}_{\Delta,\phi^2}(0; p; \ldots),$$

(4.23)

as q goes to Euclidean infinity. Doing things in the other order, we find

$$\left(\mu\frac{\partial}{\partial\mu}+\beta\frac{\partial}{\partial\lambda}-(n+2)\gamma\right)\Gamma^{(n+2)}\left(\frac{p+q}{2},\frac{p-q}{2},\dots\right)$$

$$=\left(\mu\frac{\partial}{\partial\mu}+\beta\frac{\partial}{\partial\lambda}-(n+2)\gamma\right)f(q^2)\Gamma^{(n)}_{\phi^2}(p;\dots)$$

$$=f(q^2)\Gamma^{(n)}_{\Delta,\phi^2}(0;p;\dots)$$

$$+\Gamma^{(n)}_{\phi^2}(p;\dots)\left(\mu\frac{\partial}{\partial\mu}+\beta\frac{\partial}{\partial\lambda}-2\gamma+\gamma_{\phi^2}\right)f(q^2). \tag{4.24}$$

Two things equal to the same thing are equal to each other; thus

$$\left(\mu\frac{\partial}{\partial\mu}+\beta\frac{\partial}{\partial\lambda}-2\gamma+\gamma_{\phi^2}\right)f(q^2)=0. \tag{4.25}$$

But this is a homogeneous differential equation, of the same form as the renormalization group eq. (4.1)! Thus, by exactly the same arguments as we used in Sect. (4.2), we find that as q^2 goes to Euclidean infinity,

$$f(q^2)\sim\left(\frac{q^2}{\mu^2}\right)^{[2\gamma(\lambda_1)-\gamma_{\phi^2}(\lambda_1)]/2}$$

$$\sim\left(\frac{q^2}{\mu^2}\right)^{[2d(\phi)-d(\phi^2)]/2}, \tag{4.26}$$

verifying Wilson's rule, and showing that the dimensions that occur in Wilson's rule are the same as those that give the exponents in the deep-Euclidean asymptotic form.

5   Conclusions and questions

We have gone through a lot of complicated analysis, so perhaps I should end by summarizing our main results:

(1) The formal theory of broken scale invariance is a pack of lies, hopelessly afflicted with anomalies. In one case only (the low-energy theorems associated with scale invariance), the true equations have been found that replace the false equations. These are the Callan–Symanzik equations. They are simple and elegant, but practically useless for doing low-energy phenomenology, because they involve derivatives with respect to coupling constants.

(2) Under a very strong assumption (that the physical coupling constant is a zero of the function β that appears in the Callan–Symanzik equations), the Callan–Symanzik equations become equivalent to the naive low-energy theorems, except that the dimensions of fields, are, in general, anomalous. Nothing is known about what happens, under this assumption, to the rest of the formal theory. In particular, nothing is known about

what happens to the formal low-energy theorems that follow from broken conformal invariance.[20]

(3) Under a much weaker assumption (the existence of a zero in β, not necessarily at the physical coupling constant), Green's functions show simple scaling behaviour in the deep Euclidean region, with anomalous dimensions. So do the coefficient functions that occur in Wilson's operator product expansion.

There is obviously much to be done. Here are some live questions:

(1) Perturbation theory tells us nothing about the zeros of β. Is there some other way to investigate this problem? Do the axioms of field theory say anything? Is there any way to get some sort of sum rule for β that can be evaluated experimentally?

(2) Is there any simple and compact way, analogous to the Callan–Symanzik equations, of expressing the anomalies in the low-energy theorems associated with broken conformal invariance? If there is, what do such expressions tell us about asymptotic behaviour?

(3) Is there any way of saving PCDC?

(4) Can the methods we have been playing with here, or extensions of them, be used to get information about the light-cone operator product expansion? Less abstractly, how do we make contact with the experimental data on deep inelastic electron–nucleon scattering?

None of these are easy questions, but progress in this field has been surprisingly rapid in the last few years. Maybe we will get the answers sooner than we expect.

Notes and references

1. Notation: Greek indices run from 0 to 3, with 0 denoting the time axis. Latin indices run from 1 to 3. Space-time points are labelled in three ways: $x \equiv x^\mu \equiv (x^0, \mathbf{x})$. $\partial_\mu = \partial/\partial x^\mu$. The signature of the metric tensor is $(+ - - -)$.
2. For a proof, see, for example, S. Coleman and R. Jackiw: *Ann. of Physics.* **67**, 552 (1971).
3. I should emphasize that when I say these results are perfectly general, I mean only that they can be established in the most general case by formal manipulations of the Lagrangian. Such manipulations are obviously invalid in a quantum field theory, where the divergences that occur when two fields approach the same space-time point make even *defining* the Lagrangian a delicate matter. Nevertheless, the Ward identities are frequently valid, at least to all orders in renormalized perturbation theory. Later on, we will discuss when they can be trusted.
4. Equation (2.7) should be read as follows: Under the transformation labelled by the parameter α, the space-time point x goes into $e^\alpha x$.
5. This, and all other statements in this section, are derived in tedious detail in C. Callan, S. Coleman and R. Jackiw: *Ann. of Phys.*, **59**, 42 (1970). This paper also contains references to the earlier literature on conformal invariance.

6. This equation shows very clearly the lack of connection between scale dimension and dimension in the sense of dimensional analysis. In the sense of dimensional analysis, σ has the dimensions of mass, like any other scalar field; however, because of its inhomogeneous transformation law, its scale dimension is not even defined.

7. Good papers to read are J. Ellis: *Nucl. Phys.* B 22, 478 (1970), B 26, 537 (1971), and the lectures of B. Zumino at the 1970 Brandeis Summer School in *Lectures on Elementary Particles and Quantum Field Theory*, edited by S. Deser *et al.* (M.I.T. Press 1970).

8. I remind you that a one-particle-irreducible Green's function is the sum of all connected Feynman diagrams that cannot be cut in two by breaking a single internal line. By convention the diagrams are evaluated with no propagators on the external lines. Thus, $\Gamma^{(2)}$ is the inverse propagator; $\Gamma^{(3)}$ is the proper three-line vertex (zero in the theory (3.1), by parity), $\Gamma^{(4)}$ is the two-into-two off-mass-shell scattering amplitude, less the pole terms (which are absent anyway in (3.1)), etc.

9. S. Weinberg: *Phys. Rev.* 118, 838 (1960).

10. An instructive example is a recent computation of T. Appelquist and J. Primack: *Phys. Rev.* D4, 2454 (1971). In vector-meson–nucleon theory, these authors isolate those diagrams which give the dominant asymptotic behaviour for the nucleon electromagnetic form factor, in each order of perturbation theory (the 'leading logarithms'). They then sum up these leading terms, and find an expression proportional to $\exp[-c \ln^2 s]$, where c is a positive constant, proportional to the square of the coupling constant. This goes to zero more rapidly than any powers of s; that is to say, the sum of the leading terms is highly damped, unlike the individual leading terms, which grow like powers of logarithms. Also, and even more discouraging, in the asymptotic region the sum of the leading terms is much less important than any one of the terms that were neglected (the non-leading terms) in the original summation.

11. C. G. Callan: *Phys. Rev.*, D2, 1541 (1970); K. Symanzik: *Comm. Math. Phys.* 18, 227 (1970).

12. Unfortunately, for the Ward identities at arbitrary momentum transfers, and, particularly unfortunately for the low-energy theorems that follow from broken conformal invariance, the situation remains obscure.

13. See Chapter 4 in this book.

14. M. Gell-Mann and F. E. Low: *Phys Rev.* 95, 1300 (1954); N. N. Bogoliubov and D. V. Shirkov: *Introduction to the Theory of Quantized Fields*, Interscience, 1959.

15. K. Wilson: *Phys. Rev.* D3, 1818 (1971). The immediate ancestors of Wilson's arguments can be found in the studies of quantum electrodynamics by M. Baker, K. Johnson and R. Willey: *Phys. Rev.* 136 B, 1111 (1964); 163, 1699 (1967); 183, 1292 (1969); D 3, 2516 (1971), and also in the original work on the renormalization group by Gell-Mann and Low[14]; however, I believe that the application of these ideas to strong-interaction scaling first appeared in Wilson's work.

16. See, e.g., the lectures of G. Preparata in *Properties of the Fundamental Interactions.* (Editrice Compositori, Bologna, 1973).

17. K. Wilson: *Phys. Rev.* 179, 1499 (1969).

18. K. Symanzik: *Comm. Math. Phys.* 23, 49 (1971).

19. C. Callan *Phys. Rev.* D 5, 3205 (1972). W. Zimmerman: in *Lectures on Elementary Particles and Quantum Field Theory*, edited by S. Deser *et al.*, M.I.T. Press., (1970).

20. It has recently been shown by B. Schroer: *Lettere al Nuovo Cimento* 2, 867 (1971) that these are also free of anomalies, at a zero of β.

4

Renormalization and symmetry:
a review for non-specialists
(1971)

1 Introduction

I suppose that as good a way as any of explaining the contents of this lecture is to explain the title. By 'renormalization' I mean the removal of infinities for Feynman amplitudes, in perturbation theory, for Lagrangian field theories with polynomial interactions. In particular non-perturbative renormalization (the work of Jaffe, Glimm, etc.) is outside the scope of this lecture, as are the properties of non-polynomial interactions (the work of Efimov, Salam, Lehmann, etc.). By 'renormalization and symmetry' I mean that we will be concerned not only with the renormalization of scattering amplitudes, but also with the renormalization of the matrix elements of conserved and partially conserved currents. In particular, we will discuss some fairly recent results of Symanzik, Benjamin Lee, Preparata, Weisberger, and others. By 'a review for non-specialists' I mean that I hope that this talk will be intelligible to people who can do nothing more complicated than remove the divergences from the self-energy of the electron.

Since renormalization theory has a well-deserved reputation for complexity, it is obvious that I will be able to do all this in a single lecture only by cheating. To be precise, I will explain a very powerful theorem due to Klaus Hepp, but not prove it (this is the cheat); then I will show how a wide variety of results can be obtained from this master theorem by elementary methods.[1]

2 Bogoliubov's method and Hepp's theorem

For simplicity, we will restrict ourselves to field theories involving spin-zero and spin-one-half fields only, which we will call Bose and Fermi fields, respectively. We will write the Lagrangian for such a theory in the

99

form

$$\mathscr{L} = \mathscr{L}_0 + \sum_i \mathscr{L}_i,\tag{1}$$

where $\mathscr{L}_0$ is a sum of free Lagrangians of standard form, one for each field, and each $\mathscr{L}_i$ is a monomial in the fields and their derivatives. For future use, it will be convenient to establish some notation, and denote by f_i the number of Fermi fields in $\mathscr{L}_i$, by b_i the number of Bose fields, and by d_i the number of derivatives. Thus, for example, the ps-ps meson–nucleon interaction

$$g\bar{\psi}\gamma_5\psi\phi,$$

has $f = 2, b = 1$, and $d = 0$, while the ps-pv interaction

$$f\bar{\psi}\gamma_\mu\gamma_5\psi\partial^\mu\phi,$$

has $f = 2, b = 1$, and $d = 1$.

If we attempt to calculate scattering amplitudes with such a Lagrangian, following the conventional Feynman rules, we soon encounter divergent diagrams, that is to say, infinite Feynman integrals. I will assume that we have cut off the theory in some way (say, by modifying the propagators) so that instead of divergent amplitudes we have cutoff-dependent ones. The renormalization procedure of Bogoliubov[2] consists of adding to the Lagrangian extra terms, the so-called renormalization counter-terms, whose function is to cancel the cutoff-dependence of the amplitude. First I will explain how these extra terms are constructed; later I will explain their physical meaning.

To explain the construction, three definitions are needed:

(1) One-particle-irreducible diagrams. A Feynman diagram is said to be one-particle-irreducible (abbreviated IPI) if it is connected and cannot be disconnected by cutting any one internal line. Fig. 1 shows three Feyn-

Fig. 1

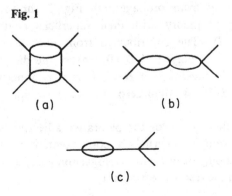

(a)   (b)

(c)

man diagrams in ϕ^4 theory. The first two are IPI; the third is not. (If the horizontal line is cut, the diagram falls into two pieces.)

(2) **Taylor expansions about the point zero.** A Feynman amplitude with n external lines is a function of $n - 1$ independent four-momenta. Furthermore, if there are no massless particles in the theory (as we shall assume from now on) it is an analytic function of these momenta in some neighbourhood of the point zero, the point where *all* external momenta vanish. Thus, it may be expanded in a Taylor series in these variables. For example, the third-order vertex diagram of ps-ps meson–nucleon theory, shown in Fig. 2, has an expansion of the form

$$a\gamma_5$$
$$+ b\gamma_5\gamma_\mu p^\mu + c\gamma_5\gamma_\mu p'^\mu$$
$$+ d\gamma_5 p^2 + e\gamma_5 p'^2 + f\gamma_5 p \cdot p'$$
$$+ \cdots,$$

where a, b, c, etc. are constants. The term on the first line is called a term of zeroth order, those on the second line terms of first order, those on the third line terms of second order, etc.

Fig. 2

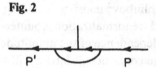

(3) **Superficial degree of divergence.** A Feynman amplitude is, in general, a multiple integral. The superficial degree of divergence of such an integral is the difference between the number of momenta in the numerator of the integral (arising from loop integration variables and from explicit momenta at vertices due to derivative interactions) and the number of momenta in the denominator (arising from propagators). Fig. 3 shows three Feynman diagrams from ϕ^4 theory, with their superficial degrees of divergences (denoted by D). The contribution from numerator and denominator are separately displayed. If $D = 0$, we say the diagram is superficially logarithmically divergent, if $D = 1$, that it is superficially linearly divergent, etc. If D is less than zero, we say it is superficially convergent.

Fig. 3(c) demonstrates the reason for the pejorative adjective 'superficial'. Although the diagram is superficially convergent, it is in fact divergent; the integration along the lower loop is logarithmically divergent no matter what happens in the rest of the diagram.

Fig. 3

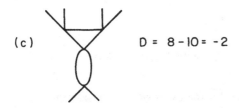

(a) $D = 4 - 4 = 0$

(b) $D = 8 - 6 = 2$

(c) $D = 8 - 10 = -2$

It will be convenient later to have a general expression for the superficial degree of divergence of a connected Feynman diagram. For such a diagram let

 B be the number of external boson lines,

 IB be the number of internal boson lines,

 F be the number of external fermion lines,

 IF be the number of internal fermion lines, and

 n_i be the number of vertices of the ith type, i.e. those that come from the ith term in the Lagrangian (1).

There is an elementary relation between these numbers. Since a vertex of the ith type has b_i boson line ends attached to it, and since every internal boson line has two ends attached to vertices and every external boson line has one, we can readily deduce that

$$B + 2(IB) = \sum n_i b_i,$$

'the law of conservation of boson ends'. By the same reasoning, we can deduce 'the law of conservation of fermion ends',

$$F + 2(IF) = \sum n_i f_i.$$

It is also elementary to compute the superficial degree of divergence:

$$D = \sum n_i d_i + 2(IB) + 3(IF) - 4 \sum n_i + 4.$$

The five terms in this formula have the following origins. (1) Every derivative in an interaction puts a momentum in the numerator of the Feynman integral. (2) Every internal boson line puts four integration momenta in the numerator and two propagator momenta in the denominator. (3) Every internal fermion line puts four integration momenta in the numerator and one in the denominator. (4) Every vertex has a four-dimensional delta-function attached to it, which, upon integration, cancels four integration momenta, (5) except for one delta-function that is left over to give overall four-momentum conservation.

Putting all of this together, we find that

$$D = -B - \tfrac{3}{2}F + 4 + \sum n_i \delta_i, \tag{2}$$

where δ_i, 'the index of divergence of $\mathscr{L}_i$', is given by

$$\delta_i = b_i + \tfrac{3}{2}f_i + d_i - 4. \tag{3}$$

It is worth remarking that, for the cases we are considering,

$$\delta_i = \dim \mathscr{L}_i - 4, \tag{4}$$

where $\dim \mathscr{L}_i$ is the dimension of $\mathscr{L}_i$, in the usual sense of dimensional analysis, in units of mass. (This is, however, special to the theories we are considering; eq. (4) is not true, for example, for the interactions of a vector meson coupled to a non-conserved current.)

This completes our three definitions (plus one long digression). We are now in a position to state the renormalization prescription of Bogoliubov.[2] As advertised, this is an iterative procedure; as we calculate in perturbation theory, to each order we change the Lagrangian, adding to it extra terms. The procedure is as follows:

(1) Calculate in perturbation theory until you encounter an IPI diagram whose superficial degree of divergence, D, is greater than or equal to zero.

(2) Add to the Lagrangian extra terms (the counterterms) chosen to precisely cancel, to this order, all terms in the Taylor expansion of this diagram of order D or less.*

As an example of this procedure, let us consider $\lambda\phi^4$ theory, for which the Lagrangian (1) is

$$\mathscr{L} = \frac{1}{2}\partial_\mu\phi\partial^\mu\phi - \frac{1}{2}\mu^2\phi^2 - \frac{\lambda}{4!}\phi^4. \tag{5}$$

In order λ^2, we encounter the divergent diagrams 3(a) and 3(b). For the first of these, $D = 0$, for the second $D = 2$. Thus we change the Lagrangian

---

* Please note that it follows from this and Eqs. (2) and (3) that the counterterms induced have index of divergence, δ, less than or equal to the sum of the indices of divergence of the interactions occurring in the diagram. This observation has been stuck in a footnote because it is not important now, but it will be useful later.

that do not enter into (4.2), the large variable s and the dimensionless kinematic variables.

Just to check that we have not made a sign error, let us evaluate (4.7) under the assumption that we are actually at a zero of β, as conjectured at the end of Sect. 3. In this case, the solution of the ordinary differential equation is trivial, $\lambda' = \lambda$, and (4.7) reduces to

$$\Gamma_{as}^{(n)} = s^{(4-n)/2} \left(\frac{s}{\mu^2}\right)^{-n\gamma(\lambda)/2} f^{(n)}\left(\lambda, \frac{p_i \cdot p_j}{s}\right). \tag{4.8}$$

This is the correct answer in this special case; everything scales as it would naively, except that the scale dimension is anomalous.

4.2   *The return of scaling in the deep Euclidean region*

The final result of the last section, Eq. (4.7) evidently has a lot of content. It tells us that the asymptotic form of a Green's function in the deep Euclidean region, which *a priori* could depend in an arbitrary way on the dimensionless variables λ and $\ln s/\mu^2$, in fact depends in an arbitrary way only on a certain function of these variables, λ', which is, in turn, completely determined in terms of the function $\beta(\lambda)$ by Eqs. (4.5) and (4.6). This is useful information as it stands; for example, it has been used to enormously simplify the summation of leading logarithms on the deep Euclidean region. However, as was pointed out by Wilson,[15] if we make some (apparently very mild) additional assumptions, we get a surprisingly greater amount of information.

These additional assumptions are: (1) The functions $f^{(n)}$ appearing in Eq. (4.7) are continuous functions of the variable λ'. Likewise, β and γ are continuous functions of the variable λ. This is certainly mild, and is trivially true in any finite order of perturbation theory, in which these functions are polynomials in these variables. (2) The function $\beta(\lambda)$ has a zero some place along the positive real axis. It is very difficult to say whether this is a plausible or implausible assumption from perturbation theory alone, since any finite order of perturbation theory merely tells us the behaviour of β near the origin, and gives us no information about the presence of zeros away from the origin. It is certainly much weaker than the conjecture we toyed with at the end of Sect. 3, that β had a zero *and* that this zero was the physical coupling constant.

The assumed behaviour of β is shown in Fig. 3. The quadratic zero at the origin is not part of the assumption; that can be established by direct perturbative computation (see eq. (3.27)). The zero at λ_1 is an assumption. The two subsequent zeros at λ_2 and λ_3 are not necessary to the argument, but they lead to interesting results if they are assumed; I have put them in

(5) by adding to it extra terms

$$\mathscr{L} \to \mathscr{L} - \frac{A_2}{4!}\,\phi^4 + \frac{1}{2}\,B_2 \partial_\mu \phi \,\partial^\mu \phi - \frac{1}{2}\,C_2 \phi^2. \tag{6}$$

(The subscript 2 is to remind you that these terms are of second order in λ; they are also cutoff-dependent, but that is not important at the moment). The A_2 term is chosen to cancel the zeroth-order term in the Taylor expansion of 3(a); the B_2 and C_2 terms to cancel the zeroth and second order terms in the Taylor expansion of 3(b). (There is no need for a first-order counterterm because Lorentz-invariance forbids a first-order term in the Taylor expansion.)

(3) Continue computing, now using the corrected Lagrangian.

Theorem (Hepp).[3] This procedure eliminates all divergences. That is to say, the resultant perturbation expansion is independent of the cutoff in the limit of infinite cutoff.[4]

For the moment, I would like you to think of this purely as a mathematical theorem about Feynman expansions; we will try to understand its physical significance shortly. However, there is one point I would like to make now – we can already begin to see why it is the superficial degree of divergence, rather than the true degree of divergence, that is important. Remember the order λ^4 diagram 3(c), which has $D = -2$. By our prescription, even though this diagram is in fact divergent, it does not induce a counterterm. We can now see the reason for this: there is another diagram of order λ^4, shown in Fig. 4, where the heavy dot is the A_2 term in eq. (6), the counterterm that was added to the Lagrangian in order λ^2. This diagram *automatically* cancels the divergence of Fig. 3(c). Speaking very roughly, we only need new counterterms at a given order of perturbation theory to take care of new divergences; old divergences, divergences caused by lower-order diagrams hiding inside higher-order ones, as 3(a) is hiding inside 3(c), are taken care of by old counterterms.

Fig. 4

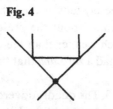

3   Renormalizable and non-renormalizable interactions

Let us look a little more closely at the theory defined by eq. (5). We have already classified the counterterms that arise in order λ^2; what happens in an arbitrary order of perturbation theory? Equations (2) and (3) give us the answer; for the special case of a ϕ^4 interaction, eq. (3) becomes

$$\delta = 4 + 0 + 0 - 4 = 0,$$

and eq. (4) becomes

$$D = 4 - B.$$

Thus the only superficially divergent diagrams are those with B equal to two or four (diagrams with odd numbers of external lines vanish because of the symmetry of the Lagrangian under $\phi \to -\phi$), and their superficial degrees of divergence are the same in a general order as in second order. The only effect of renormalization, to any order, is to change (5) into

$$\mathscr{L} = \frac{1}{2}(\partial_\mu \phi)^2 - \frac{1}{2}\mu^2\phi^2 - \frac{\lambda}{4!}\phi^4 - \frac{A}{4!}\phi^4 + \frac{B}{2}(\partial_\mu \phi)^2 - \frac{1}{2}C\phi^2, \quad (7)$$

where the constants A, B, and C are power series in λ, with coefficients that are, in general, cutoff-dependent. We can now see the physical meaning of the renormalization procedure; for if we define

$$\phi_u = (1 + B)^{\frac{1}{2}}\phi, \tag{8a}$$

$$\mu_0^2 = (\mu^2 + C)(1 + B)^{-1}, \tag{8b}$$

$$\lambda_0 = (\lambda + A)(1 + B)^{-2}, \tag{8c}$$

then we may rewrite the Lagrangian (7) as

$$\mathscr{L} = \frac{1}{2}(\partial_\mu \phi_u)(\partial^\mu \phi_u) - \frac{1}{2}\mu_0^2\phi_u^2 - \frac{\lambda_0}{4!}\phi_u^4.$$

This is of the same form as our starting Lagrangian (5), except that the coefficients have been changed. The field ϕ_u is called the unrenormalized field; it obeys canonical commutation relations, but has cutoff-dependent matrix elements (because of the cutoff-dependent quantity B in eq. (8a)). The quantities μ_0 and λ_0 are called the bare mass and bare coupling constants. Thus, for this theory the content of Hepp's theorem is that if we choose the bare mass and coupling constants in an appropriate cutoff-dependent fashion, and rescale the fields in an appropriate cutoff-dependent way, all the divergences disappear, order by order, in perturbation theory. A Lagrangian that has this property is said to be renormalizable.

The field ϕ and the quantities μ and λ are not the renormalized field, mass, and coupling constants as usually defined; this is because they are defined in terms of Green's functions at the point zero, rather than at some astutely chosen mass-shell point. However, they are cutoff-independent parameters that characterize the theory; the usual parameters can be computed in terms of them to any order of perturbation theory, and, if one wishes, these expressions can be inverted in the standard way to obtain a perturbation theory in terms of the usual parameters. For our purposes they are more convenient than the usual parameters because it is easier to do power series expansions about the point zero than about mass-shell points, where one has to worry about possible singularities of Feynman integrals. To distinguish them from the usual parameters we will refer to μ and λ as the intermediate mass and coupling constants; likewise, we will refer to the Green's functions associated with the field ϕ as intermediate Green's functions.

Not all Lagrangians are renormalizable. For example, if we had added a ϕ^5 interaction to our starting Lagrangian, this would have δ equal to one, and the renormalization procedure would inexorably add to the Lagrangian, as we computed higher and higher orders of perturbation theory, higher and higher order monomials in the field and its derivatives. Once such expressions appear in the counterterms, there is no physical reason to exclude them from the starting Lagrangian. (After all, it was only mathematical convenience that made us choose the point zero for our renormalization prescription; if we had chosen a different point (or even a separate point for every Green's function, or different points in different orders of perturbation theory) we would have obtained different values for the coefficients of the counterterms.) Thus we would be led to a theory with an infinite number of parameters. Such theories are called non-renormalizable.

4   Symmetry and symmetry-breaking: Symanzik's rule.

We are now ready to begin pulling interesting results out of Hepp's theorem. I would like to begin by taking the observation made in a footnote to Sect. 2 and raising it to the dignified status of a

Lemma. The counterterms induced by a given Feynman diagram have index of divergence, δ, less than or equal to the sum of the indices of divergence of all the interactions in the diagram.

I would also like to adopt a somewhat more stringent definition of renormalizability than usual: I will call a Lagrangian renormalizable only if all the counterterms induced by the renormalization procedure can be absorbed into a redefinition of the parameters in the Lagrangian. Thus, by this strict test, the theory of a single nucleon field interacting with a single pseudoscalar meson field through Yukawa coupling, $\bar{\psi}\gamma_5\psi\phi$, is *not* renormalizable, because renormalization induces a ϕ^4 counterterm, not present in the original Lagrangian. However, the same theory, *with a ϕ^4 interaction in the original Lagrangian, is* renormalizable, because now all the counterterms are of the same form as terms originally present.

With this definition, we can now state our:

First result. Given a set of spin-zero and spin-one-half fields, the most general Lagrangian constructed from this set containing all terms with δ less than or equal to zero (equivalently, with dimension less than or equal to four) is renormalizable.

This is a trivial consequence of the Lemma.

Second result. If we restrict the Lagrangians defined in the first result to only contain parity-conserving terms, they are still renormalizable. Likewise, if we restrict them to preserve some internal symmetry, such as isotopic spin, they are still renormalizable.

This is also trivial. Unless we have been so stupid as to introduce parity violation into our cutoff procedure, Feynman diagrams computed from a parity-conserving Lagrangian will be parity-conserving. Thus, they will have no parity-violating terms in their Taylor expansions about the point zero, and hence no parity-violating counterterms will be induced by the renormalization procedure. Ditto for internal symmetries. (In fact, ditto for chiral symmetries, such as those of the σ-model, although here one must be more clever than usual to construct a cutoff procedure that does not break the symmetry.)[5]

Third result. (Symanzik's rule for symmetry-breaking):[6] If we generalize the preceding set of Lagrangians to include symmetry-breaking terms, but only with dimensions less than or equal to n, where n is either 3, 2, or 1, they are still renormalizable.

Although this is our first 'new' (1970) result, it is also trivial.[7] The symmetric terms in the Lagrangian have $\delta \leqslant 0$; the symmetry-breaking terms have $\delta \leqslant n - 4 < 0$. A symmetry-breaking counterterm can arise only from a diagram that involves at least one symmetry-breaking interaction. By the Lemma, this must also have $\delta \leqslant n - 4$.

Thus, for example, if, in the standard isospin-symmetric theory of pions and nucleons, we choose to break isospin only by giving the charged and neutral pions different masses, then renormalization will not force us to change our intention and also introduce symmetry-breaking Yukawa couplings. Remember, though, that we are speaking here of the intermediate coupling constants. The physical renormalized coupling constants *do* display the effects of symmetry-breaking; the new terms we have added to the Lagrangian *do* affect the three-particle Green's functions. It is just that these effects are not divergent, and hence do not require counterterms. If we look at the equations that define the bare masses and coupling constants, discussed in the preceding section, we see that another way of stating this result is to say that the constraint that the internal symmetry be broken only by the bare masses, while the bare coupling constants remain symmetric, does not introduce any divergences. (Unfortunately it is the opposite case – equal bare masses but asymmetric coupling (to electromagnetism) – that is of greatest physical interest, for this is the problem of the electromagnetic mass differences within isotopic multiplets. Alas, we have to go, one way or another, beyond conventional renormalized field theory, to solve this problem.)

The most important special case of Symanzik's rule is the renormalization of the outstanding example of a Lagrangian field theory obeying PCAC, the σ model. This can be characterized as the theory of the interactions of pions, sigmas (scalar isoscalar mesons) and nucleons, such that the chiral symmetry group $SU_2 \times SU_2$ is broken only by terms of dimension one (i.e., linear in the σ field). Symanzik's rule then immediately says that this model is renormalizable.[5]

5   Symmetry and symmetry-breaking: currents

Field theories with internal symmetries have the famous feature of possessing conserved currents, and frequently the matrix elements of these currents are objects of great physical interest (e.g. electromagnetic form factors). These currents are typically bilinear forms in *unrenormalized* fields and their derivatives. Thus, one would naively expect them to be doubly divergent – divergent because the unrenormalized fields are themselves divergent, and divergent also because we are bringing two fields together at the same space-time point. Thus the following result is as surprising as it is beautiful:

Fourth result. In a renormalizable field theory with internal symmetry, the matrix elements of the conserved currents associated with the symmetry are cutoff-independent in every order of perturbation theory.[8]

To prove this result we shall need two pieces of information. Firstly, we need to know how to compute the Green's functions for one current and a string of fields in a Lagrangian way, so we can apply Hepp's theorem, which is about Lagrangians. Fortunately, there is a standard trick for doing this: let j_μ be the current, and let $\mathscr{I}_\mu(x)$ be an arbitrary c-number function of space and time. Change the Lagrangian of the theory by adding to it an extra term:

$$\mathscr{L} \to \mathscr{L} + \mathscr{I}_\mu(x) j^\mu. \qquad (9)$$

Compute all Green's functions to first order in the added term, and then functionally differentiate with respect to $\mathscr{I}_\mu$. The result is the Green's function with a current inserted.

Secondly, we need the Ward identities for conserved currents. I have discussed these in some detail in my other lectures at this school. Here we need the Ward identities only in the somewhat sketchy form depicted in Fig. 5. The blob on the left is a Green's function for one current and a string of Bose and Fermi fields, represented by the solid lines, without and with arrows. The current is represented by a wiggly line; it carries momentum k, and vector index μ. The right-hand side of the Ward identity is some linear combination of Green's functions without a current, represented by the blob on the right.

Fig. 5

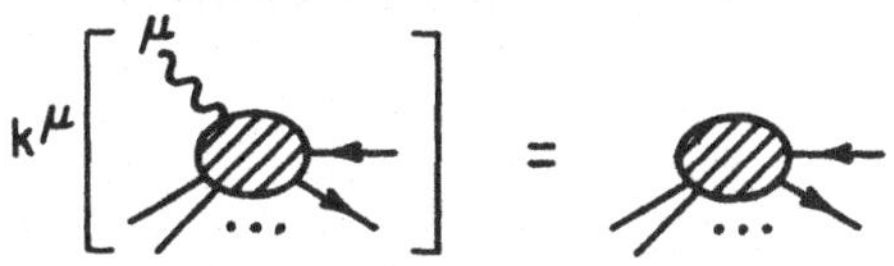

A crucial property of this equation is that it involves the same number of fields on the right as on the left, and is therefore true for any normalization of the fields that respects the internal symmetry. In particular, it is true for the intermediate fields we have been using, and therefore, for this choice of fields, the right-hand side of the Ward identity is independent of the cutoff in the limit of large cutoff. This is the essential fact we will use in the sequel.

Now let us begin counting divergences. Let us, in the manner of eq. (9), add to the Lagrangian an extra term

$$\mathscr{L} \to \mathscr{L} + \mathscr{S}_\mu^{(x)} \left(\sum_{i,j} \alpha_{ij} \bar{\psi}_i \gamma^\mu \psi_j + \sum_{i,j} \beta_{ij} \psi_i \gamma^\mu \gamma^5 \psi_j \right.$$
$$\left. + \sum_{i,j} \gamma_{ij} \phi_i \partial^\mu \phi_j + \sum_i \varepsilon_i \partial^\mu \phi_i \right), \qquad (10)$$

where the αs, βs, γs, and εs are numerical coefficients, and the sums run over all the Fermi (or Bose) fields in the theory.[9] The interactions that give us the Green's functions for the conserved currents are certainly of this form, with special choices for the numerical coefficients. However, for the moment, let us consider a general interaction of the form (10), without asking whether or not it is associated with a conserved current. Now let us follow the renormalization procedure for this new interaction (but only going to first order in $\mathscr{S}_\mu$). Since (10) is the most general Lorentz-covariant interaction linear in $\mathscr{S}_\mu$ and of dimension three or less, the counterterms induced will also be of the form (10). That is to say, starting with any interaction of the form (10), we can obtain a cutoff-independent interaction, (i.e. one that leads to cutoff-independent Green's functions; the actual numerical coefficients in the interaction will be, of course, cutoff-dependent) order by order in renormalized perturbation theory, by appropriately adjusting the numerical coefficients.

We can choose a certain subset of these interactions – say, those that are generated by starting with interactions (10) for which all but one of the numerical coefficients vanishes – as a linearly independent set. Then any interaction of the form (10) is a linear combination of these with some coefficients. The value of these coefficients is completely determined by certain terms in the Taylor expansions of certain Green's functions about the point zero. The relevant Green's functions, and their expansions, are shown in Fig. 6, where the latin letters label the fields. The one-to-one correspondence between these coefficients and the terms in (10) is evident.

Fig. 6

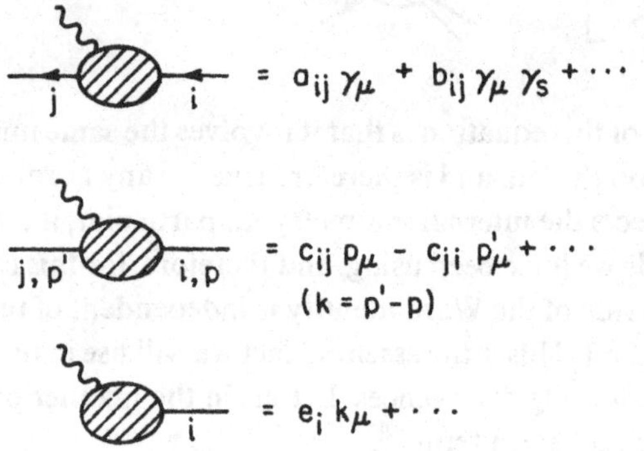

$$= a_{ij}\,\gamma_\mu + b_{ij}\,\gamma_\mu\,\gamma_5 + \cdots$$

$$= c_{ij}\,p_\mu - c_{ji}\,p'_\mu + \cdots$$
$$(k = p' - p)$$

$$= e_i\,k_\mu + \cdots$$

If, for the particular case of the Green's functions of a conserved current, we can show that these expansion coefficients are cutoff-independent, we will have shown that these Green's functions are linear combinations of cutoff-independent Green's functions with cutoff-independent coefficients, and we will have the desired result. But this is just where the Ward identities come in, for they tell us that k^μ dotted into the expansions shown in Fig. 6 must be cutoff-independent, and it is trivial to check that this is enough to tell us that the coefficients themselves are cutoff-independent.

(Please note that if we had had to go to higher orders in the Taylor expansion, the Ward identities would not have been sufficient. For example, they tell us nothing about the coefficient of the following term which can occur in the expansion of the second line of Fig. 6:

$$p_\mu k^2 - k_\mu p \cdot k,$$

because k^μ dotted into this expression vanishes. We need the divergence-counting of the renormalization procedure to tell us that all possible divergences are controlled if we can control only a few terms in the Taylor expansion. Only then can we use the Ward identities to control those terms.)

This completes the proof of our fourth result.

It is now fairly trivial to get a generalization.

Fifth result. The matrix elements of internal symmetry currents are cutoff-independent even if the symmetry is broken, provided: (1) it is broken in the manner described in the third result, that is to say, by terms of dimension three or less; and (2) the theory possesses no Bose fields with the same internal-symmetry transformation properties as the symmetry-breaking terms of dimension three. (This is a slight generalization of a result of Preparata and Weisberger.[10]

Here we proceed just as we did when establishing the third result. We treat the symmetry-breaking as a perturbation, and ask if it can introduce new divergences into current Green's functions – that is to say, whether it can induce new counterterms in the interaction (10). Since the symmetry-breaking has $\delta \leqslant -1$, and since (10) has dimension three or less, these new counterterms, if they exist, must be of dimension two or less. Thus, they must be proportional to the gradient of a Bose field. But such terms are excluded by hypothesis (2) above; they have the wrong internal-symmetry transformation properties.

Notes and references

1. Sometimes I will make further cheats. I will warn you about them in notes like this.
2. N. N. Bogoliubov and D. V. Shirkov: *Introduction to the Theory of Quantized Fields* (Interscience, 1959), especially Chapter IV and references contained therein.
3. K. Hepp: *Comm. Math. Phys.* 1, 95 (1965).
4. A cheat: we will treat this theorem as if it had been proved for general cutoff procedures; in fact it has been proved only for a restricted class of cutoffs.
5. Cheating again! Here I am blatantly ignoring the fact that the σ-model displays the Goldstone phenomenon, and that we are, therefore, not perturbing about the solution with manifest symmetry, but the one with a Goldstone boson. This cheat is not so bad, though. What we are really interested in is whether the counterterms spoil the Ward identities of chiral symmetry; these are independent of whether we are in the manifest-symmetry mode or in the Goldstone mode. See B. W. Lee, *Nucl. Phys.* B9, 649 (1969).
6. K. Symanzik in *Fundamental Interactions at High-Energies*, ed. by A. Perlmutter *et al.* (Gordon and Breach, 1970)
7. After it was done first by Symanzik.
8. Remember, we are discussing theories without vector mesons. The result is not true if the theory contains vector mesons with the same quantum numbers as the conserved currents, as does quantum electrodynamics.

9. As the γ_5 may indicate, these arguments work for chiral symmetries as well as for internal symmetries in the more usual sense. This may disturb those of you who know that the Ward identities for chiral theories sometimes contain anomalies, but don't worry – those with only one current have no anomalies, and those are the only ones we are using.
10. G. Preparata and W. Weisberger: *Phys. Rev.* **175**, 1973 (1968).

5

Secret symmetry: an introduction to spontaneous symmetry breakdown and gauge fields
(1973)

1   Introduction

Here are some long-standing problems in particle theory:

(1) How can we understand the hierarchical structure of the fundamental interactions? Are the strong, medium strong (i.e. SU(3)-breaking), electromagnetic, and weak interactions truly independent, or is there some principle that establishes connections between them?

(2) How can we construct a renormalizable theory of the weak interactions, one which reproduces the low-energy successes of the Fermi theory but predicts finite higher-order corrections?

(3) How can we construct a theory of electromagnetic interactions in which electromagnetic mass differences within isotopic multiplets are finite?

(4) How can we reconcile Bjorken scaling in deep inelastic electroproduction with quantum field theory? The SLAC–MIT experiments seem to be telling us that the light-cone singularities in the product of two currents are canonical in structure; ordinary perturbation theory, on the other hand, tells us that the canonical structure is spoiled by logarithmic factors, which get worse and worse as we go to higher and higher orders in the perturbation expansion. Are there any theories of the strong interactions for which we can tame the logarithms, sum them up and show they are harmless?

Enormous progress has been made on all of these problems in the last few years. There now exists a large family of models of the weak and electromagnetic interactions that solve the second and third problem, and we have discovered a somewhat smaller family of models of the strong interactions that solve the fourth problem. As we shall see, the structure of these models is such that we are beginning to get ideas about the solution to the (very deep) first problem; connections are beginning to appear in unexpected places, and an optimist might say that we are on the road to the first truly unified theory of the fundamental interactions. All of these marvelous developments are based upon the ideas of spontaneous symmetry breakdown and gauge fields, the subjects of these lectures.

Honesty compels me to moderate the sales-pitch of the last paragraph by pointing out that there is a fifth long-standing problem with which these theories have not yet made contact:

(5) How do we explain experiments?

We can see the reason for this embarrassing lacuna if we think a little bit more closely about the second problem, constructing a renormalizable theory of the weak interactions. At the moment, there is a plethora of such theories; they all predict that higher-order weak effects are finite, and they all predict that they are small. To find which, if any, of these theories is correct requires precision measurements of higher-order weak effects (preferably purely leptonic ones, so the strong interactions don't corrupt our predictions); these are hard to come by. Phrased another way, the Fermi theory is obviously dead wrong, because it predicts infinite higher-order corrections, but it is experimentally nearly perfect, because there are few experiments for which lowest-order Fermi theory is inadequate. Likewise for electromagnetic mass differences within isotopic multiplets: to make the differences finite, we need only to tame the high-energy behaviour of self-mass integrals; to actually compute them, though, we have to know the integrals at all energies, including the low-energy region where the strong interactions are dominant (and incalculable).

These lectures are intended as an introduction to the basic ideas of spontaneous symmetry breakdown and gauge fields, not as a survey of all the work done to date, and there are some important topics that I will not discuss at all. In particular, I will not touch at all upon the important subject of model-building; indeed, in order to simplify my examples as much as possible, I will barely mention theories involving fermions at all. Also, although I will try and make the renormalizability of the theories we discuss plausible, I will have no time to go into the guts of the renormalization problem, and therefore will say nothing about the beautiful dimensional regularization procedure of Veltman and 't Hooft, nor about the non-Abelian generalizations of the Ward identities of quantum electrodynamics, the Slavnov identities.[1]

The organization of these lectures is as follows: Section 2 is a discussion of spontaneous symmetry breakdown, Goldstone bosons, gauge fields, and the Higgs phenomenon in the simplest context, that of classical field theory. Section 3 shows how these ideas can be extended to quantum field theory in such a way that the classical reasoning of the previous section becomes the first term in a systematic quantum expansion. The important concept of the effective potential makes its first appearance here, and its properties are discussed at length. However, an important part of the quantization program is postponed: the quantization of gauge fields. This gaping hole in the arguments of Section 3 is filled in the next two sections. Section 4 is an introduction to functional integration as a method of quantization, and Section 5 is an application of this method to gauge fields, following the ideas of Faddeev and Popov. I have tried to make Section 4 as self-contained as possible, so it may be useful to the reader who wants to learn functional integration, even if he is uninterested in the other topics of these lectures. Section 6 takes off in a new direction and explores the asymptotic properties of gauge field theories. It includes a brief review of the renormalization group.

I have learned much from conversations with Ludwig Faddeev, Howard Georgi, Sheldon Glashow, Jeffrey Goldstone, David Gross, Benjamin Lee, David Politzer, Gerard 't Hooft, Tini Veltman, Erick Weinberg, Steven Weinberg, and Frank Wilczek. Many authors who have made major contributions to this subject (including a large subset of the above) are inadequately represented in the references at the end of these lectures, because of my eccentric choice of topics and methods of approach; to these I apologize, as I do to those whom I have omitted through ignorance.[1]

2 Secret symmetries in classical field theory

2.1 *The idea of spontaneous symmetry breakdown*

In general, there is no reason why an invariance of the Hamiltonian of a quantum-mechanical system should also be an invariance of the ground state of the system. Thus, for example, the nuclear forces are rotationally invariant, but this does not mean that the ground state of a nucleus is necessarily rotationally invariant (i.e. of spin zero). This is a triviality for nuclei, but it has highly non-trivial consequences if we consider systems which, unlike nuclei, are of infinite spatial extent. The standard example is the Heisenberg ferromagnet, an infinite crystalline array of spin-$\frac{1}{2}$ magnetic dipoles, with spin–spin interactions between nearest neighbors such that neighboring dipoles tend to align. Even though the Hamiltonian is rotationally invariant, the ground state is not; it is a state in which all the dipoles are aligned in some arbitrary direction, and is infinitely degenerate for an infinite ferromagnet. A little man living inside such a ferromagnet would have a hard time detecting the rotational invariance of the laws of nature; all his experiments would be corrupted by the background magnetic field. If his experimental apparatus interacted only weakly with the background field, he might detect rotational invariance as an approximate symmetry; if it interacted strongly, he might miss it altogether; in any case, he would have no reason to suspect that it was in fact an exact symmetry. Also, the little man would have no hope of detecting directly that the ground state in which he happens to find himself is in fact part of an infinitely degenerate multiplet. Since he is of finite extent (this is the technical meaning of 'little'), he can only change the direction of a finite number of dipoles at a time; but to go from one ground state of the ferromagnet to another, he must change the directions of an infinite number of dipoles – an impossible task.

At least at first glance, there appears to be nothing in this picture that can not be generalized to relativistic quantum mechanics. For the Hamiltonian of a ferromagnet, we can substitute the Hamiltonian of a quantum field theory; for rotational invariance, some internal symmetry; for the ground state of the ferromagnet, the vacuum state; and for the little man, ourselves. That is to say, we conjecture that the laws of nature may possess symmetries which are not manifest to us because the vacuum state is not invariant under them.[2] This situation is usually called 'spontaneous breakdown of symmetry'. The terminology is slightly deceptive, because the symmetry is not really broken, merely hidden, but we'll use it anyway.

We will begin by investigating spontaneous symmetry breakdown in the case of classical field theory. For simplicity, we will restrict ourselves to theories involving a set of n real scalar fields, which we will assemble into a real n-vector, ϕ, with Lagrange density[3]

$$\mathscr{L} = \frac{1}{2}(\partial_\mu \phi) \cdot (\partial^\mu \phi) - U(\phi), \tag{2.1}$$

where U is some function of the ϕs, but not of their derivatives. We will treat these theories purely classically, but use quantum-mechanical language; thus, we will call the state of lowest energy 'the vacuum', and refer to the quantities which characterize the spectra of small oscillations about the vacuum as 'particle masses'. For any of these theories, the energy density is

$$\mathscr{H} = \frac{1}{2}(\partial_0 \phi)^2 + \frac{1}{2}(\nabla \phi)^2 + U(\phi). \tag{2.2}$$

Thus the state of lowest energy is one for which the value of ϕ is a constant, which we denote by $\langle \phi \rangle$. The value of $\langle \phi \rangle$ is determined by the detailed dynamics of the particular theory under investigation, that is to say, by the location of the minimum (or minima) of the potential U. Sticking to our policy of using quantum language, we will call $\langle \phi \rangle$ 'the vacuum expectation value of ϕ'.

Within this class of theories, it is easy to find examples for which symmetries are either manifest or spontaneously broken. The simplest one is the theory of a single field for which the potential is

$$U = \frac{\lambda}{4!} \phi^4 + \frac{\mu^2}{2} \phi^2, \tag{2.3}$$

where λ is a positive number and μ^2 (despite its name) can be either positive or negative. This theory admits the symmetry

$$\phi \rightarrow -\phi. \tag{2.4}$$

If μ^2 is positive, the potential is as shown in Fig. 1. The vacuum is at $\langle \phi \rangle$ equals zero, the symmetry is manifest, and μ^2 is the mass of the scalar meson. If μ^2 is negative, though, the situation is quite different; the potential is as shown in Fig. 2. In this case, it is convenient to introduce the quantity

$$a^2 = -6\mu^2/\lambda, \tag{2.5}$$

and to rewrite the potential as

$$U = \frac{\lambda}{4!} (\phi^2 - a^2)^2, \tag{2.6}$$

plus an (irrelevant) constant. It is clear from this formula, and also from the figure, that the potential now has two minima, at $\phi = \pm a$. Because of the symmetry (2.4), which one we choose as the vacuum is irrelevant to the resulting physics; however, whichever one we choose, the symmetry is spontaneously broken. Let us choose $\langle \phi \rangle = a$. To investigate physics

Fig. 1

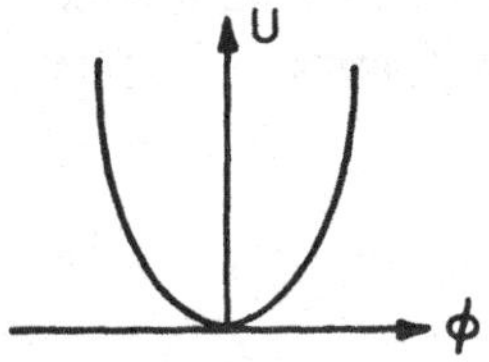

Fig. 2

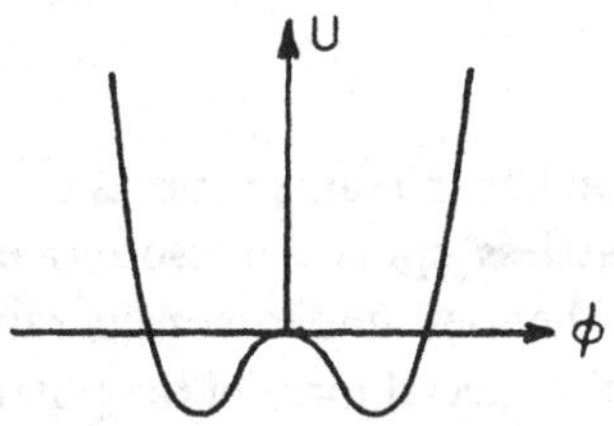

about the asymmetric vacuum, let us define a new field

$$\phi' = \phi - a. \tag{2.7}$$

In terms of the new ('shifted') field,

$$U = \frac{\lambda}{4!} (\phi'^2 + 2a\phi')^2$$

$$= \frac{\lambda}{4!} \phi'^4 + \frac{\lambda a}{6} \phi'^3 + \frac{\lambda a^2}{6} \phi'^2. \tag{2.8}$$

We see that the true mass of the meson is $\lambda a^2/3$. Note that a cubic meson self-coupling has appeared as a result of the shift, which would make it hard to detect the hidden symmetry (2.4) directly.

2.2   *Goldstone bosons in an Abelian model*

A new phenomenon appears if we consider the spontaneous breakdown of continuous symmetries. Let us consider the theory of two scalar fields, A and B, with

$$U = \frac{\lambda}{4!} [A^2 + B^2 - a^2]^2. \tag{2.9}$$

This theory admits a continuous group of symmetries isomorphic to the two-dimensional rotation group, SO(2):

$$A \to A \cos \omega + B \sin \omega,$$
$$B \to -A \sin \omega + B \cos \omega. \tag{2.10}$$

The minima of the potential lie on the circle

$$A^2 + B^2 = a^2. \tag{2.11}$$

Just as before, which of these we choose as the vacuum is irrelevant, but whichever one we choose, the SO(2) internal symmetry is spontaneously broken. Let us choose

$$\langle A \rangle = a, \quad \langle B \rangle = 0. \tag{2.12}$$

As before, we shift the fields,

$$\phi' = \phi - \langle \phi \rangle, \tag{2.13}$$

and find

$$U = \frac{\lambda}{4!} (A'^2 + B'^2 + 2aA')^2. \tag{2.14}$$

Expanding this, we see that the A-meson has the same mass as before, but the B-meson is massless. Such a massless spinless meson is called a Goldstone boson;[4] for the class of theories under consideration, its appearance does not depend at all on the special form of the potential U, but is a consequence only of the spontaneous breakdown of the continuous SO(2) symmetry group (2.10).

To show this, let us introduce 'angular variables',

$$A = \rho \cos \theta,$$
$$B = \rho \sin \theta. \tag{2.15}$$

In terms of these variables, (2.10) becomes

$$\rho \rightarrow \rho$$
$$\theta \rightarrow \theta + \omega, \tag{2.16}$$

and the Lagrange density becomes

$$\mathscr{L} = \frac{1}{2}(\partial_\mu \rho)^2 + \frac{1}{2} \rho^2 (\partial_\mu \theta)^2 - U(\rho). \tag{2.17}$$

In terms of these variables, SO(2) invariance is simply the statement that U does not depend on θ. The transformation to angular variables is, of course, ill-defined at the origin, and this is reflected in the singular form of the derivative part of the Lagrange density (2.17). However, this is of no interest to us, since we wish to do perturbation expansions not about the origin, but about an assumed asymmetric vacuum. With no loss of generality, we can assume this vacuum is at $\langle \rho \rangle = a$, $\langle \theta \rangle = 0$. Introducing shifted fields as before,

$$\rho' = \rho - a,$$
$$\theta' = \theta, \tag{2.18}$$

we find

$$\mathscr{L} = \frac{1}{2}(\partial_\mu \rho')^2 + \frac{1}{2}(\rho' + a)^2(\partial_\mu \theta')^2 - U(\rho' + a). \tag{2.19}$$

It is clear from this expression that the θ-meson is massless, just because the θ-field enters the Lagrangian only through its derivatives.

This can also be seen purely geometrically, without writing down any formulae. If the vacuum is not invariant under SO(2) rotations, then there is a curve passing through the vacuum along which the potential is constant; this is the curve of points obtained from the vacuum by SO(2) rotations – in terms of our variables, the curve of constant ρ. If we expand the potential around the vacuum, no terms can appear involving the variable that measures displacement along this curve – the θ variable. Hence we always have a massless meson.

2.3 *Goldstone bosons in the general case*

This argument can easily be generalized to the spontaneous breakdown of a general continuous internal symmetry group. I will give the generalization using somewhat more mathematical apparatus than is really necessary, in order to establish some notation that will be useful to us later on. Let us assume that we have a set of n real fields, ϕ, such that the potential is invariant under a group of transformations

$$\phi \to e^{T_a \omega^a}\phi,\tag{2.20}$$

where the Ts are a set of N real antisymmetric matrices, the group generators, the ωs are arbitrary real parameters, and the sum over repeated indices is implied. The associated infinitesimal transformations are

$$\delta\phi = T_a \delta\omega^a \phi.\tag{2.21}$$

Since the Ts are group generators, they obey the relations

$$[T_a, T_b] = c_{abc}T_c,\tag{2.22}$$

where the cs are the structure constants of the group. If we choose the Ts to be orthonormal (in the trace norm), then c^{abc} is completely antisymmetric. Invariance of the Lagrange density (2.1) implies that

$$U(\phi) = U(e^{T_a \omega^a}\phi).\tag{2.23}$$

Now let us consider the subgroup of (2.20) that leaves $\langle\phi\rangle$, the minima of U, invariant. Depending on the structure of U, this may be anything from the trivial identity subgroup (all symmetries spontaneously broken) to the full group (no symmetries spontaneously broken). In any case, though, we can always choose our group generators such that this subgroup is generated by the first M generators, where $N \geqslant M \geqslant 0$. In equations,

$$T_a\langle\phi\rangle = 0, \quad a \leqslant M.\tag{2.24}$$

By definition, the remaining $(N - M)$ generators do not leave $\langle\phi\rangle$ invariant; thus we have, passing through $\langle\phi\rangle$, an $(N - M)$-dimensional surface of constant U. Thus, by the same arguments as before, the theory must contain $(N - M)$ massless spinless mesons, one for each spontaneously broken infinitesimal symmetry. (Note that I say 'spinless', not 'scalar' or 'pseudoscalar'. The mesons may be either scalar or pseudoscalar, depending on the parity-transformation properties of the spontaneously broken generators; they may even have no well-defined parity at all, if parity is itself spontaneously broken, or if the original Lagrangian is not parity conserving.)

These mesons are called Goldstone bosons, and what we have proved in the preceding paragraph is a special case of Goldstone's theorem.[4] The theorem can be proved in much greater generality: given a field theory obeying the usual axioms (Lorentz invariance, locality, Hilbert space with positive-definite inner product, etc.), if there is a local conserved current (the axiomatic version of the statement that the Lagrangian is invariant under some continuous transformation) such that the space integral of its time component does not annihilate the vacuum state, then the theory necessarily contains a massless spinless meson, with the same internal-symmetry and parity properties as the time component of the current.[5]

At first glance, Goldstone's theorem seems to be a killing blow to the idea that spontaneous breakdown (at least of continuous symmetries) is at work in the real world, for there is not a smidgen of experimental evidence for the existence of massless spinless mesons. However, there is one loophole: there do exist perfectly respectable field theories which do not obey the usual axioms. These are gauge field theories, of which quantum electrodynamics is the most familiar. There is no gauge in which quantum electrodynamics obeys all the axioms simultaneously; if we quantize in a covariant gauge, the theory contains states of negative norm, associated with the longitudinal photons; if we quantize in a gauge in which the theory has only states of positive norm, such as radiation gauge, the theory is not covariant. We will now investigate this loophole in more detail.

2.4 The Higgs phenomenon in the Abelian model

I will begin by reviewing the minimal-coupling prescription of ordinary quantum electrodynamics, and its connection with gauge invariance. Let ϕ be a set of fields (not necessarily real and spinless), with dynamics determined by a Lagrange density, $\mathscr{L}(\phi, \partial_\mu \phi)$. Let $\mathscr{L}$ be invariant under a one-parameter group of transformations,

$$\phi \to e^{iQ\omega}\phi, \tag{2.25}$$

where Q is a Hermitian matrix, called the charge matrix. (Conventionally, a set of complex basis fields of definite charge is chosen, so that Q is diagonal. However, for our purposes, it will be more convenient to choose a real set of fields, so that iQ is a real antisymmetric matrix, like the Ts in Eq. (2.20).) The associated infinitesimal transformation is

$$\delta\phi = iQ\phi\delta\omega. \tag{2.26}$$

Now let us consider transformations of the same form as Eq. (2.26), but with $\delta\omega$ space-time dependent (gauge transformations). Our theory is not invariant under these transformations, since

$$\delta(\partial_\mu \phi) = iQ(\partial_\mu \phi)\delta\omega + iQ\phi\partial_\mu(\delta\omega), \tag{2.27}$$

and the second term spoils the invariance. We can take care of this, though, by enlarging the theory and introducing a new field, A_μ, the gauge field, that transforms according to

$$\delta A_\mu = -\frac{1}{e}\partial_\mu(\delta\omega), \tag{2.28}$$

where e is a free parameter, called the electric charge. If we now define

$$D_\mu \phi = \partial_\mu \phi + ieQA_\mu \phi, \tag{2.29}$$

then

$$\delta D_\mu \phi = iQ\phi \delta \omega, \tag{2.30}$$

and

$$\mathcal{L}(\phi, D_\mu \phi) \tag{2.31}$$

is gauge invariant. $D_\mu \phi$ is called the gauge-covariant derivative, or sometimes just the covariant derivative. Of course, the expression (2.31) by itself can not be the total Lagrange density for a physically interesting theory; it contains no terms proportional to the derivatives of A_μ, so if we vary it with respect to A_μ we obtain, not true equations of motion, but equations of constraint. To make the gauge field a true dynamical variable, we must add a term involving derivatives; the simplest gauge-invariant choice is a term proportional to $(F_{\mu\nu})^2$, where

$$F_{\mu\nu} = \partial_\mu A_\nu - \partial_\nu A_\mu. \tag{2.32}$$

By convention, A_μ is normalized such that the final Lagrange density is

$$-\frac{1}{4}(F_{\mu\nu})^2 + \mathcal{L}(\phi, D_\mu \phi). \tag{2.33}$$

This is just the usual Lagrange density of minimally-coupled electrodynamics, and it has the usual physical interpretation (charged particles, massless photons, etc.), if the dynamics of the ϕ-fields are such that the symmetry (2.25) does not suffer spontaneous breakdown. But what happens if the symmetry is spontaneously broken, as in (2.10)?

This question is most easily answered if we use the angular variables defined by Eq. (2.15). We can avoid some tedious algebra by observing that Eq. (2.29) can be rewritten as

$$D_\mu \phi = \partial_\mu \phi + eA_\mu \frac{\delta \phi}{\delta \omega}. \tag{2.34}$$

In this form, it can be directly applied to the angular variables. From Eq. (2.16), it follows that

$$D_\mu \rho' = \partial_\mu \rho',$$

and

$$D_\mu \theta' = \partial_\mu \theta' + eA_\mu. \tag{2.35}$$

Applying this to Eq. (2.19) we obtain

$$\mathcal{L} = \frac{1}{4}(\partial_\mu A_\nu - \partial_\nu A_\mu)^2 + \frac{1}{2}(\partial_\mu \rho')^2$$

$$+ \frac{1}{2}(\rho' + a)^2(\partial_\mu \theta + eA_\mu)^2 - U(\rho' + a). \tag{2.36}$$

It is hard to directly read off the predictions of this expression for small oscillations about the vacuum, because of the presence of quadratic cross terms, terms proportional to $A_\mu \partial^\mu \theta'$. However, these can be eliminated by introducing the new variable

$$c_\mu = A_\mu + e^{-1} \partial_\mu \theta. \tag{2.37}$$

In terms of this,

$$\mathscr{L} = -\frac{1}{4} (\partial_\mu c_\nu - \partial_\nu c_\mu)^2 + \frac{1}{2} (\partial_\mu \rho')^2 + \frac{e^2}{2} (\rho' + a)^2 (c_\mu)^2 - U(\rho' + a).$$

$$\tag{2.38}$$

Since the quadratic part of the Lagrangian is now in diagonal form, we can read off the eigenmodes for small vibrations about the ground state, or, in the quantum language we have been using, the particle spectrum. We see that there is a massive scalar meson associated with the ρ'-field, whose mass depends on the form of U. There is also a massive vector meson associated with the c-field, with mass given by

$$m_C^2 = e^2 a^2. \tag{2.39}$$

But the Goldstone boson, the θ-field, has completely disappeared! This seems a little less preposterous if we count degrees of freedom. A massive vector meson has three degrees of freedom, the three spin states of a spin-one particle, while a massless vector meson has only two, the two helicity states of the photon. What has happened is that the two degrees of freedom of the massless gauge field and the one degree of freedom of the Goldstone boson have combined together to make the three degrees of freedom of the c-field. The vector meson has eaten the Goldstone boson and grown heavy.

This magic trick was discovered by Peter Higgs, and is called the Higgs phenomenon. (Actually, the terminology is unfair, since the phenomenon was discovered independently by several other investigators, but we will use it anyway, since it is awkward to talk of the Brout–Englert–Guralnik–Hagen–Higgs–Kibble phenomenon.)[6] We can gain further insight into the Higgs phenomenon if we remember the motivation for the minimal-coupling prescription – gauge invariance.

Gauge invariance tells us that our theory is invariant under transformations of the form

$$\theta \to \theta + \omega, \tag{2.40}$$

with ω an *arbitrary* function of space and time. In particular, this means we can choose ω to be minus θ, that is to say, pick our gauge in such a way that the θ-field is identically zero. The reason the Goldstone boson disappears in the gauge-invariant theory is that it was never there in the first place; the degree of freedom that would be associated with the Goldstone boson is a mere gauge phantom, an object that can be gauged away, like a longitudinal photon.

It is now clear how to extend the Higgs phenomenon to a general internal symmetry group, like (2.20). We merely have to add extra degrees of freedom (gauge fields) to promote the whole internal symmetry group to a gauge group. If we can do this, then we can always gauge away the degrees of freedom that would correspond to Goldstone bosons, and kill the Goldstone bosons before they are born. To carry out this scheme, though, we need first to develop the theory of gauge fields for general internal symmetry groups.

2.5   *Yang–Mills fields and the Higgs phenomenon in the general case*

How do we make a general internal symmetry group a gauge group? We will follow closely our discussion of electromagnetism. We begin with a theory that is invariant under transformations of the form (2.21),

$$\delta\phi = T_a \delta\omega^a \phi. \tag{2.21}$$

Now let us consider transformations of the same form, but with $\delta\omega^a$ space-time dependent. Our theory is not invariant under these transformations, since

$$\delta(\partial_\mu\phi) = T_a \delta\omega^a \partial_\mu\phi + T_a(\partial_\mu\delta\omega^a)\phi, \tag{2.41}$$

and the second term spoils the invariance. We will try to take care of this by introducing a set of N gauge fields, A_μ^a, one for each group generator, and defining the covariant derivatives

$$D_\mu\phi = \partial_\mu\phi + g T_a A_\mu^a \phi. \tag{2.42}$$

where g, like e, is a free parameter. (For the moment, we will postpone the question of whether we can choose different gs for different gauge fields.) We wish to define the transformation properties of the gauge fields such that

$$\delta(D_\mu\phi) = T_a \delta\omega^a D_\mu\phi. \tag{2.43}$$

It is easy to see that this implies that

$$\delta A_\mu^a = c^{abc}\delta\omega^b A_\mu^c - \frac{1}{g}\partial_\mu\delta\omega^a, \tag{2.44}$$

where the cs are the structure constants of the group, defined in Eq. (2.22). (Both terms in this expression are easy to understand. The second term is a trivial generalization of the electromagnetic gauge transformation, Eq. (2.28). The first term is necessary to insure the invariance of the gauge-field couplings under space-time *independent* transformations; it states that, under such transformations, the gauge fields transform like the group generators. (E.g., if the gauge group is isospin, the gauge fields must form an isovector.)) It follows from Eq. (2.43) that

$$\mathscr{L}(\phi, D_\mu \phi) \tag{2.45}$$

is gauge invariant.

It is a bit harder to see what is the generalization of the free electromagnetic Lagrange density, $(F_{\mu\nu})^2$. The trick is to observe that, for electromagnetism

$$(D_\mu D_\nu - D_\nu D_\mu)\phi = iQ F_{\mu\nu}\phi \tag{2.46}$$

From this equation, the gauge invariance of $F_{\mu\nu}$ follows directly. In our case,

$$(D_\mu D_\nu - D_\nu D_\mu)\phi = T_a F^a_{\mu\nu}\phi, \tag{2.47}$$

where

$$F^a_{\mu\nu} = \partial_\mu A^a_\nu - \partial_\nu A^a_\mu + g c^{abc} A^b_\mu A^c_\nu. \tag{2.48}$$

From Eq. (2.46), it follows directly that $F^a_{\mu\nu}$ is, not gauge-invariant, but gauge-covariant,

$$\delta F^a_{\mu\nu} = c^{abc}\delta\omega^b F^c_{\mu\nu}. \tag{2.49}$$

However, the quadratic form $(F^a_{\mu\nu})^2$ is gauge-invariant, and therefore the generalization of the electromagnetic Lagrange density (2.33) is

$$-\frac{1}{4}(F^a_{\mu\nu})^2 + \mathscr{L}(\phi, D_\mu \phi). \tag{2.50}$$

The first Lagrange density of this type (for the special case of the isospin group) was constructed by Yang and Mills; for this reason non-Abelian gauge fields are frequently called Yang–Mills fields.[7]

Note that for non-Abelian gauge fields, in contrast to electromagnetism, there is a non-trivial interaction even in the absence of the ϕ-fields, because of the non-linear form of $F^a_{\mu\nu}$. There is a good physical reason for this, which is most easily seen by going to a particular example. Let us imagine that the gauge group is isospin. Just as the photon couples to every field that carries non-zero charge, so the I_z gauge meson, for example, must couple to every field that carries non-zero I_z. But among these fields are the other two members of the isotriplet of gauge fields. (It is for precisely the same reason that gravitation is inherently non-linear; the gravitational field couples to everything that carries energy density, including the gravitational field itself.)

Now let us return to the postponed question of whether we can have different coupling constants for different gauge fields. If the gauge group is simple (like SU(2) or SU(3)), the generators of the group, and therefore the gauge fields, transform irreducibly under the action of the group; therefore they must all have the same coupling constant. However, if the gauge group is a product of simple factors (like SU(2)⊗SU(2)), then the generators of different factors never mix with each other under the action of the group, and the associated gauge fields can have different coupling constants. Thus there are as many independent coupling constants as there are simple factors in the gauge group, and Eq. (2.44), for example, should properly be written as

$$\delta A_\mu^a = c^{abc} \delta \omega^b A_\mu^c - \frac{1}{g_a} \partial_\mu \delta \omega^a. \tag{2.44'}$$

(no sum on a), where g_a can take on different values for gauge fields associated with different factor groups.

Now that we have developed the classical theory of non-Abelian gauge fields, let us apply it to spontaneous symmetry breakdown. Since the entire internal symmetry group has been promoted to a gauge group, we can always choose our gauge such that the degrees of freedom that would become Goldstone bosons disappear. From our experience with the Abelian model, we would expect the gauge fields associated with the spontaneously broken symmetries to acquire masses. It is easy to see that the only relevant part of the Lagrange density (2.1) is the derivative term

$$\mathcal{L} = \frac{1}{2} (\partial_\mu \phi) \cdot (\partial^\mu \phi) + \dots. \tag{2.51}$$

In the presence of the gauge fields this becomes

$$\mathcal{L} = \frac{1}{2} (\partial_\mu \phi + g_a A_\mu^a T_a \phi) \cdot (\partial^\mu \phi + g_b A_\mu^b T_b \phi) + \dots. \tag{2.52}$$

When we shift the fields, this generates a mass term

$$\mathcal{L} = (g_a A_\mu^a T_a \langle \phi \rangle) \cdot (g_b A_\mu^b T_b \langle \phi \rangle) + \dots. \tag{2.53}$$

Note that gauge fields associated with symmetries that are not spontaneously broken, that is to say, those for which

$$T_a \langle \phi \rangle = 0, \tag{2.54}$$

remain massless. Thus, if we wish to have a theory of this type with a realistic particle spectrum, the entire gauge group must be spontaneously broken, except for a one-parameter subgroup. We identify this subgroup with electric charge, and the corresponding gauge field with the only observed massless vector meson, the photon.

2.6 *Summary and remarks*

(1) We have discovered a large family of field theories that display spontaneous breakdown of internal symmetries. If the spontaneously

broken symmetry is discrete, this causes no problems; however, if the symmetry is continuous, symmetry breakdown is associated with the appearance of Goldstone bosons. This can be cured by coupling gauge fields to the system and promoting the internal symmetry group to a gauge group; the Goldstone bosons then disappear and the gauge mesons acquire masses. It is pleasant to remember that, at the times of their inventions, both the theory of non-Abelian gauge fields and the theory of spontaneous symmetry breakdown were thought to be theoretically amusing but physically untenable, because both predicted unobserved massless particles, the gauge mesons and the Goldstone bosons. It was only later that it was discovered that each of these diseases was the other's cure.

(2) Everything we have done so far has been for classical field theory. One of the main tasks before us is to see to what extent the apparatus of this section can be extended into the quantum domain. We shall see that, at least for weak couplings, it survives substantially unchanged; in particular, all of the equations we have derived can be reinterpreted as the first terms in a systematic quantum expansion.

(3) We have not touched at all on theories with fermions. It is trivial that if we couple fermions to the scalar-meson systems we have discussed, either directly (through Yukawa couplings) or indirectly (through gauge field couplings), then the shift in the scalar fields will induce an apparent symmetry-violating term in the fermion part of the Lagrangian. A more interesting question is whether spontaneous symmetry breakdown can occur in a theory without fundamental scalar fields. For example, perhaps bilinear forms in Fermi fields can develop symmetry-breaking vacuum expectation values all by themselves. I will have nothing to say about this possibility here, not because it is not important, but because so little is known about it.[8] (There is one exactly soluble model without fundamental scalars that displays the full Goldstone–Higgs phenomenon. This is the Schwinger model, quantum electrodynamics of massless fermions in two-dimensional space-time.)[9]

(4) It is important to realize that we can make the effects of spontaneous symmetry breakdown as large or as small as we want, by appropriately fudging the parameters in our models. Thus, in the real world, some of the spontaneously broken symmetries of nature may be observed as approximate symmetries in the usual sense, and others may be totally inaccessible to direct observation. Also, of course, there is no objection to exact or approximate symmetries of the usual kind coexisting with spontaneously broken symmetries. Presumably symmetries such as nucleon number conservation, neither broken nor coupled to a massless gauge meson, are of this sort.

(5) All of this is very pretty, but what does it buy us? What is the practical use of the idea of spontaneous symmetry breakdown, even by the generous standards of practicality current among high-energy theoreticians? The answer to this question will be given in the next section, when we leave classical physics and turn to quantum field theory.

3   Secret renormalizability
3.1   *The order of the arguments*
We are going to plunge immediately into the study of spontaneous symmetry breakdown in quantum field theory, despite the fact that we know nothing of the properties of quantum non-Abelian gauge fields, even in the absence of spontaneous symmetry breakdown. Logically, this is not a good order in which to do things, but I would like to get to the heart of the matter as soon as possible. Thus, if you have a critical disposition, you should assume in this section that I am talking about symmetry breakdown in the presence of at most some Abelian gauge fields, and you should ignore my occasional remarks about the non-Abelian case. In any case, we will quantize non-Abelian gauge fields later on.

In this section, we will first review the elements of renormalization theory, without worrying about spontaneous symmetry breakdown. Then we will develop a formalism for handling symmetry breakdown, without worrying about renormalization. Finally, we will bring the two strands of argument together.

3.2   *Renormalization reviewed*[10]
In any non-trivial quantum field theory, divergent integrals appear in the perturbation expansion for the Green's functions. Renormalization is a procedure for removing these divergences, order by order in perturbation theory, by adding extra terms, called counterterms, to the Lagrangian that defines the theory. For example, let us consider the expansion of the proper four-point-function (i.e. the off-mass-shell scattering amplitude) in the theory defined by

$$\mathscr{L} = \frac{1}{2}(\partial_\mu \phi)^2 - \frac{1}{2}\mu^2\phi^2 - \frac{\lambda}{4!}\phi^4. \tag{3.1}$$

The first few terms in this expansion are shown in Fig. 3. All the graphs except the first correspond to divergent Feynman integrals, If we cut off the integrations at some large momentum, Λ, we obtain

$$\Gamma^{(4)} = -\lambda + a\lambda^2 \ln \Lambda + \lambda^2 f, \tag{3.2}$$

where a is a finite (i.e. cutoff-independent in the limit of large cutoff)

Fig. 3

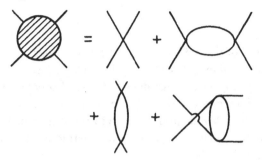

constant and f is a finite function of the external momenta. We now change the theory, by adding an extra term (the counterterm) to $\mathscr{L}$:

$$\mathscr{L} \to \mathscr{L} - \frac{a\lambda^2}{4!} \ln \Lambda \phi^4. \tag{3.3}$$

The divergent term in Eq. (3.2) is now cancelled, and the Green's function is rendered finite. Of course, the extra term in the Lagrangian must be taken into account as an internal vertex when we compute to yet higher orders, but to this order at least, everything is OK.

It turns out that the obvious generalization of this idiotically simple manipulation gets rid of all the infinities for any field theory with polynomial interactions, to any order in perturbation theory. (I ask you to take this statement, and the ones that will follow it, on trust; they are true, but very difficult to prove.) Furthermore, it is possible to give a general rule for the counterterms that occur in each order of perturbation theory. For simplicity, I will begin by giving this rule and explaining its consequences for theories involving scalar (or pseudoscalar – parity conservation will not be assumed) and Dirac bispinor fields only. Let us write the Lagrange density of our theory in the form

$$\mathscr{L} = \mathscr{L}_0 + \sum \mathscr{L}_i, \tag{3.4}$$

where $\mathscr{L}_0$ is the standard free Lagrange density, and each $\mathscr{L}_i$ is a monomial in the fields and their derivatives. To each of these terms, let us assign a dimension, d_i, according to the rule that the dimension of a scalar field is one, of a Dirac field 3/2, and of a derivative operator, one. Thus, in Eq. (3.1), the three terms have dimensions four, two, and four respectively. (Note that we do not count dimensionful parameters, like μ^2, in computing these dimensions.) Then, to any given order of perturbation theory, all divergences can be canceled with counterterms, themselves polynomials in the fields and their derivatives, whose dimensions obey the inequality

$$(d-4)\leqslant\sum n_i(d_i-4),$$ (3.5)

where n_i is the number of times $\mathscr{L}_i$ occurs in the given order.

Of course, not all counterterms allowed by the inequality (3.5) are necessary. For example, the Lagrange density (3.1) is Lorentz invariant, parity invariant, and invariant under the internal symmetry $\phi \rightarrow -\phi$. Thus, unless we are so foolish as to use a cutoff procedure which breaks these symmetries, we need never worry about counterterms which are not invariant under them.

Let us check (3.5) against our sample computation. For the Lagrange density (3.1), there is only one interaction, and its dimension is four. Thus, to $O(\lambda^2)$, the order to which we worked, the right-hand side of the inequality is zero, and there are only three counterterms of appropriate dimensions and symmetry properties:

$$\mathscr{L} \rightarrow \mathscr{L} + \frac{1}{2} A(\partial_\mu \phi)^2 - \frac{1}{2} B\phi^2 - \frac{1}{4!} C\phi^4,$$ (3.6)

where A, B, and C are cutoff-dependent. We only saw the last of these in our sample computation, but the other two are also needed in this order, to cancel the infinities in the second order self-energy (Fig. 4).

Fig. 4

But these are not only the only counterterms to second order, they are the only ones *to general order*, because no matter how many interactions we sum up, the right-hand side of the inequality is still zero. (The new interactions induced by the counterterms themselves do not affect this argument; their dimensions are also less than or equal to four.) But these three counterterms are of the same form as the three terms in the original Lagrangian; thus they can be thought of as simply readjustments of the parameters in the original theory. (More precisely, the A term can be absorbed in a rescaling of ϕ; the B and C terms are then corrections to the mass and coupling constant.)

A theory which has this property, for which all the counterterms induced by renormalization are of the same form as terms in the original Lagrangian, is said to be renormalizable. Phrased another way, a renormalizable theory is one for which all cutoff-dependence can be removed from the

Green's functions by rescaling the fields and choosing the parameters of the theory in appropriate cutoff-dependent ways. Renormalizable theories are a very small subset of the set of all quantum field theories one can write down. (Although they may exhaust the set of theories that make sense.) For example, it is clear from our inequality (or from direct computation) that any theory involving an interaction of dimension greater than four is nonrenormalizable. However, not all theories with only interactions of dimension four or less are renormalizable. For example, the theory of mesons and nucleons interacting only through a Yukawa coupling, $\bar{\psi}\gamma_5\psi\phi$, is not renormalizable, for this interaction induces a ϕ^4 counterterm, not present in the original theory. On the other hand, the same theory with both Yukawa *and* ϕ^4 interactions *is* renormalizable. (This is a somewhat stricter definition of renormalizability than the one in common use. Most people define renormalizable to mean that there are only a finite number of counterterms induced, whether or not they were all present in the original Lagrangian.)

I have said only that the counterterms are to be chosen to cancel the infinities. This obviously leaves them undetermined, in each order, up to finite additions. For renormalizable theories, these ambiguities are usually resolved by a set of equations, called renormalization conditions, which define the scales of fields and values of renormalized masses and coupling constants in terms of Green's functions evaluated at some conventionally chosen point in momentum space. Exactly how we choose these conventions will not be relevant to our immediate purposes. For nonrenormalizable theories, in the common sense (i.e. those with an infinite number of counterterms), there are an infinite number of free parameters, which is why these theories are commonly (and properly, I think) considered disgusting.

Until now, I have said nothing about vector fields. The rules I gave for assigning dimensions to fields were in fact derived from the high-energy behavior of free propagators, as one might expect, since these are obviously the properties that control the divergences of Feynman integrals. Thus, although the dimension of a massive vector field is one, in the normal sense of dimensional analysis, its propagator is

$$-i\,\frac{g_{\mu\nu} - k_\mu k_\nu/\mu^2}{k^2 - \mu^2}. \tag{3.7}$$

Because of the second term, this grows at high momentum like the propagator for the gradient of a scalar field, an object of dimension two, and our dimension-counting formula, (3.5), breaks down. In fact, most interactions of a massive vector field are nonrenormalizable. However, if the massive vector field is coupled to a conserved current, as if it were an Abelian gauge field, then we can shuffle variables to rewrite the theory in such a way that the propagator is

$$-i \frac{g_{\mu\nu} - k_\mu k_\nu / k^2}{k^2 - \mu^2}. \tag{3.8}$$

This grows just like a scalar propagator, so the dimension-counting procedure is good again. For a true (massless) Abelian gauge field, the theory may also be quantized in such a way that the propagator is of the form (3.8), (with μ^2 zero, of course). Thus, here also the dimension-counting procedure is good, as I trust you know from your experience with quantum electrodynamics. We shall see in Section 5 that this can also be done (with some complications) for non-Abelian gauge fields (but here only for the massless case).

However, even for quantum electrodynamics, dimension-counting is not sufficient to establish renormalizability. For example, Eq. (3.5) will certainly allow an $(A_\mu)^4$ counterterm (dimension four), but if we really had to introduce such a term into the Lagrangian, it would be a disaster – it would destroy gauge invariance. In QED, one shows such a term can not occur by a complicated sequence of arguments. (1) The theory is cut off in a cunning way that does not destroy gauge invariance. (2) Gauge invariance is used to establish relations between Green's functions, Ward identities. (3) The Ward identities are used to show that the possible gauge-noninvariant counterterms are not necessary. The same sequence of steps can be carried through for non-Abelian gauge theories, but the arguments are much more complicated; I will not have time to cover them in these lectures, and must refer you to the literature.[11]

3.3   *Functional methods and the effective potential*[12]

I would now like to put aside renormalization for the moment, and begin a new line of development, one that will lead (after an orgy of formalism) to a method for treating spontaneous symmetry breakdown in quantum field theory. For simplicity, in explaining the formalism, I will restrict myself to the theory of a single scalar field, ϕ, whose dynamics are described by a Lagrange density, $\mathscr{L}(\phi, \partial_\mu \phi)$. The generalization to more complicated cases is trivial. Let us consider the effect of adding to the Lagrange density a linear coupling of ϕ to an external source, $J(x)$, a c-number function of space and time:

$$\mathscr{L}(\phi, \partial_\mu \phi) \rightarrow \mathscr{L} + J(x)\phi(x). \tag{3.9}$$

The connected generating functional, $W(J)$, is defined in terms of the transition amplitude from the vacuum state in the far past to the vacuum state in the far future, in the presence of the source $J(x)$,

$$e^{iW(J)} = \langle 0^+|0^-\rangle_J. \tag{3.10}$$

We can expand W in a functional Taylor series

$$W = \sum_n \frac{1}{n!} \int d^4x_1 \dots d^4x_n G^{(n)}(x_1 \dots x_n)J(x_1) \dots J(x_n). \tag{3.11}$$

It is well known that the successive coefficients in this series are the connected Green's functions; $G^{(n)}$ is the sum of all connected Feynman diagrams with n external lines.

The classical field, ϕ_c, is defined by

$$\phi_c(x) = \frac{\delta W}{\delta J(x)}$$

$$= \left[\frac{\langle 0^+|\phi(x)|0^-\rangle}{\langle 0^+|0^-\rangle} \right]_J. \tag{3.12}$$

The effective action, $\Gamma(\phi_c)$, is defined by a functional Legendre transformation

$$\Gamma(\phi_c) = W(J) - \int d^4x J(x)\phi_c(x). \tag{3.13}$$

From this definition, it follows directly that

$$\frac{\delta\Gamma}{\delta\phi_c(x)} = -J(x). \tag{3.14}$$

This equation will shortly turn out to be critical in the study of spontaneous breakdown of symmetry. The effective action may be expanded in a manner similar to that of (3.11):

$$\Gamma = \sum_n \frac{1}{n!} \int d^4x_1 \dots d^4x_n \Gamma^{(n)}(x_1 \dots x_n)\phi_c(x_1) \dots \phi_c(x_n). \tag{3.15}$$

It is possible to show that the successive coefficients in this series are the IPI Green's functions[13] (sometimes called proper vertices); $\Gamma^{(n)}$ is the sum of all IPI Feynman diagrams with n external lines. (An IPI (one-particle-irreducible) Feynman diagram is a connected diagram that cannot be disconnected by cutting a single internal line. By convention, IPI diagrams are evaluated with no propagators on the external lines.) There is an alternative way to expand the effective action: Instead of expanding in powers of ϕ_c, we can expand in powers of momentum (about the point where all external momenta vanish). In position space, such an expansion looks like

$$\Gamma = \int d^4x[-V(\phi_c) + \tfrac{1}{2}(\partial_\mu\phi_c)^2 Z(\phi_c) + \cdots]. \tag{3.16}$$

$V(\phi_c)$ – an ordinary function, not a functional – is called the effective potential. By comparing the expansions (3.15) and (3.16), it is easy to see that the nth derivative of V is the sum of all IPI graphs with n vanishing external momenta. In tree approximation (that is to say, neglecting all diagrams with closed loops), V is just the ordinary potential, the object we called U in Section 2.

The usual renormalization conditions of perturbation theory can be expressed in terms of the functions that occur in (3.15). For example, if we define the squared mass of the meson as the value of the inverse propagator at zero momentum, then

$$\mu^2 = \frac{d^2 V}{d\phi_c^2}\bigg|_0. \tag{3.17a}$$

Likewise, if we define the four-point function at zero external momenta to be the coupling constant, λ, then

$$\lambda = \frac{d^4 V}{d\phi_c^4}\bigg|_0. \tag{3.17b}$$

Similarly, the standard condition for the normalization of the field becomes

$$Z(0) = 1. \tag{3.17c}$$

We are now ready to apply this apparatus to the study of spontaneous symmetry breaking. Let us suppose our Lagrange density possesses an internal symmetry, like the classical field theories of Section 2. Then, spontaneous symmetry breaking occurs if the quantum field ϕ develops a nonzero vacuum expectation value, even when the source $J(x)$ vanishes. From Eqs. (3.12) and (3.14) this occurs if

$$\frac{\delta \Gamma}{\delta \phi_c} = 0, \tag{3.18}$$

for some non-zero value of ϕ_c. Further, since we are typically only interested in cases where the vacuum expectation value is translationally invariant (that is to say, we are not interested in the spontaneous breakdown of momentum conservation), we can simplify this to

$$\frac{dV}{d\phi_c} = 0, \tag{3.19}$$

for some non-zero value of ϕ_c. The value of ϕ_c for which the minimum occurs, which we denote by $\langle\phi\rangle$, is the expectation value of ϕ in the new (asymmetric) vacuum.

To explore the properties of the spontaneously broken theory, we define a new quantum field with vanishing vacuum expectation value,

$$\phi' = \phi - \langle\phi\rangle. \tag{3.20}$$

This generates a corresponding redefinition of the classical field,

$$\phi'_c = \phi_c - \langle \phi \rangle, \tag{3.21}$$

from which it immediately follows that the actual mass, coupling constant, etc. are computable from equations exactly like the Eqs. (3.17), except that the derivatives are evaluated at $\langle \phi \rangle$, rather than at zero. Thus, we have recreated the entire structure of our study of spontaneous symmetry breakdown in classical field theory. The only difference is that, instead of working with the classical potential U, we work with the effective potential V.

3.4 The loop expansion

Unfortunately, except for trivial models, we do not know the effective potential; to calculate it requires an infinite summation of Feynman diagrams, a task beyond our computational abilities. Thus, it is important to know a sensible approximation method for V. I shall now attempt to show that one such sensible method is the loop expansion: first summing all diagrams with no closed loops (tree graphs), then those with one closed loop, etc. Of course, each stage in this expansion also involves an infinite summation, but, as we shall see, this summation is trivial.

Let us introduce a parameter a into our Lagrange density, by defining

$$\mathscr{L}(\phi, \partial_\mu \phi, a) \equiv a^{-1} \mathscr{L}(\phi, \partial_\mu \phi). \tag{3.22}$$

We shall now show that the loop expansion is equivalent to a power-series expansion in a. Let P be the power of a associated with any graph. Then it is easy to see that

$$P = I - V, \tag{3.23}$$

where I is the number of internal lines in the graph and V is the number of vertices. This is because the propagator, being the inverse of the differential operator occurring in the quadratic terms in $\mathscr{L}$, carries a factor of a, while every vertex carries a factor of a^{-1}. (Note that it is important that we are dealing with IPI graphs, for which there are no propagators attached to external lines.) On the other hand, the number of loops, L, is given by

$$L = I - V + 1. \tag{3.24}$$

This is because the number of loops in a diagram is equal to the number of independent integration momenta; every internal line contributes one integration momentum, but every vertex contributes a δ function that reduces the number of independent momenta by one, except for one δ function that is left over for overall energy–momentum conservation.

Combining Eqs. (3.23) and (3.24), we find that

$$P = L - 1,$$ (3.25)

the desired result.

The point of this analysis is not that the loop expansion is a good approximation scheme because a is a small parameter; indeed, a is equal to one. (However, it is certainly no worse than ordinary perturbation theory for small coupling constants, since the set of graphs with n loops or less certainly includes, as a subset, all graphs of nth order or less in the coupling constants.) The point is, rather, since the loop expansion corresponds to expansion in a parameter that multiplies the total Lagrange density, it is unaffected by shifts of fields, and by the redefinition of the division of the Lagrangian into free and interacting parts associated with such shifts.[14]

Thus we have a systematic expansion procedure, in any order of which we can apply the methods of Section 2. Further, the first term in the expansion of V is the classical potential, U, the negative sum of all non-derivative terms in the Lagrange density. Thus, we have not only justified in the quantum world many of the classical methods of Section 2, we have justified many of the actual computations of Section 2. They should be reliable in the corresponding quantum field theories for the usual conditions under which we expect diagrams with closed loops to be negligible, that is to say, for small coupling constants.

3.5   *A sample computation*

To put some flesh on this dry formalism, let us compute the effective potential for the theory of a single scalar field with Lagrange density

$$\mathscr{L} = \tfrac{1}{2}(\partial_\mu \phi)^2 - U(\phi),$$ (3.26)

where U is a polynomial, not necessarily of renormalizable type. As stated, in the zero loop approximation,

$$V = U(\phi_c).$$ (3.27)

Now let us turn to the one-loop approximation. Since the one-loop approximation does not depend on how we break the Lagrangian into free and interacting parts, let us take only the first term in (3.26) as the free Lagrange density, and all of U (including possible mass terms) as the interaction. All the one-loop graphs are then shown in Fig. 5. The black dot stands for a sum of terms with zero, one, two, etc. external lines, arising from terms in U of second, third, fourth, etc. order in ϕ. (Terms linear in ϕ do not contribute to IPI one-loop diagrams.) Each of these external lines carries zero external momentum and a factor of ϕ_c. Thus, the value of the

Fig. 5

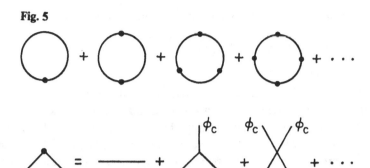

vertex in Fig. 5 is

$$i\frac{d^2U}{d\phi^2}\bigg|_{\phi=\phi_c} = iU''(\phi_c). \tag{3.28}$$

(The i is just the usual i from Dyson's formula.) For example, if we take the U of our old Abelian model, Eq. (2.6), then

$$U''(\phi_c) = \frac{\lambda}{6}(3\phi_c^2 - a^2). \tag{3.29}$$

Every line carries the usual massless propagator,

$$\frac{i}{k^2 + i\varepsilon}, \tag{3.30}$$

where k is the momentum going around the loop. Thus, the sum of all the graphs in Fig. 5 corrects Eq. (3.27) in the following way:

$$V = U + i\int \frac{d^4k}{(2\pi)^4} \sum_{n=1}^{\infty} \frac{1}{2n}\left(\frac{U''(\phi_c)}{k^2 + i\varepsilon}\right)^n. \tag{3.31}$$

Two factors in this expression require further explanation. (1) The i in front is just a reflection of the i in the definition of W, Eq. (3.10). (2) The $1/2n$ is a combinatoric factor; rotating or reflecting the n-dot graph does not lead to a new contraction in the Wick expansion, and therefore the $1/n!$ in Dyson's formula is incompletely cancelled.

It is easy to sum this infinite series. Aside from an irrelevant (divergent) constant, the answer is

$$V = U + \frac{1}{2}\int \frac{d^4k}{(2\pi)^4} \ln(k^2 + U''(\phi_c) - i\varepsilon), \tag{3.32}$$

where I have rotated the integral into Euclidean space in the standard way, but unconventionally have not dropped the $i\varepsilon$. (The reason for this eccentricity will become clear shortly.) The integral is divergent; if the integration is cut off at some large momentum Λ, we obtain

$$V = U + \frac{\Lambda^2}{32\pi^2} U'' + \frac{(U'')^2}{64\pi^2} \left(\ln \frac{U'' - i\varepsilon}{\Lambda^2} - \frac{1}{2} \right), \qquad (3.33)$$

plus an irrelevant constant.

The distinction between renormalizable and non-renormalizable interactions emerges very clearly in this computation. If U is a quartic polynomial (the renormalizable case), then we can remove all the cutoff-dependence from Eq. (3.33) by adding counterterms to the Lagrangian which are themselves at most quartic polynomials, and which can therefore be interpreted as corrections to the parameters in the original Lagrangian. For example, for our old Abelian model, we obtain in this way

$$V = \frac{\lambda}{4!} (\phi_c^2 - a^2)^2 + \frac{\lambda^2}{2304\pi^2} (3\phi_c^2 - a^2)^2 \ln(3\phi_c^2 - a^2 - i\varepsilon) + b\phi_c^4 + c\phi_c^2,$$

$$(3.34)$$

where b and c are finite constants, undetermined until we state our renormalization conventions, the conditions that define the renormalized parameters of the theory, and fix the finite parts of the counterterms.[15] (Note that it is a good thing that we retained the $i\varepsilon$, for the argument of the logarithm can become negative, and the $i\varepsilon$ is needed to tell us the sign of the imaginary part of V. We will return to this point later.)

On the other hand, if U is of quintic order or higher (the non-renormalizable case), the counterterms we must add are of yet higher order, and we are launched on the unending escalation of ambiguities that characterizes non-renormalizable theories. (A technical point: as in all renormalization schemes, the counterterms added in first order are to be considered as quantities of first order in the relevant expansion parameter. In our case, this is the (suppressed) loop-counting parameter, a, of Eq. (3.22). Thus, if we go to higher loops, the counterterms introduced at this stage are to be counted as one-loop internal parts, despite the fact that they are represented graphically by simple point vertices.)

3.6   The most important part of this lecture

The significant feature of the computation we have just done is that we needed to invoke no more counterterms than would have been required if there had not been spontaneous symmetry breakdown; the ultraviolet divergences of the theory respect the symmetry of the Lagrangian, even if the vacuum state does not. That this occurred in our specific computation should be no surprise; our entire formalism has been constructed so this is what happens in any computation. For $\mathcal{L}$, the Lagrange density in Eq. (3.9), is the *total* Lagrange density for the theory. In particular, this means that it contains *all* the counterterms needed to eliminate *all* ultraviolet divergences. None of the subsequent manipulations in Section 3.3 involve any integrations over internal momenta, and therefore none of them can introduce new ultraviolet divergences.

This point is important enough to be worth stating again in a slightly different way. We have developed the theory of spontaneous symmetry breakdown in quantum field theory in such a way that we remove all the ultraviolet divergences from the theory *before* we shift the fields. Before we shift the fields, everything is manifestly symmetric under the full internal symmetry group of the theory; therefore there is no way in which asymmetric counterterms can arise.

Once more, with feeling: *the divergence structure of a renormalizable field theory is not affected by the occurrence of spontaneous symmetry breakdown.* This simple observation is the most important part of this lecture. It is the secret of the construction of renormalizable theories of the weak interactions. These theories are apparently non-renormalizable, for they involve massive vector mesons (the W-bosons) coupled to non-conserved currents. However, this is only an appearance; in actuality, the Lagrangians of these theories involve only massless gauge fields coupled minimally to conserved currents, and are perfectly renormalizable. The mass of the vector mesons and the non-conservation of the currents are a result of spontaneous symmetry breakdown.

Likewise, we see how to construct theories in which mass differences within an isotopic multiplet are finite. We begin with a theory in which the photon is part of a set of gauge mesons that couple in an isospin-symmetric way. In such a theory, one needs only isosinglet mass counterterms. Spontaneous symmetry breakdown now occurs; the friends of the photon become massive; but there is still no need for an isospin-breaking mass counterterm.

At the end of Section 2, I asked, 'What does it buy us?' We now have the wonderful answer: secret symmetry buys us secret renormalizability.

3.7 *The physical meaning of the effective potential*

In classical field theory, the ordinary potential, $U(\phi)$, is an energy density; it is the energy per unit volume for that state in which the field assumes the value ϕ. I will now show that, in quantum field theory, the effective potential, $V(\phi_c)$, is also an energy density; it is the expectation value of the energy per unit volume in a certain state for which the expectation value of the field is ϕ_c.[16] An immediate consequence of this is that, if V has several local minima, it is only the absolute minimum that corresponds to the true ground state of the theory, the state of lowest energy. As a byproduct, we will obtain the essential clue to the meaning of the mysterious imaginary part of V which appeared in our sample computation.

We begin the proof by expanding $W(J)$, defined in Eq. (3.10), in the same way we expanded Γ in Eq. (3.16):

$$W = \int d^4x \{ -\mathcal{E}(J) + \tfrac{1}{2}(\partial_\mu J)^2 X(J) + \cdots \}. \tag{3.35}$$

Now let us consider a $J(x)$ which has a constant value, which we denote by J, throughout a box of side L, during a time T, and which goes to zero smoothly outside this space-time region. Under these conditions, for very large L and T, the first term in Eq. (3.35) is the dominant one, and

$$e^{iW} = \langle 0^+ | 0^- \rangle \approx e^{-iL^3 T \mathcal{E}(J)}. \tag{3.36}$$

What has happened physically is that, throughout the box, we have smoothly changed the Hamiltonian density of the theory:

$$\mathcal{H} \rightarrow \mathcal{H} - J\phi. \tag{3.37}$$

Thus we would expect the ground state of the theory, within the box, to go adiabatically into the ground state of the theory with the additional term in is Hamiltonian density. This state would evolve in time according to the Schrödinger equation; since it is a ground state, this means that it simply develops a phase. When we turn off the perturbation, the state goes adiabatically back to the ground state of the unperturbed theory, but the phase remains. Thus, $\mathcal{E}(J)$ is the energy per unit volume of the ground state of the perturbed Hamiltonian. (Of course, level crossing might take place. To be precise, we should say not 'the ground state' but 'that stationary state of the perturbed theory that is obtained from the ground state of the unperturbed theory by adiabatically turning on the perturbation'.)

I will now begin an independent line of argument, which, when combined with the above observation, will yield the desired result. For notational simplicity, I will construct this argument for ordinary quantum mechanics, not for field theory, so we will speak of energies, rather than of energy densities; the proper generalization will be obvious. Let us remember the ancient Rayleigh–Ritz variational problem: to construct a state $|a\rangle$ that is a stationary state of the quadratic form

$$\langle a|H|a\rangle, \tag{3.38}$$

under the constraint that the norm of the state be one,

$$\langle a|a\rangle = 1. \tag{3.39}$$

This problem is traditionally solved by Lagrange multipliers; one introduces a Lagrange multiplier, called E, and varies without constraint the form

$$\langle a|(H-E)|a\rangle. \tag{3.40}$$

In this way one obtains

$$(H-E)|a\rangle=0. \tag{3.41}$$

Hence, $|a\rangle$ is an eigenstate of H with energy E.

Now let us consider a slight variation of this problem. We add an extra equation of constraint

$$\langle a|A|a\rangle=A_c, \tag{3.42}$$

where A is some Hermitian operator and A_c some number. We must now introduce two Lagrange multipliers, which I will call E and J, and vary without constraint

$$\langle a|(H-E-JA)|a\rangle. \tag{3.43}$$

We thus obtain

$$(H-E-JA)|a\rangle=0. \tag{3.44}$$

Hence $|a\rangle$ is an eigenstate of the perturbed Hamiltonian, $H-JA$, and E is its energy. Of course, this gives us E as a function of J, and we are really interested in how things depend, not on J, but on A_c. The connection between these two quantities is easily obtained by a standard formula of first-order perturbation theory,

$$A_c=\langle a|A|a\rangle=-\frac{dE}{dJ}. \tag{3.45}$$

Hence the quantity we originally set out to make stationary is given by

$$\langle a|H|a\rangle=E+JA_c=E-J\frac{dE}{dJ}. \tag{3.46}$$

It can hardly have escaped you that (with the obvious substitution of energy densities for energies and ϕ for A) this is precisely the chain of manipulations that led to the definition of the effective potential. Thus we have found the physical meaning of the effective potential;

$$V(\phi_c)=\langle a|\mathcal{H}|a\rangle, \tag{3.47}$$

for a state $|a\rangle$ such that

$$\delta\langle a|\mathcal{H}|a\rangle=0, \tag{3.48}$$

under the constraints

$$\langle a|a\rangle=1, \tag{3.49a}$$

and

$$\langle a|\phi|a\rangle=\phi_c. \tag{3.49b}$$

We can check this interpretation in another way, by reducing the four dimensions of space-time to one. The Lagrange density then becomes the

Lagrangian for a particle of unit mass, ϕ becomes x, the position of the particle, and $U(\phi)$ becomes $U(x)$, the potential in which the particle moves. Eq. (3.32) becomes

$$V = U + \frac{1}{2} \int \frac{d\omega}{2\pi} \ln(\omega^2 + U'' - i\varepsilon)$$

$$= U + \tfrac{1}{2}(U'' - i\varepsilon)^{\frac{1}{2}}. \tag{3.50}$$

This has a direct physical interpretation: classically, the particle sits in a minimum of the potential, and its energy is the value of the potential at the minimum. To get the first quantum correction to this picture, we approximate the potential near the minimum by a harmonic oscillator potential, and add the zero-point energy of the oscillator; this is the second term in Eq. (3.50).[17]

Once we know V is an energy density, we can understand the meaning of its imaginary part. When we follow an energy level as we change the parameters of a theory, it may often happen that, at a certain point, the energy level becomes unstable; at this moment the energy acquires a negative imaginary part, equal in magnitude to half the probability of decay per unit time. This can also be seen from our earlier discussion of $\mathscr{E}(J)$ in terms of the adiabatic turning-on of a perturbation. If the ground state of the unperturbed system adiabatically moves into an unstable state of the perturbed system, it will decay, and

$$\langle 0^+ | 0^- \rangle = \exp[-iL^3 T \mathscr{E}(J)], \tag{3.51}$$

will be a number with modulus less than one. Of course, for a system of infinite spatial extent, one should not speak of decay probability per unit time, any more than one speaks of energy; one speaks of decay probability per unit time per unit volume, just as one speaks of energy density. Thus the imaginary part of the effective potential is to be interpreted as half a decay probability per unit time per unit volume. (Note that the $i\varepsilon$ in Eq. (3.32) insures that the imaginary part is negative, as it must be if this interpretation is to be consistent.)[18]

3.8   Accidental symmetry and related phenomena

Like all perturbative expansions, the loop expansion is trustworthy only for small dimensionless coupling constants. For small coupling constants, one usually expects higher terms in a perturbation expansion to be small compared to lower terms. This is indeed the case in our sample computation; for the Abelian model, for example, the zero-loop effective potential is of order λ, and the one-loop correction is of order λ^2. Nevertheless, there are important cases in which the one-loop corrections are more important than the tree graphs, and play the dominant role in determining the structure of spontaneous symmetry breaking.

This is because our theory may contain interactions that do not appear at all in the zero-loop approximation to the effective potential, such as Yukawa couplings or gauge-field couplings. We have not yet explicitly computed any graphs involving closed loops of virtual fermions or gauge particles, but it is obvious that their magnitude depends only on the magnitude of the Yukawa or gauge coupling constants. Since these are independent parameters of the theory, it is always possible to choose them so the one-loop graphs are more important than the zero-loop graphs, even if all coupling constants are small. Thus, for example, in the Abelian gauge model of Section 2.4, closed loops of virtual photons turn out to make a contribution to V of order e^4. (See the Appendix for the computation.) This is more important than the zero-loop effective potential if e^4 is much greater than λ, which can happen even if e and λ are both much less than one.

There are even cases in which the one-loop effective potential is important whatever the relative magnitude of the dimensionless coupling constants. This is most easily explained with a specific example. Consider an SO(3) quintuplet of scalar mesons, which we denote by ϕ^a, where a runs from 1 to 5. The transformation properties of these fields can be most simply expressed if we assemble them into a real traceless symmetric 3×3 matrix, which we denote by ϕ. Under an SO(3) transformation, characterized by a rotation matrix R,

$$\phi \to R\phi R^T. \tag{3.52}$$

In addition, we will assume invariance under the discrete symmetry

$$\phi \to -\phi. \tag{3.53}$$

Thus we can only have quadratic and quartic self-couplings. The only invariant quadratic form is

$$\mathrm{Tr}\,\phi^2 = \sum (\phi^a)^2. \tag{3.54}$$

There are apparently two possible quartic couplings, $\mathrm{Tr}\,\phi^4$ and $(\mathrm{Tr}\,\phi^2)^2$; however, these are related by the tracelessness of ϕ:

$$\mathrm{Tr}\,\phi^4 = \frac{1}{2}(\mathrm{Tr}\,\phi^2)^2 = \frac{1}{2}\left[\sum (\phi^a)^2\right]^2. \tag{3.55}$$

As it is clear from the right-hand sides of these equations, both of these terms are invariant under a larger symmetry group than SO(3), to wit, SO(5). Thus, the constraints of renormalizability (no higher than fourth-order interactions) have forced the scalar meson self-interaction, and therefore the zero-loop effective potential, to be invariant under a larger symmetry group than we started out with. This phenomenon has been dubbed accidental symmetry by Weinberg.[19] However, if the scalar mesons are coupled to a triplet of gauge fields, the gauge interaction is not forced to be (indeed, can not be) SO(5)-invariant; however, it also does not appear in the zero-loop approximation for V.

Thus, if we attempted to analyze this model in the zero-loop approximation, we would be in the soup for two reasons: (1) We would have too rich a set of vacua – an SO(5) family instead of just an SO(3) one. (2) Even if we miraculously picked the right vacuum from this over-rich set, we would find some massless scalars that were only SO(5) Goldstone bosons, and not SO(3) ones. (Weinberg calls these pseudo-Goldstone bosons.) To find the right vacuum, and to give a mass to the pseudo-Goldstone bosons, it is necessary to compute the effects of gauge-field loops.

3.9   An alternative method of computation

In these lectures I have stressed a method of computation in which we first compute higher-order corrections, and then shift the fields. To be honest, I must tell you that most workers in this field prefer to do things in the other order. They rewrite the Lagrangian of the theory in terms of shifted fields

$$\phi = \phi' + \langle \phi \rangle. \tag{3.56}$$

This gives them a Lagrangian with an extra free parameter for each spinless field (the value of the shift). These are fixed at some stage in the computation by demanding that the vacuum expectation values of the shifted fields vanish,

$$\langle 0|\phi'|0 \rangle = 0. \tag{3.57}$$

In other words, all IPI graphs with only one external line (tadpole graphs) should sum to zero.

This is just as good a way of doing things as the way I have explained; it is equivalent to computing directly the derivative of V and demanding that it vanish, without bothering to compute V first. The only reason I have developed the theory in the way I have is a pedagogical one; in the alternative method of development, it is not so easy to see that spontaneous symmetry breakdown does not lead to asymmetric counterterms. (The only case I can think of in which our method would be clearly superior would be for a theory in which V had two local minima; in this case, we would need to know the value of V in order to determine which of them was the absolute minimum, the true vacuum.)

4 Functional integration (vulgarized)

4.1 *Integration over infinite-dimensional spaces*

Functional integration is a method for defining and manipulating integrals over function spaces, that is to say, over infinite-dimensional spaces, in the same way the ordinary integral calculus enables us to define and manipulate integrals over finite-dimensional spaces. It is useful in theoretical physics because it is possible to represent the generating functional of a quantum field theory as a functional integral. Such a representation has many virtues; from our point of view, the chief of these is that this makes it especially easy to see how the theory changes if we make non-linear transformations on its fundamental dynamical variables. The larger the set of physically interesting nonlinear transformations, the more useful is the functional-integral representation; thus it is most useful in studying non-Abelian gauge theories.

This lecture will be devoted to explaining functional integration and its connection with field theory. Our approach will be, from a mathematical viewpoint, despicable. Nothing will be proved; everything will be done by analogy, formal manipulation of ill-defined (and sometimes divergent) quantities, and handwaving. I hope that this will at least give you an idea of what is going on and teach you to manipulate functional integrals; if you want a deeper understanding, you must go elsewhere.[20]

We begin with a very simple one-dimensional integral, the Gaussian integral,

$$\int dx\, e^{-\frac{1}{2}ax^2} = (2\pi/a)^{\frac{1}{2}}, \tag{4.1}$$

where a is a positive real number. By analytic continuation, the formula is also true for complex a whenever the integral is defined, that is to say, whenever a has a positive real part. Eq. (4.1) can readily be generalized to n-dimensional space. We will call a vector in such a space, x. We will denote the usual inner product of two such vectors, x and y, by (x, y). Then, if A is a real symmetric positive-definite matrix,

$$\int d^n x\, e^{-\frac{1}{2}(x,Ax)} = (2\pi)^{n/2}(\det A)^{-\frac{1}{2}}, \tag{4.2}$$

as can easily be seen by diagonalizing A. As before, this formula is also true if A is a complex symmetric matrix with positive-definite real part, by analytic continuation.

To keep from continually writing πs and ns, we define

$$(dx) = d^n x (2\pi)^{-n/2}. \tag{4.3}$$

Thus, Eq. (4.2) becomes

$$\int (dx)e^{-\frac{1}{2}(x,Ax)} = (\det A)^{-\frac{1}{2}}. \tag{4.4}$$

If we can integrate Gaussians, we can integrate exponentials of general quadratic forms. Let

$$Q(x) = \tfrac{1}{2}(x, Ax) + (b, x) + c \tag{4.5}$$

where **b** is some vector and c is a number. Let $\bar{x}$ be the minimum of Q,

$$\bar{x} = -A^{-1}b. \tag{4.6}$$

Then

$$Q(x) = Q(\bar{x}) + \frac{1}{2}(x - \bar{x}, A[x - \bar{x}]), \tag{4.7}$$

and

$$Q(\bar{x}) = -\frac{1}{2}(b, A^{-1}b) + c. \tag{4.8}$$

Whence,

$$\int (dx)e^{-Q(x)} = e^{-Q(\bar{x})}(\det A)^{-\frac{1}{2}}. \tag{4.9}$$

Once we have Eq. (4.9), we can do the integral of any polynomial times the exponential of a quadratic form, just by differentiating with respect to **b**,

$$\int (dx)P(x)e^{-Q(x)} = P\left(-\frac{\partial}{\partial b}\right) \int (dx)e^{-Q(x)}. \tag{4.10}$$

It will be convenient later to also have formulae for integrating over an n-dimensional complex vector space, not in any contour-integral sense, but merely in the sense of integrating separately over imaginary and real parts. We will denote the usual Hermitian inner product in such a space by (z^*, w), and the $2n$-dimensional real integration (with appropriate factors of π inserted) by $(dz^*)(dz)$. Then if A is a positive-definite Hermitian matrix,

$$\int (dz^*)(dz)e^{-(z^*, Az)} = (\det A)^{-1}, \tag{4.11}$$

as can easily be seen by diagonalizing A. Note the change in the power of the determinant. This is because each eigenvalue of A contributes twice to the integral, once from the integration over the real part of z, and once from the integration over the imaginary part. The missing $\frac{1}{2}$ in the exponential is just a matter of convention; its effects are absorbed in the definition of $(dz^*)(dz)$. From this formula equations analogous to those we derived before follow directly; I will not bother to write them out explicitly.

Now comes the great leap of faith: there is nothing in our integration formulae that refers explicitly to the dimension of the vector space; therefore we boldly extend them to infinite-dimensional vector spaces. Let me be a bit more precise about how this is done, using Eq. (4.4) as an example. Given a quadratic form, (x, Ax), defined by a linear operator, A, on an infinite-dimensional real Hilbert space, we first restrict the form to some finite-dimensional subspace. On this finite-dimensional subspace, both sides of Eq. (4.4), the integral and the determinant, are well-defined. We then let the finite-dimensional subspace grow, until, in the limit, it becomes the whole space. More precisely, we consider an increasing sequence of finite-dimensional subspaces such that their union contains a dense set of vectors. This limit defines both the infinite-dimensional integral and the infinite determinant. It is a deep problem to determine for what operators A the limits exist and are independent of the sequence of subspaces, but it is not one I will worry about here. We will assume in our manipulations that expressions like (4.4) are well-defined whenever we need them.

The infinite-dimensional spaces we will be most concerned with will be spaces of functions, for example, the space of functions of a single real variable. This special case has an unnecessary, but traditional, special notation associated with it. The vectors in the space are traditionally denoted not by x, as we have been doing, but by some symbol that makes their nature as functions manifest, e.g., by $q(t)$, where t is the real variable. The inner product is written as

$$(q, q) = \int dt [q(t)]^2. \tag{4.12}$$

Also, a function from the vector space to the real or complex numbers is called a functional, and derivatives, like those appearing in Eq. (4.10), are called variational derivatives, and denoted by expressions like $\delta/\delta q(t)$, rather than $\partial/\partial x$. In field-theoretical applications, we will consider spaces of functions of four-dimensional space-time, usually denoted by expressions like $\phi(x)$, where x is a space-time point. In this case,

$$(\phi, \phi) = \int d^4x [\phi(x)]^2, \tag{4.13}$$

and Eq. (4.4) would be written as

$$\int (d\phi) e^{-\frac{1}{2}(\phi, A\phi)} = (\det A)^{-\frac{1}{2}}. \tag{4.14}$$

In the cases that will most concern us, A will be an integral or differential operator.

4.2   Functional integrals and generating functionals

There are a large number of cases in which the generating functionals of quantum theories can be written as functional integrals. I will begin with an especially simple case, that of a single scalar field with non-derivative self-interaction. As in Section 3.3, let us write the Lagrange density for such a theory in the presence of an external c-number source, $J(x)$,

$$\mathcal{L} = \tfrac{1}{2}(\partial_\mu \phi)^2 - \tfrac{1}{2}\mu^2\phi^2 + \mathcal{L}'(\phi) + J(x)\phi. \tag{4.15}$$

Here $\mathcal{L}'$ is the interaction, some polynomial function of ϕ. Let us consider this as a classical Lagrange density for a c-number field, and let us construct the classical action integral

$$S(\phi, J) = \int d^4x\, \mathcal{L}. \tag{4.16}$$

S is a functional of the two c-number fields, ϕ and J. In Section 3.3, we also introduced the generating functional for the quantum theory, $\exp[iW(J)]$, defined as the sum of all vacuum-to-vacuum graphs in the presence of the source J. I will now demonstrate the following remarkable connection between the quantum generating functional and the classical action integral:

$$e^{iW(J)} = N \int (d\phi) e^{iS(\phi, J)}, \tag{4.17}$$

where N is a normalization factor, chosen such that W vanishes when J vanishes. Eq. (4.17) is a version of Feynman's sum over histories; a quantum transition amplitude is obtained by summing over all possible classical histories of the system. As it stands, Eq. (4.17) is ill-defined, even by our sloppy standards; the integrand is an awful oscillating object, nothing like the nicely damped Gaussians of Section 4.1. This problem is remedied by stating that the generating functional on the left-hand side of Eq. (4.17) is that of *Euclidean* Green's functions, and the functional integral is to be evaluated for fields in *Euclidean* space that vanish at infinity.

This prescription requires some explanation: Feynman amplitudes are defined, to begin with, for real external three-momenta and real external energies. However, we can analytically continue them to imaginary energies, by simultaneously rotating all energies (internal as well as external) by $\pi/2$ in the complex energy plane. It is trivial to verify that no singularities of the Feynman integral are encountered in the course of this rotation. (The analytic continuation can also be proved without recourse to perturbation theory, but this is the easiest way to see that it is possible.)

Thus we arrive at Euclidean momentum space – real three-momenta and imaginary energies. For any Euclidean momentum k_μ, we define the real variable k_4 by

$$k_0 = ik_4. \tag{4.18}$$

Thus,

$$k^2 = -k_E^2, \tag{4.19}$$

where

$$k_E^2 = \mathbf{k}^2 + k_4^2, \tag{4.20}$$

the standard Euclidean square of a vector. Also,

$$d^4k = id^4k_E \tag{4.21}$$

Euclidean position-space Green's functions are defined by analytically continuing the Fourier transforms of momentum-space Green's functions. So that the Fourier exponential factor, $\exp(ik \cdot x)$, will not blow up and spoil the continuation, we must rotate x_0 through minus $\pi/2$ at the same time we rotate k_0 through plus $\pi/2$. Thus we obtain

$$x_0 = -ix_4, \tag{4.22}$$
$$d^4x = -id^4x_E, \tag{4.23}$$

etc. Thus, for example, the Feynman propagator for a free scalar field of mass μ,

$$\Delta_F(x) = \int \frac{d^4k}{(2\pi)^4} e^{-ik \cdot x} \frac{i}{k^2 - \mu^2 + i\varepsilon}, \tag{4.24}$$

becomes, in Euclidean space,

$$\Delta_E(x) = \int \frac{d^4k_E}{(2\pi)^4} e^{-ik \cdot x} \frac{1}{k_E^2 + \mu^2}. \tag{4.25}$$

Note that there is no need to retain the $i\varepsilon$ in Euclidean space. It will be important to us shortly that this function obeys

$$(-\Box_E^2 + \mu^2)\Delta_E(x) = \delta^{(4)}(x), \tag{4.26}$$

where

$$\Box_E^2 = \nabla^2 + \partial_4^2. \tag{4.27}$$

Since the integrand in Eq. (4.25) has no pole, Δ_E is the unique solution to Eq. (4.26); this is in contrast to the situation in Minkowski space, where the corresponding equation has many solutions, and the $i\varepsilon$ is needed to resolve the ambiguity.

Let us now turn to the verification of the functional-integral formula, Eq. (4.17). I will begin with the case of a free field, $\mathscr{L}' = 0$. The Minkowski-space generating functional is

$$e^{iW} = \exp\left[-\frac{1}{2} \int d^4x\, d^4y\, J(x)\Delta_F(x-y)J(y) \right].$$ (4.28)

Hence, the Euclidean generating functional is

$$\exp\left[\frac{1}{2} \int d^4x_E\, d^4y_E\, J(x)\Delta_E(x-y)J(y) \right].$$ (4.29)

This takes care of the left-hand side of Eq. (4.17). As for the right-hand side,

$$iS = i \int d^4x \left(\frac{1}{2} [(\partial_0\phi)^2 - (\nabla\phi)^2 - \mu^2\phi^2] + J\phi \right)$$

$$= -\int d^4x_E \left(\frac{1}{2} [(\partial_4\phi)^2 + (\nabla\phi)^2 + \mu^2\phi^2] - J\phi \right).$$ (4.30)

(I emphasize that this is not an analytic continuation, just a formal substitution. We are not proving that one well-defined object is an analytic continuation of another; we are *defining* the functional integrand.) Thus the functional integral is of the form (4.9), with

$$A = -\Box_E^2 + \mu^2,\ b = -J,\ c = 0.$$ (4.31)

Hence,

$$N \int (d\phi)e^{iS} = N(\det A)^{-\frac{1}{2}}e^{\frac{1}{2}(J, A^{-1}J)}.$$ (4.32)

We can now determine the normalization factor, N,

$$N = (\det A)^{\frac{1}{2}}.$$ (4.33)

This saves us the trouble of computing the determinant. (This is a good thing, because, in cold fact, the determinant is divergent.) Thus we obtain our final answer for the integral,

$$\exp\left[\frac{1}{2}(J, A^{-1}J) \right] = \exp\left[\frac{1}{2} \int d^4x_E\, d^4y_E\, J(x)\Delta_E(x-y)J(y) \right].$$

(4.34)

This is in agreement with Eq. (4.29); in this case, at least, the functional integral has reproduced the generating functional, as promised.

If we had attempted to evaluate the integral directly in Minkowski space, using (erroneously) the integral formulae of Section 4.1, we would have arrived at a similar result, except that A would have been the Klein–Gordon operator. We would then have been stymied, for we would not have known what Green's function to use for A^{-1}. The Euclidean calculation contains no such ambiguity; the right answer (Feynman's iε rule) comes about automatically as a consequence of our prescription for continuing back into Minkowski space, *after* we have done the integration. Thus, Euclidean integration is not just a mathematical nicety, but is essential if we are to obtain an unambiguous answer. From now on, I will not explicitly do the continuations into Euclidean space and out again, but simply write my integrals as if they were to be done in Minkowski space, as in Eq. (4.17). You should always remember, though, that this is just a notational convention; really we are always integrating over Euclidean fields.

Now for the interacting case. Still being slapdash, I will ignore all questions of divergences, cutoffs, and renormalizations, and simply write down Dyson's formula for the generating functional:

$$e^{iW} = N'T \left\langle 0 \left| \exp\left[i \int (\mathcal{L}'(\phi_1) + J\phi_1)d^4x \right] \right| 0 \right\rangle, \tag{4.35}$$

where $|0\rangle$ is the bare vacuum, ϕ_1 is the interaction-picture field, T is the time-ordering symbol, and N' is a normalization factor, chosen as before. This can be written as

$$N' \exp\left[i \int d^4y \mathcal{L}'\left(-i\frac{\delta}{\delta J(y)} \right) \right] \left\langle 0 \left| \exp\left[i \int J\phi_1 d^4x \right] \right| 0 \right\rangle$$

$$= N' \exp\left[i \int d^4y \mathcal{L}'\left(-i\frac{\delta}{\delta J(y)} \right) \right] \exp[iW_0(J)], \tag{4.36}$$

where W_0 is the generating functional for the free field. Now for the functional integral. We split the action into two parts,

$$S = \int d^4x \mathcal{L}'(\phi) + S_0(\phi, J), \tag{4.37}$$

where S_0 is the action for the free field (including the source term), the quantity denoted by S in Eq. (4.27). In the spirit of Eq. (4.11),

$$N \int (d\phi)e^{iS} = N \exp\left[i \int d^4y \mathcal{L}'\left(-i\frac{\delta}{\delta J(y)} \right) \right] \int (d\phi)e^{iS_0}$$

$$= N \exp\left[i \int d^4y \mathcal{L}'\left(-i\frac{\delta}{\delta J(y)} \right) \right] e^{iW_0}, \tag{4.38}$$

by our preceding evaluation. Things equal to the same thing are equal to each other. Q.E.D.

This result generalizes immediately to a theory involving several scalar fields

$$\mathcal{L} = \tfrac{1}{2}(\partial_\mu \phi^a)(\partial^\mu \phi^a) - U + J_a(x)\phi^a, \tag{4.39}$$

where U depends on the fields but not their derivatives, and the sum on repeated indices is implied. In this case, Eq. (4.17) becomes

$$e^{iW} = N \int \prod_a (d\phi^a) e^{iS}. \tag{4.40}$$

Likewise, descending from four dimensions to one, we see that for the parallel system in particle mechanics,

$$L = \tfrac{1}{2}(\dot{q}^a)^2 - U(q^a) + J_a(t)q^a, \tag{4.41}$$

a similar formula applies,

$$e^{iW} = N \int \prod_a (dq^a) e^{iS}. \tag{4.42}$$

Of course, in this case, the action is just a single integral, not a quadruple one,

$$S = \int dt \, L. \tag{4.43}$$

We can also go backwards, from 'particles' to fields, by letting the index a run over an infinite range, and identifying the qs with the Fourier components of the fields at fixed time. Thus, Eq. (4.42) is in fact more general than Eq. (4.40); it involves no conditions on the Lorentz transformation properties of the dynamical variables, merely a condition on the way in which their time derivatives enter the Lagrangian. For this reason, I will in the future use 'particle' language when the discussion is general, and return to field language only for special cases.

4.3   Feynman rules

Let us return for a moment to the case of a single scalar field with non-derivative interactions. Eq. (4.38) gives a formal expression for the functional integral in this case, but, if the interaction is non-trivial, it is impossible to turn this into an explicit closed form. However, it is perfectly feasible to evaluate it perturbatively, by expanding in powers of the interaction. Such an expansion gives the ordinary Feynman rules.

This can be seen most easily with the aid of a functional identity. I will first state and prove this identity for finite-dimensional real vector spaces, and then, as usual, extend it to function spaces. Let $F(x)$ and $G(x)$ be any two numerical-valued functions on a vector space; then

$$F\left(-i\frac{\partial}{\partial x}\right)G(x) = G\left(-i\frac{\partial}{\partial y}\right)F(y)e^{i(x,y)}\big|_{y=0}. \tag{4.44}$$

The identity is most easily proved by Fourier analysis, that is to say, by taking F and G to be plane waves,

$$F = e^{i(a,x)}, \quad G = e^{i(b,x)}, \tag{4.45}$$

with **a** and **b** fixed vectors. Then

$$e^{(\mathbf{a},\partial/\partial \mathbf{x})}e^{i(\mathbf{b},\mathbf{x})}=e^{i(\mathbf{b},\mathbf{x}+\mathbf{a})}. \tag{4.46}$$

and

$$e^{(\mathbf{b},\partial/\partial \mathbf{y})}e^{i(\mathbf{x}+\mathbf{a},\mathbf{y})}=e^{i(\mathbf{x}+\mathbf{a},\mathbf{y}+\mathbf{b})}. \tag{4.47}$$

This proves Eq. (4.44).

Extending this to a function space, and applying it to Eq. (4.38), we obtain

$$e^{iW}=N\exp\left[\frac{1}{2}\int d^4y\,d^4x\,\Delta_F(x-y)\frac{\delta}{\delta\phi(x)}\frac{\delta}{\delta\phi(y)}\right]$$
$$\times \exp\left[i\int d^4x(\mathcal{L}'(\phi)+J\phi)\right]\Bigg|_{\phi=0} \tag{4.48}$$

Here I have made the obvious substitutions of J for $\mathbf{x}$, ϕ for $\mathbf{y}$, and variational derivative for ordinary derivative. Eq. (4.48) is manifestly the Feynman rules for the vacuum-to-vacuum matrix element. (If it is not manifest to you, I suggest that you compute the first few terms in the expansion for a ϕ^4 interaction.) Note that diagrams occur in the expansion in which two fields from the same interaction vertex are linked by a propagator; the functional integral does not normal-order the interaction for us. If we wish to treat normal-ordered interactions, we must do the normal-ordering by hand, by inserting explicit counterterms into the interaction.

This argument can immediately be extended to the general case, to give 'Feynman rules' for perturbatively evaluating a functional integral of the form

$$\int \prod_a (dq^a)e^{iS}, \tag{4.49}$$

where

$$S=S_0+S', \tag{4.50}$$

and

$$S_0=-\tfrac{1}{2}(q^a,A_{ab}q^b). \tag{4.51}$$

Here A is a linear operator (independent of the qs) with positive-definite real part (after the rotation to imaginary time has been performed), and S' is an arbitrary polynomial functional of the qs, possibly involving source terms. Then, just as above, we can develop a diagrammatic expansion for the integral in powers of S', exactly like Feynman rules. Every power of S' is represented by a vertex, and the propagator, $D_F^{ab}(t,t')$, is the solution of

$$A_{ab}D_F^{bc}(t,t')=-i\delta_a^c\delta(t-t'). \tag{4.52}$$

Any ambiguity in solving this equation is to be resolved by rotating to imaginary time.

Note that if S' contains derivatives of the qs, these will just become derivatives of propagators in the expansion. The familiar problem of pushing time derivatives of quantum fields through a time-ordering operator, the problem that makes perturbation theory for derivative interactions such a combinatoric nightmare, has no counterpart here, for we have no time-ordering operator and no quantum fields, just an integral over c-number fields.

Thus, for any theory, if we can write the generating functional in the form (4.49), we can just read off the Feynman rules from S' in the most naive way, replacing every derivative of a field with a momentum factor, etc., without making any mistakes. Unfortunately, at the moment, the only theories for which we can write the generating functional in the form (4.49) are those without any derivatives in the interaction, so this observation is without immediate use. However, it will become very useful shortly.

4.4   Derivative interactions

There is a large class of theories with derivative interactions for which it is possible to write a functional-integral representation of the generating functional. These are theories where the Lagrangian is no more than quadratic in time derivatives,

$$L = \tfrac{1}{2}\dot{q}^a K_{ab}\dot{q}^b + L_a\dot{q}^a - U, \tag{4.53}$$

where K, L, and U are functions of the qs. The only restriction I will place on these functions is that K be invertible, so that the equation for the canonical momenta,

$$p_a = K_{ab}\dot{q}^b + L_a, \tag{4.54}$$

can be solved for the $\dot{q}$s and the Hamiltonian constructed,

$$H = \tfrac{1}{2}p_a(K^{-1})^{ab}p_b + \cdots, \tag{4.55}$$

where the triplet dots indicate terms of first and zeroth order in the ps.

For these theories, the appropriate generalization of our earlier result, Eq. (4.42), turns out to be

$$e^{iW} = N \int \prod_a (dq^a)[\det K]^{\frac{1}{2}} e^{iS}. \tag{4.56}$$

In this equation, K is to be interpreted as a linear operator on the function space, and the integral is to be interpreted in the same way our earlier (Gaussian) integrals were interpreted. Everything is to be restricted to a finite-dimensional subspace, the integral is to be done over that subspace, and the limit is to be taken. I do not know of any short argument for this formula, and have to refer you to the literature for a proof.[21] However, I can try and make it plausible to you by showing that it obeys some simple consistency checks. (1) If K is independent of the qs, and L vanishes, this reduces to the previous case. The determinant can then be pulled out of the integral and absorbed by the normalization factor, reproducing Eq. (4.42). (2) If K is independent of the qs, but L does not vanish, then, by our earlier remarks, the Feynman rules are the naive ones, with the derivative in the interaction becoming a factor of momentum at the vertex. This may be a familiar result to you if you have ever gone through the derivation of the Feynman rules for ps-pv meson–nucleon theory, or the electrodynamics of charged scalar bosons. (3) If K does depend on the qs, things are not so simple. This may be familiar to you if you followed the discussion in the literature a few years ago about the Feynman rules for chiral Lagrangians.

(4) Finally, a Lagrangian of the form (4.48) becomes one of the same form if we change coordinates. To be more precise, let us trade the qs for new variables, which we denote by $\bar{q}^a$. Then

$$L = \tfrac{1}{2}\dot{q}^a K_{ab}\dot{q}^b + \cdots = \tfrac{1}{2}\dot{\bar{q}}^a \bar{K}_{ab}\dot{\bar{q}}^b + \cdots, \tag{4.57}$$

where

$$\bar{K}_{ab} = \frac{\partial q^d}{\partial \bar{q}^a} K_{cd} \frac{\partial q^d}{\partial \bar{q}^b}. \tag{4.58}$$

This takes care of the transformation of the Lagrangian, but we still have to change variables in the functional integral. As always, we will figure out how to do this by going back to the finite-dimensional case. Suppose, in a finite dimensional space, we change from coordinates x to coordinates $\bar{x}$. Even though $\bar{x}$ may be a non-linear function of x, $\partial x/\partial \bar{x}$ is a linear operator (an $n \times n$ matrix, where n is the dimension of the space), and has a determinant. The change-of-variables formula is the familiar Jacobian formula,

$$(dx) = (d\bar{x}) \det (\partial x/\partial \bar{x}). \tag{4.59}$$

As always, we simply extend this to the infinite-dimensional case, obtaining

$$[\det K]^{\frac{1}{2}} \prod_a (dq^a) = [\det K]^{\frac{1}{2}} \prod_a (d\bar{q}^a)\det(\partial q/\partial \bar{q})$$
$$= \prod_a (d\bar{q}^a)[\det \bar{K}]^{\frac{1}{2}}. \tag{4.60}$$

Thus, Eq. (4.56) is independent of our choice of coordinates.

Eq. (4.56) is sometimes written in 'Hamiltonian form',[22]

$$e^{iW} = N \int \prod_a (dp^a)(dq_a)e^{iS}, \tag{4.61}$$

where S is, as usual, the integral of the Lagrangian, but the Lagrangian is written as a function of the ps and qs, considered as independent variables,

$$L = p^a \dot{q}_a - H.$$ (4.62)

Formally, it is easy to see that this is equivalent to our earlier formula, by explicitly doing the integral over the ps. This is an integral of the exponential of a quadratic form, so it can be done with Eq. (4.9). We see that we get a determinant in front, just the one we need. In addition, in the exponential, the ps are replaced by their values at the point where S is stationary with respect to variations of the ps. This means that we must solve the equations

$$\dot{q}^a = \frac{\partial H}{\partial p_a}$$ (4.63)

But this just reverses the standard passage from the Lagrangian to the Hamiltonian, and recreates the Lagrangian in its original form, as a function of the qs and $\dot{q}$s.

The Hamiltonian form of the functional integral must be taken with a grain of salt. Unlike the Lagrangian form, the derivative terms do not become nicely damped exponentials when we rotate to imaginary time; they stay oscillating. Thus the Hamiltonian integral is much less well-defined than the Lagrangian one. Indeed, one can show that not even the most ingenious mathematician can make it well-defined; it is possible to find examples for which the value one assigns to (4.61) depends on whether one integrates first over the ps or first over the qs. (A simple one is $H = p^2 + q^2 + \lambda p^2 q^2$; the differences arise in perturbation theory in order λ^2.) However, there is nothing wrong with (4.61) as long as you remember that always attached to it is the rule: first integrate over the ps formally, then rotate to imaginary times.

4.5   Fermi fields

Everything we have done until now has been for Bose fields. What about Fermi fields? For Bose fields, we found that the generating functional could be represented as an integral over ordinary c-number fields, the classical limits of Bose fields. By analogy, we would expect that the generating functional for a theory involving Fermi fields could be written as an integral over the classical limits of Fermi fields, *anticommuting c-number fields*. Anticommuting c-numbers are notoriously objects that make strong men quail; fortunately, we will be able to circumvent the problem of defining functional integrals involving them.

Suppose we were able to define a functional integral over Fermi fields. What sort of integrals would we want to evaluate? In any theory we are interested in, the Fermi fields enter the Lagrangian at most quadratically. Thus, if we denote the Fermi field(s) by η and the conjugate field(s) by η^*, the part of the action involving Fermions is of the form

$$S_f = (\eta^*, A\eta). \tag{4.64}$$

Here A is typically the sum of two terms: a constant term, from the free fermion Lagrangian, and a term involving Bose fields, from the couplings to spinless mesons and/or gauge fields. For the moment, let us consider S_f to be the total action, and the Bose fields referred to above as external fields. (We can always integrate over them later; we know how to integrate over Bose fields.) If we were able to define a functional integral over Fermi fields, we would like to prove that

$$\langle 0^+|0^-\rangle = e^{iW} = N \int (d\eta^*)(d\eta)e^{iS_f}. \tag{4.65}$$

Now let us consider the identical integral with η a complex Bose field. In this case, we know how to do the functional integral, by Eq. (4.11),

$$\int (d\eta^*)(d\eta)e^{i(\eta^*, A\eta)} = [\det(iA)]^{-1}. \tag{4.66}$$

We also know how to directly evaluate W, by perturbation theory. W is the sum of all connected Feynman graphs. For an action of the form (4.64), these are just single-closed-loop graphs, like those drawn in Fig. 5 (except that here the lines should have arrows on them, because the field is complex). What happens to the perturbation expansion if we replace bosons by fermions? The only difference is that there is a factor of minus one for every closed Fermi loop. Every graph that contributes to W has one and only one closed loop; therefore, W is replaced by minus W, or, equivalently, the inverse determinant in Eq. (4.66) is replaced by the determinant.

Thus, we would get the right answer if

$$\int (d\eta^*)(d\eta)e^{i(\eta^*, A\eta)} = \det(iA), \tag{4.67}$$

up to a constant factor, which we can always absorb in the normalization constant, N. Therefore, we *define* the left-hand side of this equation to be equal to the right-hand side. This is a poor substitute for a deep theory of integration over anticommuting c-numbers, but it does give up a compact expression (the determinant) for a sum over Fermi closed loops, and it will turn out that this is all we will need for our purposes.

4.6   *Ghost fields*

We left the theory of derivative interactions in poor shape. It is true that we had an expression for the generating functional, Eq. (4.56), but it was not in the form of an integral of an exponential; there was a determinant sitting in front. Therefore, we could not use Eq. (4.56) to develop a diagrammatic perturbation expansion of the integral. We can now use our knowledge of Fermi fields to get the determinant up into the exponential. For, if we introduce a set of complex Fermi variables, η^a, and denote by $K^{\frac{1}{2}}$ the matrix square-root of K, then

$$[\det K]^{\frac{1}{2}} = \int (d\eta^*)(d\eta)e^{i(\eta^{a^*}, K_{ab}^{\frac{1}{2}}\eta^b)}, \tag{4.68}$$

up to a multiplicative constant, which can always be absorbed in the normalization factor, N. The ηs are called ghost variables (in the field-theory case, ghost fields). They are not true dynamical variables of the system, simply devices for getting a determinant up into an exponential.

Thus, the Feynman rules for the theory can be read off from an 'effective Lagrangian',

$$L_{\text{eff}} = L + L_{\text{g}}, \tag{4.69}$$

where L_{g}, the ghost Lagrangian, is given by

$$L_{\text{g}} = \eta^{*a} K_{ab}^{\frac{1}{2}} \eta^b. \tag{4.70}$$

It is instructive to work out in detail a field-theoretic example. Let us consider the theory of a free field coupled to an external source,

$$\mathscr{L} = \tfrac{1}{2}(\partial_\mu \phi)^2 - \tfrac{1}{2}\mu^2\phi^2 + J\phi. \tag{4.71}$$

Let us make a change of variables to a new field, A, defined by

$$\phi = A + \tfrac{1}{2}gA^2. \tag{4.72}$$

where g is a constant. (This transformation is not invertible, but that shouldn't worry us; we're only going to do perturbation theory, and (4.72) is invertible near $\phi = 0$.) In terms of A, the Lagrange density is given by

$$\mathscr{L} = \tfrac{1}{2}(\partial_\mu A)^2(1 + gA)^2 - \tfrac{1}{2}u^2 A^2(1 + \tfrac{1}{2}gA)^2 + JA(1 + \tfrac{1}{2}gA). \tag{4.73}$$

Thus we apparently have a very complicated interaction, with g some sort of coupling constant. Of course, this interaction is just an illusion; the vacuum-to-vacuum matrix element must be the same as in our original theory. However, this is not the answer you will get if you just read the Feynman rules naively out of (4.73). The right Feynman rules are obtained from an effective Lagrange density

$$\mathscr{L}_{\text{eff}} = \mathscr{L} + \mathscr{L}_{\text{g}}, \tag{4.74}$$

where

$$\mathscr{L}_g = \eta^* \eta (1 + gA). \tag{4.75}$$

The unphysical nature of the ghost fields is doubly clear from this expression. (1) The ghost fields are spinless fields obeying Fermi statistics. (2) The ghost propagator has no momentum dependence; it is a constant, i.

I recommend that you compute a few things to low orders of perturbation theory, using this effective Lagrange density, to convince yourself that everything works out as it should. A good starting point is the one-point function (tadpole) to order g. This should vanish. Does it?

5 The Feynman rules for gauge field theories

5.1 *Troubles with gauge invariance*

The quantization of gauge field theories is notoriously tricky. We can get an idea of the problem if we look at the simplest gauge-invariant field theory, electromagnetism.

$$\mathscr{L} = -\tfrac{1}{4}(\partial_\mu A_\nu - \partial_\nu A_\mu)^2 + \mathscr{L}'. \tag{5.1}$$

Let us try and derive the Feynman propagators for A_μ by straightforwardly applying the methods of Section 4, without worrying about whether electromagnetism is in fact in the class of theories we discussed there. The computation is simplified by splitting the field into (four-dimensional) transverse and longitudinal parts

$$A_\mu = A_\mu^T + A_\mu^L$$
$$= (P_{\mu\nu}^T + P_{\mu\nu}^L) A^\nu. \tag{5.2}$$

where the Ps are the transverse and longitudinal projection operators; in Fourier space they are given by

$$P_{\mu\nu}^T = g_{\mu\nu} - k_\mu k_\nu / k^2, \quad P_{\mu\nu}^L = k_\mu k_\nu / k^2. \tag{5.3}$$

(Remember, we are secretly doing all our computations in Euclidean space, so there is no ambiguity in dividing by k^2.) Then it is easy to see that

$$S = \int d^4x [\tfrac{1}{2}(\partial_\mu A_\nu^T)^2 + \mathscr{L}']. \tag{5.4}$$

We obtain the propagators for the transverse and longitudinal parts of the field by our standard formulae; thus

$$D_{\mu\nu}^F = \left(g_{\mu\nu} - \frac{k_\mu k_\nu}{k^2} \right) \left(-\frac{i}{k^2} \right) + \frac{k_\mu k_\nu}{k^2} \left(\frac{i}{0} \right). \tag{5.5}$$

The second term is obviously unacceptable; something has gone wrong.

This debacle can be explained in two ways, either from Feynman's sum over histories or from conventional canonical quantization. (1) Sum-over-histories explanation: Feynman says that to compute a transition amplitude you must sum over all possible histories of the system. This is normally what the functional integral does for us. However, in a gauge theory, summing over all gauge fields, $A_\mu(x)$, sums over each history an infinite number of times, because fields that are connected by a gauge transformation do not represent different histories but a single history. No wonder we got divergent nonsense! (2) Canonical explanation: To canonically quantize a dynamical system, you have to find a set of initial-value variables, ps and qs, which are complete, in the sense that their values at time zero determine the values of the dynamical variables at all times. It is only in this case that the imposition of canonical commutators at time zero will determine commutators at all times and define a quantum theory. In a gauge theory, this can never be done, because you can always make a gauge transformation that vanishes at time zero but does not vanish at some other time; thus you can never find a complete set of initial-value variables. To quantize a gauge theory, you must first pick a gauge, impose some condition that eliminates the freedom to make gauge transformations. Then, if you are clever and/or lucky in your choice of gauge, you may be able to canonically quantize the theory. Of course, physical quantities are gauge-invariant, and therefore should not depend on what gauge you pick for quantization, but this always has to be proved explicitly in every particular case. We worried about none of this; no wonder we get divergent nonsense!

Both of these explanations emphasize gauge invariance as the critical feature. I personally prefer the second to the first; the injunction to sum over histories seems to me to be incomplete, for it does not tell us what measure to use when the sum is continuous, and, as we saw in our study of derivative interactions, this is not a trivial question. However, in the first part of our investigation, I will accept a quantization method invented by Fadeev and Popov, which is inspired by the first viewpoint. Later on, I will justify the Faddeev–Popov method by appealing to canonical quantization. (Please do not think I am being original in this last step; I learned the canonical justification from Faddeev.)

5.2   *The Faddeev–Popov Ansatz*

As usual, I will begin by discussing finite-dimensional integrals and later extend the results to function spaces. Our model of the function space of a gauge field theory will be a space of $n + m$ real variables, which we denote collectively by z. We will also denote the first n of these variables by x, and the last m by y. The xs will be our finite-dimensional model of the gauge-independent variables (in electrodynamics, A_μ^T) and the ys of the gauge-dependent variables (A_μ^L, in electrodynamics). We will also have a model of a gauge-invariant action, a function $S(z)$, which is independent of the y-variables,

$$\frac{\partial S}{\partial y} = 0. \tag{5.6}$$

We wish to define a (finite-dimensional model of the) generating functional that avoids the divergence problems we would encounter if we integrated over all the zs. This is easy; we just integrate over the xs only, and define

$$e^{iW} = \int (dx) e^{iS}. \tag{5.7}$$

(We suppress the normalization factor for the moment.) This can also be written as

$$e^{iW} = \int (dz) e^{iS} \delta(y). \tag{5.8}$$

Here $\delta(y)$ is an m-dimensional δ-function, normalized such that

$$\int g(y)(dy)\delta(y) = g(0), \tag{5.9}$$

for any function g. Eq. (5.8) says that we integrate along the surface $y = 0$. Of course, since nothing depends on the ys, we could just as well integrate along an arbitrary surface, defined by

$$y = f(x), \tag{5.10}$$

where f is an m-vector, a set of m functions of the xs. We then obtain

$$e^{iW} = \int (dz) e^{iS} \delta(y - f(x)). \tag{5.11}$$

We may not be given the surface in the form (5.10), but as the solution to some set of equations,

$$F(z) = 0, \tag{5.12}$$

where F is again an m-vector, a set of m functions of the zs. It is easy to rewrite the integral in a form appropriate to this description of the surface,

$$e^{iW} = \int (dz) e^{iS} \det(\partial F/\partial y) \delta(F(z)). \tag{5.13}$$

Note that, because of the presence of the δ-function, we need only evaluate the determinant on the surface. I emphasize that Eq. (5.13) defines the same expression as Eq. (5.7), and is completely independent of our choice of the functions F.

I will now state the Faddeev–Popov[23] quantization procedure. Let S be the action integral for a theory involving m fields, $\phi^a(x)$ (not necessarily all scalar). Let S be invariant under some group of gauge transformations, parametrized by a set of n real functions, $\omega^b(x)$. For such a theory, a 'gauge' is defined to be a set of n equations

$$F^b(x) = 0. \qquad (5.14)$$

where the Fs are functions of the ϕs, possibly differential or even non-local, such that, given any $\phi^a(x)$, there is one and only one gauge transformation that makes Eq. (5.14) true. For electrodynamics, an example of a gauge is radiation gauge, $\nabla \cdot \mathbf{A} = 0$. Another example is Lorentz gauge, $\partial_\mu A^\mu = 0$. (You may object that in this case the gauge transformation is not unique. This is true in Minkowski space, but remember that we are always secretly working in Euclidean space.) According to Fadeev and Popov, the theory is now quantized by declaring that

$$e^{iW} = N \int \prod_a (d\phi^a) e^{iS} \det\left(\frac{\partial F^b}{\partial \omega^c}\right) \prod_b \delta(F^b). \qquad (5.15)$$

where $F^a = 0$ is some gauge. This is the functional analogue of Eq. (5.13). The δ-function in Eq. (5.15) is a δ-function on function space, a δ-functional if you will; it obeys the equation

$$\int (d\phi) G(\phi)\delta(\phi) = G(0), \qquad (5.16)$$

for any functional G. We will call Eq. (5.15) the Fadeev–Popov Ansatz.

Remarks. (1) The choice of gauge in the Faddeev–Popov Ansatz is equivalent to the choice of surface in the finite-dimensional integral we discussed earlier. Thus, whether the Ansatz is true or false, it is at least self-consistent; it is independent of the choice of gauge. (2) Thus, to verify the Ansatz, it suffices to verify it for just one gauge. If it is true in one gauge, it is true in any other. (3) The gauge-independence of the Ansatz depends on the action being gauge-independent. Thus, the action can not contain source terms coupled linearly to the gauge fields. However, it can contain source terms coupled to gauge-invariant objects, like $(F^a_{\mu\nu})^2$, for example. Phrased in another way, the Ansatz only gives us gauge-invariant expressions for gauge-invariant Green's functions. Since the standard wisdom is that in a gauge theory only gauge-invariant quantities are physical observables. this is no great restriction. Also, once we have settled down in some fixed gauge, there is no objection to computing non-gauge-invariant objects, like gauge-field propagators, as a preliminary step in the computation of gauge-invariant objects. (4) I remind you that everything we are doing is on a purely formal level; we are ignoring complications that may arise as a result of ultraviolet divergences. Therefore, everything we do should be taken as merely heuristic, to be checked later by more careful analysis. The manipulation of functional integrals is more efficient than other formal methods of treating gauge theories, but it is no more rigorous.

5.3   *The application of the Ansatz*

We will begin with the simplest gauge theory, electrodynamics. Since gauge transformations for this theory are parametrized by only a single function, only one equation is needed to determine a gauge. We will choose a slight generalization of the Lorentz gauge,

$$F = \partial^\mu A_\mu - f(x), \tag{5.17}$$

where $f(x)$ is an arbitrary function. Under an infinitesimal gauge transformation, Eq. (2.28),

$$\delta F = -e^{-1} \Box^2 \delta \omega. \tag{5.18}$$

Thus,

$$\det(\delta F/\delta \omega) = \det(-e^{-1} \Box^2). \tag{5.19}$$

This is a constant and can be brought outside the integral and absorbed in the normalization. Thus we obtain

$$e^{iW} = N \int (dA)(d\psi)e^{iS}\delta(\partial^\mu A_\mu - f(x)) \tag{5.20}$$

where, to simplify notation, I have indicated by (dA) the integrals over all four components of the gauge fields, and by $(d\psi)$ the integrals over all other fields in the theory.

We still do not have the integral of an exponential, so it is hard to evaluate Eq. (5.20) perturbatively. This is easily rectified. Since the integral is independent of the function f, we can integrate it with any functional of f, $G(f)$, without changing the integral (except perhaps for a normalization, which can always be absorbed in N). Thus,

$$e^{iW} = N \int (dA)(d\psi)(df)e^{iS}\delta(\partial^\mu A_\mu - f)G(f)$$

$$= N \int (dA)(d\psi)e^{iS}G(\partial^\mu A_\mu) \tag{5.21}$$

In particular, if we choose

$$G(f) = e^{-\frac{i}{2\alpha}\int d^4x f^2}, \tag{5.22}$$

where α is some real number, we find

$$e^{iW} = N \int (dA)(d\psi)e^{i\left[S - \frac{1}{2\alpha}\int d^4x(\partial^\mu A_\mu)^2\right]}. \tag{5.23}$$

Thus the outcome of our application of the Ansatz is to replace the

Lagrange density of the theory by an effective Lagrange density

$$\mathscr{L}_{\text{eff}} = \mathscr{L} - \frac{1}{2\alpha} (\partial^\mu A_\mu)^2. \tag{5.24}$$

From this it is easy to compute the electromagnetic propagator, since

$$\int d^4x (\partial^\mu A_\mu)^2 = \int d^4x (\partial^\mu A_\nu^L)^2. \tag{5.25}$$

Hence, the preposterous Eq. (5.5) is replaced by

$$D_{\mu\nu} = \frac{-i}{k^2} \left[g_{\mu\nu} - \frac{k_\mu k_\nu}{k^2} + \alpha \frac{k_\mu k_\nu}{k^2} \right]. \tag{5.26}$$

Any one of these propagators should give the same results as any other in the computation of gauge-invariant objects. (I hope you know enough about electrodynamics from other sources to recognize this as a true statement.) The choice $\alpha = 1$ yields what is usually called the Feynman-gauge propagator, $\alpha = 0$ the Landau-gauge propagator, etc. For any choice of α, the propagator has the same high-energy behaviour as that of a scalar field, and therefore the dimension-counting formulae of Section 3.1 are applicable in computing counterterms.

Now let us turn to non-Abelian gauge fields. For notational simplicity, we will restrict ourselves to the case where there is only one gauge-field coupling constant. We determine a gauge as in Eq. (5.17),

$$F^a = \partial^\mu A_\mu^a - f^a, \tag{5.27}$$

where the fs are arbitrary functions. Under an infinitesimal gauge transformation, (2.44),

$$\delta F^a = g^{-1}[-\Box^2 \delta\omega^a + gc^{abc}\partial^\mu(\delta\omega^b A_\mu^c)]. \tag{5.28}$$

In contrast to the Abelian case, here the determinant is not a constant. However, just as in Section 4.6, we can write it as an integral over a set of ghost fields, scalar fields obeying Fermi statistics,

$$\det \left(\frac{\delta F^a}{\delta\omega^b} \right) = \int (d\eta^*)(d\eta) e^{iS_g}, \tag{5.29}$$

where

$$S_g = \int d^4x \, \mathscr{L}_g = \int d^4x (\partial_\mu \eta^{*a})(\partial^\mu \eta^a - gc^{abc}\eta^b A_\mu^c). \tag{5.30}$$

and we have chosen to absorb an overall factor of $\det g^{-1}$ into N. The ghost Lagrange density can be written in a compact way if we consider the ghosts as a set of fields that transform according to the adjoint representation of the group,

$$\mathscr{L}_g = \partial_\mu \eta^{*a} D^\mu \eta^a. \tag{5.31}$$

In this form it is clear that $\mathscr{L}_g$ is not gauge-invariant; of course, there is no reason why it should be, since it is derived from the (purposefully) non-gauge-invariant Eq. (5.27). In contrast to the example of Section 4.6, here the ghosts have a momentum-dependent propagator, that of a set of massless charged scalar fields, $i\delta^{ab}/k^2$. However, they still reveal their unphysical nature by being spinless particles obeying Fermi statistics.

The remainder of the development of the non-Abelian case is exactly the same as that of the Abelian case. Thus we arrive at the effective Lagrange density

$$\mathscr{L}_{eff} = \mathscr{L} + \mathscr{L}_g - \frac{1}{2\alpha}(\partial^\mu A_\mu^a)^2. \tag{5.32}$$

where α is an arbitrary real number. Note that if $\mathscr{L}$ is the Lagrange density for a renormalizable field theory minimally coupled to gauge fields, every term in this expression is an interaction of renormalizable type (dimension less than or equal to four). As I explained at the end of Section 3.1, this observation is just the first step in establishing renormalizability, but it is as far as we will have time to go here.

5.4   Justification of the Ansatz

I will now justify the Faddeev–Popov Ansatz by showing that, in a particular gauge, it is equivalent to canonical quantization. The gauge is Arnowitt–Fickler[24] gauge (sometimes called axial gauge); it is defined by

$$F^a = A_3^a = 0. \tag{5.33}$$

where the 3 indicates the third spatial component. Unlike Eq. (5.27), this is not Lorentz-covariant, so this is a terrible gauge for performing Feynman calculations; however, this is not our purpose. For simplicity, I will construct the proof for pure Yang–Mills fields, uncoupled to other fields; The generalization is straightforward.

First we must construct the Faddeev–Popov Ansatz:

$$\delta F^a = -g^{-1}\partial_3\delta\omega^a + c^{abc}\delta\omega^b A_3^c$$
$$= -g^{-1}\partial_3\delta\omega^a. \tag{5.34}$$

The second line follows from Eq. (5.33). Thus, the determinant is a constant, and can be absorbed in the normalization factor; in this gauge, *there are no ghosts*, even in the non-Abelian case. Thus, the Ansatz becomes

$$e^{iW} = N \int (dA)e^{iS} \prod_a \delta(A_3^a), \tag{5.35}$$

where (dA) indicates integration over all the As. More explicitly,

$$(dA) \prod_a \delta(A_3^a) = \prod_a (dA_1^a)(dA_2^a)(dA_0^a). \tag{5.36}$$

I remind you that

$$S = -\frac{1}{4} \int d^4 x (\partial_\mu A_\nu^a - \partial_\nu A_\mu^a + g c^{abc} A_\mu^b A_\nu^c)^2, \tag{5.37}$$

plus source terms, which I shall not bother to write explicitly.

(This is off the main line of the argument, but it is a point that may have been worrying you: the ghosts are fictitious particles, but they do have real poles in their propagators. Therefore, it seems that states involving ghosts might contribute to the absorptive parts of gauge-invariant Green's functions. This would be disturbing if it happened; fortunately, the existence of a ghost-free gauge shows that it does not.)

It will be convenient to rewrite the Ansatz in so-called first-order form,

$$e^{iW} = \int (dF)(dA) \prod_a \delta(A_3^a) e^{iS'}, \tag{5.38}$$

where

$$S' = \int d^4 x [-\tfrac{1}{2}(F_{\mu\nu}^a)^2 + \tfrac{1}{4} F^{\mu\nu a}(\partial_\mu A_\nu^a - \partial_\nu A_\mu^a + g c^{abc} A_\mu^b A_\nu^c)], \tag{5.39}$$

and (dF) denotes integration over all the Fs. The integral over the Fs is trivial and obviously reproduces Eq. (5.35). S' is also equivalent to S in the normal sense of Lagrangian dynamics; if we vary S' with respect to the Fs and As independently, we get the same equations of motion we obtain by varying S with respect to the As alone. That S' is a good action in both these senses is no coincidence; it is a consequence of the integration formula (4.9). If a dynamical variable appears in the action at most quadratically, and if the coefficient of the quadratic term is a constant, then integrating over the variable is the same as eliminating it from the action by using the Euler–Lagrange equations.

So much for the Ansatz; now let us turn to canonical quantization. Again, we will use S', the first-order action, and work in Arnowitt–Fickler gauge, setting A_3^a equal to zero. Let us write Eq. (5.39) in such a way that the dependence on various tensor components is explicit:

$$\mathcal{L}' = -\tfrac{1}{2}(F_{\mu\nu}^a)^2 + \tfrac{1}{2} F^{ija}(\partial_i A_j^a - \partial_j A_i^a + g c^{abc} A_i^b A_j^c)$$
$$+ F^{0ia}(\partial_0 A_i^a - \partial_i A_0^a + g c^{abc} A_0^b A_i^c)$$
$$+ F^{03a}(-\partial_3 A_0^a) + F^{i3a}(-\partial_3 A_i^a). \tag{5.40}$$

where i and j run over the range 1, 2. Note the drastic simplification of the last two terms, caused by the gauge condition. We now see that canonical quantization of (5.40) is like shooting fish in a barrel: A_0^a, F^{ija}, F^{03a}, F^{13a} are constrained variables; their Euler–Lagrange equations involve no time derivatives and are, therefore, not true equations of motion but equations of constraint, fixing the constrained variables on the initial-value surface in terms of the remaining variables, A_i^a and F^{0ia}.

Let us denote the action obtained by eliminating the constrained variables by S''; it is a functional only of A_i^a and F^{0ia}. Furthermore, it is in Hamiltonian form, with the As the canonical fields and the Fs the conjugate momentum densities. Thus, we can use Eq. (4.61) to write

$$e^{iW} = N \int \prod_a (dF^{01a})(dF^{02a})(dA_1^a)(dA_2^a)e^{iS''}. \tag{5.41}$$

However, because the constrained variables enter Eq. (5.40) at most quadratically, and because the coefficients of the quadratic terms are constants, we can equally well write this as

$$e^{iW} = N \int (dF) \prod_a (dA_0^a)(dA_1^a)(dA_2^a)e^{iS'}$$

$$= N \int (dF)(dA) \prod_a \delta(A_3^a)e^{iS'}. \tag{5.42}$$

But this is the Faddeev–Popov Ansatz, Eq. (5.38). Q.E.D.

5.5 Concluding remarks

(1) The chain of arguments we have just constructed shows both the power and the limitations of functional-integral methods. Functional integration is a supplement to canonical quantization, not a replacement for it. For example, when writing down the Ansatz, I could well have multiplied the integrand by some function of $(F_{\mu\nu}^a)^2$. This would have been just as gauge-invariant, and just as plausible *a priori* as the original Ansatz. It would have been wrong, but there would be no way to tell this without appealing to canonical quantization. On the other hand, once we have justified the Ansatz by canonical quantization, we can use it to pass from one gauge to another with incomparable ease. In particular, we can use it to pass from a gauge in which canonical quantization is simple to a gauge in which the Feynman rules are simple.

(2) I have said this before, but it deserves emphasis: Everything we have done in this section is purely heuristic; we have paid no attention to the problems caused by ultraviolet divergences. Properly, everything should be redone with careful attention to cutoffs, renormalizations, etc. Such careful investigations have been done;[11] the result is that the heuristic arguments have not betrayed us: these theories are renormalizable; renormalization does not spoil gauge invariance; ghost states never contribute to the absorptive parts of gauge-invariant Green's functions; etc. There is one exception: in theories in which some of the gauge trans-

formations are chiral, the familiar Adler–Bell–Jackiw triangle anomalies can falsify our arguments. However, if the transformation properties of the Fermi fields are chosen such that there are no anomalies in the lowest-order triangle graphs with gauge currents at the vertices, then there are no anomalies anywhere, and everything is all right.[25]

(3) People are sometimes worried that the formal apparatus for treating spontaneous symmetry breakdown, explained in Section 3, is not gauge-invariant. This is true; the vacuum expectation value of a scalar field, the effective potential, indeed, even the Feynman propagators themselves, are not gauge-invariant objects. This is also irrelevant. In quantum electrodynamics, we continually do computations using non-gauge-invariant objects, like propagators, at intermediate stages. There is nothing wrong with this, as long as we are careful to express our final results in terms of gauge-invariant quantities, like masses and cross-sections. The occurrence of spontaneous symmetry breakdown does not affect this; the form of the effective potential and the location of its minimum are indeed gauge-dependent, but the values of masses and cross-sections computed with the aid of these objects are not.

(4) At the end of Section 3, I explained how many workers prefer to do computations in terms of shifted fields, defined by

$$\phi' = \phi - \langle \phi \rangle, \tag{5.43}$$

and to determine the parameters $\langle \phi \rangle$ at the end of the computation, by self-consistency. There is one awkwardness in doing things this way; the shift generates a bilinear scalar–vector coupling from the scalar Lagrange density:

$$\tfrac{1}{2}D^\mu\phi \cdot D_\mu\phi + \cdots = g\partial^\mu\phi' \cdot A_\mu^a T_a\langle\phi\rangle + \cdots \tag{5.44}$$

This coupling causes a scalar–vector mixed propagator to appear in the Feynman rules of the theory; this is no difficulty in principle, but is an annoyance in practice. Fortunately, it is possible to cancel this term by a clever choice of gauge.[26] For our gauge condition, we choose

$$F^a = \partial^\mu A_\mu^a - f^a(x) - \xi\phi' \cdot T_a\langle\phi\rangle. \tag{5.45}$$

where ξ is a number to be determined later. If we go through what should be by now familiar arguments, we obtain an effective Lagrange density of the form

$$\begin{aligned}
\mathscr{L}_{\text{eff}} = \mathscr{L} &+ \partial_\mu\eta^{*a}D^\mu\eta^a \\
&- g\xi\eta^{*a}\eta^b(T_b\langle\phi\rangle \cdot T_a\langle\phi\rangle + T_b\phi' \cdot T_a\langle\phi\rangle) \\
&- \frac{1}{2\alpha}(\partial^\mu A_\mu^a - \xi\phi' \cdot T_a\langle\phi\rangle)^2.
\end{aligned} \tag{5.46}$$

Hence, if we choose

$$\xi = \alpha g, \tag{5.47}$$

we can cancel the annoying cross terms. Note that the interactions in this Lagrange density are still of renormalizable type, dimension less than or equal to four.

6 Asymptotic freedom

6.1 *Operator products and deep inelastic electroproduction*

The topic we are now going to discuss seems, at first glance, to have very little to do with the previous lectures. It is a topic in strong-interaction physics, that of reconciling the apparent scaling in the SLAC–MIT electroproduction experiments with the predictions of quantum field theory. I will begin by summarizing very briefly the standard lore on this problem.[27]

(1) The electroproduction experiments at Stanford measure the total cross sections for the process

electron + nucleon → electron + anything,

which is, of course, the same thing as

virtual photon + nucleon → anything.

The process is therefore described by two kinematic variables: q^2, the mass of the virtual photon, a negative number, and E, the energy of the virtual photon in the lab frame. It is convenient to trade E for the dimensionless variable

$$x = -q^2/2mE, \tag{6.1}$$

where m is the nucleon mass. Elementary kinematics restricts x to be between zero and one. The nucleons in the experiment are unpolarized, while the virtual photons can be either transverse or longitudinal; thus the cross-section can be described in terms of two dimensionless invariants, $F_i(q^2, x)$, where i is 1 or 2. The Fs are called structure functions; the details of their definitions will not be relevant to our immediate purposes.

As $-q^2$ increases, the Fs quite rapidly lose their dependence on q^2; by $q^2 = -(2\,\text{GeV})^2$, the Fs appear to be functions of x alone, within experimental error. This phenomenon is called Bjorken scaling. There are two schools of thought on Bjorken scaling. (1) Bjorken scaling is a true asymptotic phenomenon. It will persist even if the range of q^2 is increased greatly. (2) SLAC energies are too small for us to believe that we are really in the asymptotic region. Bjorken scaling is some sort of low-energy epiphenomenon, and has nothing to do with true high-energy limits.

I will adopt the first position for this lecture, but you should be aware that this is just a matter of prejudice. The second position may well be correct; only future experiment can decide the question.

It will turn out to be convenient for our purposes to phrase matters in terms of the moments of the structure functions,

$$F_i^n(q^2) = \int_0^1 dx \, x^n F_i(q^2, x).$$ (6.2)

The problem is: why do these moments become constants (within experimental error) as q^2 becomes large and spacelike?

(2) The operator product expansion was invented by Wilson and proved to all orders of renormalized perturbation theory by Callan and Zimmerman. It is an asymptotic expansion for the product of two local operators as the distance between them becomes small, but for our purposes it will be most convenient to express the expansion in momentum space. Let A and B be any two local operators (renormalized polynomials in canonical fields and their derivatives) and let $|a\rangle$ and $|b\rangle$ be any two states. Then,

$$\int e^{iq \cdot x} d^4x \, \langle a|A(x)B(-x)|b\rangle$$

$$= \sum_C f_{ABC}(q) \langle a|C(0)|b\rangle.$$ (6.3)

as q goes to Euclidean infinity. The sum is over a complete set of local operators (all renormalized monomials in canonical fields and their derivatives). The expansion is useful because the rate of growth of the coefficient functions, the fs, is that given by naive dimensional analysis, modulo polynomials in $\ln q^2$. (This is true to any finite order in perturbation theory; we will later investigate whether these polynomials can pile up and change the asymptotic behavior if we sum the perturbation series.) Thus, for any given A and B in any given field theory, only a finite set of operators contributes to the leading asymptotic behaviour; higher monomials give lower powers of q^2.

I emphasize that the fs are independent of the states $|a\rangle$ and $|b\rangle$. In particular, this means the operator product expansion is unaffected by the occurrence of spontaneous symmetry breakdown. This will be important to us later.

If A, B, and C are other than Lorentz scalars, f_{ABC} will have a non-trivial tensor structure. Since f is only a function of a single four-vector, q, it is a known tensor function of q times an unknown scalar function of q^2. It will be convenient to multiply this scalar function by a power of q^2

so that it becomes dimensionless. We will call the resulting dimensionless scalar function $\bar{f}_{ABC}$.

(3) By choosing $|a\rangle$ and $|b\rangle$ to be one-nucleon states, and A and B to be electromagnetic currents, we can use the operator product expansion to get an expression for the moments of the structure functions. The calculation is straightforward, and I do not want to do it in detail here; the result is of the form

$$F_i^n(q^2) = \sum_C d_{iC}^m \bar{f}_{ABC}(q^2) \langle a|C(0)|b \rangle_R. \tag{6.4}$$

Here the ds are constant coefficients, terms that are less important by powers of q^2 than the terms retained have been dropped, the subscript R indicates a reduced (scalar) matrix element, and, in any given theory, for fixed i and n, the sum runs over only a finite set of Cs.

Thus the problem becomes: why do the $\bar{f}$s become constants (within experimental error) as q^2 becomes large and negative?

Because of the logarithmic polynomials mentioned before, this constant behavior is not an obvious prediction of field theory. To be specific, let us consider a theory in which there is only one coupling constant, like the standard quark–vector-gluon model. In this case, a perturbative expansion of one of the $\bar{f}$s typically yields an asymptotic expression like

$$\bar{f} = a_0 + a_{11}g^2\ln q^2 + a_{10}g^2 + a_{22}g^4(\ln q^2)^2 + \cdots \tag{6.5}$$

where the as are constant coefficients. Since we are interested in both large g (strong interactions) and large q^2 (asymptotic behaviour), this is worse than useless. Even for the (unrealistic) case of small g, Eq. (6.5) tells us noting about asymptotic behaviour, for the largeness of the logarithm eventually overcomes the smallness of g. The only case in which we can predict asymptotic behaviour is free field theory ($g=0$); in this case, the $\bar{f}$s are indeed constants. It is for this reason that it is sometimes said that, at high negative q^2, the effects of the interactions seem to disappear, and the theory behaves as if it were free. At the moment, this may seem to you to be an excessively dramatic way of describing Bjorken scaling; nevertheless, we shall see, for a certain class of field theories, this is exactly what happens. Before we do this though, we need to develop a systematic formalism for going beyond perturbation theory and summing up the logarithms in Eq. (6.5).

6.2   *Massless field theories and the renormalization group*[28]
       It can be shown, in any renormalizable field theory, to all orders of perturbation theory, that the asymptotic behaviour of the coefficient

functions in the operator product expansion is the same as it would be in a massless field theory. By a massless field theory I mean one that has only dimensionless coupling constants in its Lagrangian; not only are masses excluded but also interactions with dimensionful coupling constants, like cubic meson self-couplings. This is very plausible; the coefficient functions depend only on a single momentum, and this momentum is going to Euclidean infinity, getting as far as it can from the mass shell, and therefore losing all memory of the masses. (I emphasize that this does *not* mean that the structure functions themselves are the same as they would be in a massless theory. Eq. (6.4) contains not just $\bar{f}$, but also $\langle a|C|b \rangle$, and this stays on the mass shell.)

Thus we need only analyze the behavior of the $\bar{f}$s in a massless theory. A massless renormalizable field theory is parametrized by a set of renormalized dimensionless coupling constants, which I will call g^a. These may be either Yukawa coupling constants, quartic meson self-interaction constants, or gauge field coupling constants. In addition, another parameter is required to complete the description – a mass, M. This extra parameter is needed to *define* the others (and the scale of the renormalized fields).

Let me explain why this is so, using the simplest renormalizable field theory, $\lambda\phi^4$ theory, as an example. In the massive version of this theory, the renormalized coupling constant, λ, is usually defined as the value of $\Gamma^{(4)}$ on the mass shell, at the symmetry point, $s = t = u$. (Sometimes it is defined as the value of $\Gamma^{(4)}$ when all external momenta vanish, as in Eq. (3.17b), but this becomes the same definition when the mass vanishes.) Likewise, the renormalized field is defined as the field scaled in such a way that the derivative of $\Gamma^{(2)}$ is one on the mass shell (or, sometimes, at zero momentum). However, for a massless theory, these definitions are unworkable; all of the normal thresholds collapse on the renormalization point, and it is obviously bad policy to define λ as the value of a Green's function at the locus of an infinite number of singularities. The cure for this disease is simple; we define λ as the value of $\Gamma^{(4)}$ at some point in Euclidean space, where there are no singularities, even in the massless theory. For example, we would define λ as the value of $\Gamma^{(4)}$ at $s = t = u$, with all external momenta squared equal to $-M^2$. M can be anything; any M is as good as any other M, so long as it is not zero. Likewise, we could define the scale of the field by demanding that the derivative of $\Gamma^{(2)}$ be one when $p^2 = -M^2$. These definitions can be extended in an obvious way to more complicated theories with other kinds of couplings.

Thus, the parameterization of a massless field theory requires a mass, M.

But M is arbitrary; in a given physical theory, if you change the value of M, this can always be compensated for by an appropriate change in the gs and in the scales of the renormalized fields, because the only function of M is to define these quantities. Phrased in equations, if we make a small change in M,

$$M \to M(1+\varepsilon),\tag{6.6a}$$

where ε is infinitesimal, this can always be compensated for by an appropriate small change in the coupling constants

$$g^a \to g^a + \beta^a \varepsilon,\tag{6.6b}$$

and a corresponding small change in the scale of renormalized operators, e.g.

$$A(x) \to (1+\gamma_A \varepsilon)A(x).\tag{6.6c}$$

By dimensional analysis, the βs and γs can depend only on the gs,

$$\beta^a = \beta^a(g),\ \gamma_A = \gamma_A(g),\tag{6.7}$$

where, to simplify notation, a single g in the argument of a function stands for all the gs. If A is one of a set of operators that can mix with one another as a result of renormalization (as is the case, for example, with ϕ^4 and $\partial_\mu \phi \partial^\mu \phi$), Eq. (6.6c) should be replaced by a matrix equation. However, for simplicity, we will ignore this possible complication here.

The infinitesimal transformations (6.6) define a one-parameter group, called the renormalization group. All physical quantities must be invariant under this group. In particular, the $\bar{f}$s must be invariant; thus

$$\left[M\frac{\mathrm{d}}{\mathrm{d}M} + \beta^a(g)\frac{\partial}{\partial g^a} + \gamma_{ABC}(g) \right] \bar{f}_{ABC} = 0.\tag{6.8}$$

where

$$\gamma_{ABC} = \gamma_A + \gamma_B - \gamma_C.\tag{6.9}$$

Of course, similar equations can be derived for any other object in the theory, in particular, for Green's functions. Only the γ-terms depend on the object under consideration.

Since these renormalization-group equations are exactly valid, they must be valid order-by-order in renormalized perturbation theory. Thus, from perturbation expansions of Green's functions, it is possible to deduce perturbation expansions for the βs and γs. If this is done for the quark–vector-gluon model, for example, one finds that the power series for β begins with terms of order g^3, while those for either the quark or gluon γ begin with terms of order g^2. This is reasonable, because β reflects the effects of coupling-constant renormalization, which begin at order g^3, while γ reflects those of wave-function renormalization, which begin at order g^2.

6.3   Exact and approximate solutions of the renormalization group equations

The differential equations of the renormalization group are a mathematical expression of a physical triviality, that the only function of the mass M is to define the renormalized coupling constants and the scale of the renormalized fields. Nevertheless, they can, in favorable circumstances, be used to obtain highly non-trivial information about the asymptotic behaviour of the theory. The basic reason for this is simple dimensional analysis; since $\bar{f}$ is dimensionless,

$$\bar{f}_{ABC} = \bar{f}_{ABC}(Q/M, g), \tag{6.10}$$

where $Q = (-q^2)^{\frac{1}{2}}$. Thus, knowledge of the (trivial) dependence on M is equivalent to knowledge of the (non-trivial) dependence on Q.

To work this out in detail, let me assume that we know the βs and γs exactly. Then there is a standard method[29] for solving the linear partial differential equation

$$\left[M \frac{\partial}{\partial M} + \beta^a(g) \frac{\partial}{\partial g^a} + \gamma(g) \right] \bar{f}(Q/M, g) = 0, \tag{6.11}$$

where I have suppressed the ABC subscript for notational simplicity. The standard method goes in two steps. First, one constructs $g'^a(g, t)$, a set of functions of the gs and a single extra variable, t, defined as the solution to the ordinary differential equations

$$dg'^a/dt = \beta^a(g'), \tag{6.12a}$$

with the boundary condition

$$g'^a(g, 0) = g^a. \tag{6.12b}$$

Then, the general solution to Eq. (6.11) is

$$\bar{f} = F(g'(g, \ln[Q/M])) \times \exp \int_0^{\ln[Q/M]} \gamma(g'(g, t)) dt \tag{6.13}$$

where F is an arbitrary function. Thus we see the power of the renormalization group; if we know everything for all gs at $Q = M$, then we know everything for all gs at all Qs.

Unfortunately, we do not know everything for all gs. Typically, we only know the first few terms in a power series in g. Even in this case, though, it is possible to use the renormalization group to squeeze out extra information. To show how this is done, let me return to the quark–

Abelian-vector-gluon model. Here I have argued that

$$\beta(g) = bg^3 + O(g^5).$$ (6.14)

where b is a numerical coefficient. The only thing I will ask you to take on trust is that, if you actually do the relevant Feynman calculations, you will find that b is positive. Eq. (6.12) then becomes

$$dg'/dt = bg'^3 + O(g'^5).$$ (6.15)

Now let us attempt to construct an approximate solution of this equation, for small g, by ignoring the terms of order g^5. The solution is trivially obtained by quadratures,

$$\frac{1}{g'^2} = \frac{1}{g^2} - 2bt,$$ (6.16)

or

$$g'^2 = \frac{g^2}{1 - 2btg^2}.$$ (6.17)

When can we trust this approximate solution? When t gets large and positive, the approximate g' becomes large, and the terms we have neglected become comparable to the terms we have retained. For this range of t, the approximation is garbage. On the other hand, as t becomes large and negative, the approximate g' becomes smaller and smaller, and the terms we have neglected therefore become smaller and smaller than the terms we have retained. For this range of t, the approximation is wonderful.

Now, when we plug g' into Eq. (6.13), t becomes $\ln(Q/M)$. Thus our approximation gets better and better the smaller Q is. Furthermore, we can improve on it as much as we want, simply by doing more perturbation calculations to get the higher terms in the expansions of β, γ, and F.

To phrase the whole thing more generally, an ordinary perturbation expansion, like (6.5), has two conditions for its reliability, $|g| \ll 1$ and $|\ln(Q/M)| \ll 1$. The approximation scheme I have described replaces these with a single condition, $|g'| \ll 1$. This single condition may hold in regions where the logarithm is large; in the case at hand, this includes the region of arbitrarily large negative $\ln(Q/M)$.

This is marvelous stuff; the renormalization group has tamed the logarithms in Eq. (6.5). Unfortunately, this is of no physical interest, for two reasons. (1) To start the approximation, g must be small. We are interested in strong interactions. (2) We can tame the logarithms in the region of small Q, the infrared region. We are interested in large Q, the ultraviolet region. Indeed, our whole method of approach is nonsense in the region of small Q, because, when Q is small, it is no longer sensible to neglect particle masses.

Now let us do another example, pure Yang–Mills theory for some simple Lie group. Here again there is only one coupling constant, and coupling-constant renormalization begins in order g^3, so this is hardly a new example. Everything will be exactly the same as for the quark–vector-gluon model; the only possible difference can be in the value of b, the constant in Eq. (6.15). I now announce the great discovery of the last year: b is *negative*.[30] (This is true whatever the simple Lie group.)

Thus, our previous analysis is turned on its head. Large negative t is replaced by large positive t, infrared by ultraviolet. There exists a family of renormalizable field theories for which the logarithms can be tamed in the ultraviolet region! In this region, we obtain, from lowest-order perturbation theory and the renormalization group, an approximation that gets better and better as Q gets larger and larger. Furthermore, we can improve on the approximation as much as we want, simply by doing more perturbation calculations to get the higher terms in the expansions of β, γ, and F.

6.4   *Asymptotic freedom*

What we have discovered for pure Yang–Mills theory is a special case of a phenomenon called asymptotic freedom. A general renormalizable field theory is said to be asymptotically free if, for small g^a,

$$\lim_{t \to \infty} g^{a'}(g, t) = 0. \tag{6.18}$$

All my remarks for pure Yang–Mills theory carry over without alteration to a general asymptotically free theory; in particular, asymptotic behaviour for large Q is exactly computable from simple perturbation theory and the renormalization group. In principle, it is simple to test whether any given field theory is asymptotically free; all one needs to do is compute the β-functions to lowest non-vanishing order, and then solve the differential equations (6.12). In practice, the test is difficult to carry out; the computation of the β-functions is straightforward, but, in the typical case, the differential equations can not be solved analytically, and one has to resort to tedious case-by-case numerical integration. Thus, although many asymptotically free theories have been discovered, and a few general theorems have been proved, we have nothing like a complete classification of asymptotically free theories. I will tell you what is known about the classification problem shortly; first, though, I would like to convince you that asymptotic freedom offers a possible explanation of Bjorken scaling.

At first thought, this is a preposterous suggestion. Asymptotic freedom is a property of field theories for small coupling constants, and Bjorken scaling is a strong-interaction effect. Nevertheless, it is possible, with a little hand-waving, to establish a connection. For simplicity, let us consider an asymptotically free theory with only one coupling constant, g. By assumption, β is negative for small positive g. Let us denote the first positive zero of β by g_1. We certainly can not compute g_1 perturbatively; if we were asked to guess, we would probably guess that g_1 was something like 1 or π or maybe even infinity (if β has no zeros). In any case, it certainly can not be a small number; for small coupling constants, we trust perturbation theory, and perturbation theory tells us β is negative. Whatever the value of g_1, for any g less than g_1, β is negative. Therefore, if we start from such a g, and integrate

$$dg'/dt = \beta(g'), \tag{6.19}$$

g' will decrease. As we continue to integrate the equation, it will continue to decrease, until we finally reach the region of small g', where formulae like Eq. (6.17) will be valid. Thus, the asymptotic expressions derived from renormalization-group-improved perturbation theory are valid for theories defined by large coupling constants as well as small. If we are very lucky, and β has no positive zeros, they will be valid for all values of g.

The decrease from large to small g' can be quite rapid. As an example, let us take the result of the lowest-order perturbation theory, Eq. (6.16), and imagine that it is valid for large coupling constants as well as small. (I emphasize that this is undoubtedly false; I am just using it as a simple model of rapid decrease.) For a pure Yang–Mills theory with gauge group SU(3),

$$b = -11/16\pi^2. \tag{6.20}$$

Thus, Eq. (6.16) becomes

$$\left(\frac{g'^2}{4\pi}\right)^{-1} = \left(\frac{g^2}{4\pi}\right)^{-1} + \frac{11t}{2\pi}. \tag{6.21}$$

Now let us imagine that we start out with some very large value of $g^2/4\pi$, say 10^3, at $t = 0$. Then by going to $t = 1$ (that is to say, by increasing Q by a factor of e) we arrive at $g'^2/4\pi = 2\pi/11$. From this point on, the variation is quite slow; multiplying Q by e again merely halves g', and multiplication by e^2 is required to halve it again. Thus we are led to conjecture a qualitative picture in which a very large value of g' at low momentum zooms down with lightning rapidity to a small value, and then inches its way to zero.

What sort of asymptotic behaviour do we predict, once we are in the region of small g'? To evaluate Eq. (6.13), we need to know not only g', but also γ. For small g',

$$\gamma(g') = cg'^2 + o(g'^4), \tag{6.22}$$

where c is a numerical coefficient. From Eq. (6.17), for large t,

$$g'^2 \approx -1/2bt, \tag{6.23}$$

whence,

$$\gamma \approx -c/2bt. \tag{6.24}$$

Thus, the significant variation in Eq. (6.13) comes from the upper limit in the integral; for large t,

$$\bar{f} \approx K[\ln(Q/M)]^{-c/2b} \tag{6.25}$$

where K is a constant. This is *not* Bjorken scaling; the moments of the structure functions are not constants, but powers of logarithms. Nevertheless, a power of a logarithm is a very slowly varying function. I have not studied the SLAC–MIT data myself, but I am told by those who have looked at them (with optimistic eyes) that they can be fit as well with powers of logarithms as with constants.

Note that, for any given model, these powers can be computed by lowest-order perturbative calculations.[31] For example, the popular colored-quark model, with a color octet of vector gluons, is an asymptotically free theory with only one coupling constant. In this model, the moments of the isospin-odd (proton minus neutron) transverse structure function have the asymptotic form

$$F^{(n)} \propto (\ln Q)^{[-0.296\ln(n+2) + 0.051]}. \tag{6.26}$$

These are rather small powers for small moments (-0.2 for $n = 0$), and grow slowly with n, reaching -1 only for $n = 27$. Of course, since we do not know the constant coefficients of the moments, we can not reconstruct the structure functions from formulae like Eq. (6.26). However, it is easy to construct functions whose moments obey Eq. (6.26) and which display quite small deviations from scaling except for x very near to 1. (The very high moments are obviously sensitive only to the behavior of $F(q^2, x)$ in this neighborhood.)

If we accept asymptotic freedom as the explanation of Bjorken scaling, then, whatever the field theory of the strong interactions, it must be asymptotically free. (Bjorken scaling places no restrictions on the weak and electromagnetic interactions; these are negligible in the relevant energy region.) Thus, it is important to know what field theories are asymptotically free. Here is what we know now:

(1) All pure Yang–Mills theories based on groups without Abelian factors are asymptotically free.[30]

(2) Theories of non-Abelian gauge fields and Fermi multiplets are sometimes asymptotically free and sometimes not. The fermions make a positive contribution to the β-function; if the theory has too many fermions, the sign of β is reversed and asymptotic freedom is lost. 'Too many' is typically a large number. For example, if the gauge group is SU(3), sixteen triplets of fermions are not too many.[30]

(3) Much less is known about theories of non-Abelian gauge fields and scalar multiplets; typically, these theories involve a large number of quartic meson coupling constants, and this makes the investigation of the differential equations difficult. There are some theories involving scalar fields which are known to be asymptotically free.[32] At the moment, there are no known asymptotically free theories for which all the gauge mesons may be given a mass by scalar vacuum expectation values. I do not view this as a serious difficulty, for two reasons. (1) The investigation is still in its early stages; such a theory may be found next week. (2) Even if no such theory is found, we are talking about models of the strong interactions; although the couplings may become weak at large momentum, they are certainly strong at small momentum, and this is where spontaneous symmetry breakdown occurs. Therefore, symmetry breakdown might well occur non-perturbatively, as discussed in Section 2.6.

(4) Any renormalizable field theory that does not involve non-Abelian gauge fields is not asymptotically free.[33]

This last result has far-reaching consequences: if we accept asymptotic freedom as the explanation for Bjorken scaling, then the field theory of the strong interactions must be asymptotically free. If it is to be asymptotically free, then it must involve non-Abelian gauge fields. Since no-one has ever seen a massless hadron, these gauge fields must acquire masses. The only known mechanism by which gauge fields can acquire masses is through spontaneous symmetry breakdown. Thus, *the field theory of the strong interactions must be a spontaneously broken gauge field theory.*

This is a striking conclusion, suggestive of deep connections between the strong and weak interactions. It implies a complete reversal of the conventional wisdom of only a few years ago. We used to believe that at high (Euclidean) energies the weak interactions became strong; now we believe that the strong interactions become weak.

6.5   No conclusions

I know of no way to put a proper conclusion to these lectures, because I know of no way to judge the validity of the ideas we have discussed. They are certainly ideas of great beauty, and they certainly resolve many long-standing theoretical problems, but they equally certainly have not yet quantitatively confronted experiment. Spontaneously broken gauge field theories are in the uncomfortable position of SU(3) without the Gell-Mann–Okubo formula, or current algebra without the Adler–Weisberger relation. There are good reasons for this, which I explained in the Introduction, but still one can not help feeling nervous. It is very possible that this whole beautiful and complex structure will be swept into the dustbin of history by a thunderbolt from Batavia. All we can do is wait and see.

Appendix: One-loop effective potential in the general case

This appendix is a computation of the one-loop effective potential, V, for a general renormalizable field theory. Such a theory contains three types of interactions: spinless-meson self-interactions, Yukawa couplings of mesons and fermions, and gauge-field interactions. We shall see that in an appropriate gauge (Landau gauge), these three types of interactions contribute additively to the effective potential in one-loop approximation; thus,

$$V = U + V_m + V_f + V_g + V_{ct}, \tag{A.1}$$

where the first term is the zero loop effective potential, the next three terms are the contributions from the three types of interactions, and the last term is a quartic polynomial in ϕ_c, the finite residue of the renormalization counterterms, determined once we state our renormalization conditions. The method of computation will be a direct generalization of the diagrammatic summation of Section 3.[34]

(1) Spinless-meson contribution. Here the analysis is almost identical to that of Section 3; the only difference is that there may be many meson fields. Thus each internal line in the graphs of Fig. 5 carries an index a, b, etc., labeling the meson field, and the black dots represent matrices for the transition from a meson of type a to one of type b:

$$i[U''(\phi_c)]_{ab} = i\partial^2 U/\partial\phi^a\partial\phi^b\big|_{\phi=\phi_c}. \tag{A.2}$$

In computing the graphs, we must not only integrate over the internal momentum, but also sum over the internal indices. This is equivalent to multiplying the matrices around the loop and then taking the trace. Thus, from Eq. (3.33), we obtain

$$V_m = \frac{1}{64\pi^2} \text{Tr}([U''(\phi_c)]^2 \ln U''(\phi_c)). \tag{A.3}$$

(2) Fermion contribution. Here again the graphs are almost the same as in Fig. 5; the only difference is that the internal lines are fermion lines.

(Thus, you should imagine them as carrying arrows.) The relevant term in the Lagrangian is

$$\mathcal{L} = i\bar{\psi}^a \partial\!\!\!/ \psi^a + \bar{\psi}^a m_{ab}(\phi)\psi^b + \cdots. \tag{A.4}$$

Here m is the sum of two terms: a constant term (the fermion masses) and a term linear in ϕ (the Yukawa couplings). It can also be broken into two parts in a different way:

$$m = A + iB\gamma_5. \tag{A.5}$$

(I use a Hermitian γ_5.) The reality of the Lagrangian implies that A and B are Hermitian matrices. I have chosen the name m for this matrix because $m(\langle\phi\rangle)$ is the fermion mass matrix, in zero-loop approximation.

The computation can be made to look like the preceding one by grouping the terms in pairs:

$$\cdots m\,\frac{1}{p\!\!\!/}\,m\,\frac{1}{p\!\!\!/}\cdots = \cdots mm^\dagger\,\frac{1}{p^2}\cdots. \tag{A.6}$$

Now the only differences between the fermion computation and the boson one are: (1) The combinatoric factor of $\frac{1}{2}$ is missing because the lines have arrows on them, and thus the graphs are not invariant under reflections. (2) This is compensated for by the fact that the odd terms in the infinite series vanish when we take the trace on Dirac indices. (3) There is an overall Fermi minus sign. Thus we obtain

$$V_f = -\frac{1}{64\pi^2}\,\mathrm{Tr}([\,mm^\dagger(\phi_c)]^2\ln\,mm^\dagger(\phi_c)). \tag{A.7}$$

Note that here the trace is on Dirac indices as well as internal indices.

(3) *Gauge-field contribution*. If we work in a general gauge, the trilinear coupling between gauge fields and spinless mesons can lead to troublesome graphs of the form shown in Fig. 6. (Here the straight line is a spinless meson, and the wiggly line a gauge field.) However, if we work in Landau gauge,

$$D_{\mu\nu} = i\,\frac{g_{\mu\nu} - k_\mu k_\nu/k^2}{k^2} \tag{A.8}$$

Fig. 6

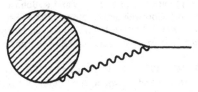

182   *Secret symmetry: spontaneous symmetry breakdown, gauge fields*

these graphs vanish. This is because the external meson carries zero momentum; the sum of the meson momenta is the same as the gauge-field momentum, and gives zero when we contract it with the propagator (A.8). Hence we need only worry about the quadrilinear coupling

$$\mathcal{L} = \cdots + \tfrac{1}{2}A_{\mu a}A_b^\mu M_{ab}^2(\phi) + \cdots, \tag{A.9}$$

where

$$M_{ab}^2 = g_a g_b (T_a \phi) \cdot (T_b \phi), \tag{A.10}$$

and g_a is the coupling constant of the ath gauge field. $M^2(\langle \phi \rangle)$ is the gauge-meson mass-squared matrix in zero-loop approximation, whence its name.

The computation is now identical with the preceding case, except that it is now gauge fields that run around the loop. Thus,

$$V_g = \frac{3}{64\pi^2} \operatorname{Tr}([M^2(\phi_c)]^2 \ln M^2(\phi_c)). \tag{A.11}$$

The factor of three comes from the trace of the propagator (A.8).

Notes and references

1. There are many excellent reviews that can be used to rectify these lapses: A. De Rujula, in *Proceedings of the First International Meeting on Fundamental Physics*; E. S. Abers and B. W. Lee, *Physics Reports* **9**, 1 (1973); J. R. Primack and H. R. Quinn (in the Proceedings of the 1973 Santa Cruz Summer School on Particle Physics); C. H. Llewellyn Smith, in *Proceedings of the 1973 Scottish Universities Summer School*, ed. R. Crawford and R. Jennings (Academic Press, 1974); M. Veltman in *Proceedings of the 1973 Bonn Conference*, (North Holland, 1974). This last contains a meticulous history.
2. This idea goes back to the classic work of Goldstone, Nambu, and Jona-Lasinio. J. Goldstone, *Nuovo Cimento* **19**, 15 (1961); Y. Nambu and G. Jona-Lasinio, *Phys. Rev.* **122**, 345 (1961); **124**, 246 (1961).
3. Notation: the signature of the metric tensor is $(+ - - -)$; $\partial_\mu = \partial/\partial x^\mu$; summation over repeated indices is always implied.
4. J. Goldstone, A. Salam, and S. Weinberg, *Phys. Rev.* **127**, 965 (1962).
5. D. Kastler, D. Robinson, and J. Swieca, *Comm. Math. Phys.* **3**, 151 (1966).
6. F. Englert and R. Brout, *Phys. Rev. Letters* **13**, 321 (1964); P. Higgs, *Phys. Letters* **12**, 132 (1964); G. Guralnik, C. Hagen and T. Kibble, *Phys. Rev. Letters* **13**, 585 (1964); P. Higgs, *Phys. Rev.* **145**, 1156 (1966); T. Kibble, *Phys. Rev.* **155**, 1554 (1967).
7. C. N. Yang and R. Mills, *Phys. Rev.* **96**, 191 (1954); R. Utiyama, *Phys. Rev.* **101**, 1597 (1956); S. Glashow and M. Gell-Mann, *Ann. Phys.* (N.Y.) **15**, 437 (1961).
8. For some tantalizing recent explorations, see R. Jackiw and K. Johnson; *Phys. Rev.* **D8**, 2386 (1973), and J. Cornwall and R. Norton, *Phys. Rev.* **D8**, 3338 (1973); J. Cornwall, *Phys. Rev.* **D10**, 500 (1974). See also R. Jackiw, in *Laws of Hadronic Structure*, 1973 Ericé Summer School, Academic Press (1975).
9. J. Schwinger, *Phys. Rev.* **128**, 2425 (1962); W. Thirring and J. Wess, *Ann. Phys.* (N.Y.) **27**, 331 (1964); J. Lowenstein and J. Swieca, *Ann. Phys.* (N.Y.) **68**, 172

(1971). This model has another amusing feature; it possesses a continuous (non-gauge) symmetry (chirality) that is not associated with a local conserved current, because of an anomaly. This symmetry is spontaneously broken but there is no Goldstone boson, because there is no conserved local current.

10. For more details (and references) see Chapter 4 in this volume.

11. Ward identities: A. Slavnov, *Theo. and Math. Phys.* **10**, 99 (1972). Gauge-invariant cutoff: G. 't Hooft and M. Veltman, *Nucl. Phys.* **44B**, 189 (1973). Renormalization: G. 't Hooft, *Nucl. Phys.* **B33**, 173 (1971); **B35**, 167 (1971); B. Lee and J. Zinn-Justin, *Phys. Rev.* **D5**, 3121, 3137, 3155 (1972). A detailed review is Abers and Lee.[1]

12. This, and much of what follows in this section, is plagiarized from S. Coleman and E. Weinberg, *Phys. Rev.* **D7**, 1888 (1973), which contains more details, references, and applications. The effective potential was introduced by Goldstone, Salam, and Weinberg,[4] and by G. Jona-Lasinio, *Nuovo Cimento* **34**, 1790 (1964).

13. This can most easily be proved with the aid of functional integration; see the discussion sections.

14. There is a parameter, which we have set equal to one by our choice of units, that enters the theory in the same way as a; this is h. Thus, it is sometimes said that the loop expansion is an expansion in powers of h. Y. Nambu, *Phys. Letters* **26B**, 626 (1966).

15. The computation of the one-loop effective potential can be done along the same lines for a general renormalizable field theory, involving arbitrary numbers of spinless mesons, fermions, and gauge fields, with hardly more labor. For the interested reader, the computation is done in the Appendix. The computation may also be done with functional integrals; this method avoids the infinite summation of diagrams. See Lee and Zinn-Justin[11] and R. Jackiw, *Phys. Rev.* **D9**, 1686 (1974).

16. This is a result of K. Symanzik, *Comm. Math. Phys.* **16**, 48 (1970).

17. This fits in with the loop expansion being an expansion in powers of h.[14] You should not allow this discussion to obscure one big difference between particle mechanics and field theory: in particle mechanics, there is no spontaneous symmetry breakdown, even for a double-welled potential of the sort shown in Fig. 2. The difference can easily be seen in the simple quadratic approximation discussed above. In particle mechanics, if we approximate the two supposed degenerate ground states by harmonic-oscillator wave functions, they have a non-zero inner product. This induces mixing which breaks the degeneracy; only one linear combination of the two states is a true ground state. It is easy to make the corresponding approximation in field theory, if we put the theory in a box of volume V. A simple computation then shows that the inner product goes to zero exponentially as the volume goes to infinity, and (at least in this approximation) the degeneracy remains.

18. This phenomenon is well-known; it occurs in the classic calculation of the effective action for electrons coupled to a constant external electromagnetic field. (W. Heisenberg and H. Euler, *Z. Physik* **98**, 714 (1936); J. Schwinger, *Phys. Rev.* **82**, 664 (1951).) This is real for a magnetic field, but imaginary for an electric field; the reason is that, in the presence of a constant electric field, the vacuum is unstable and decays into electron-positron pairs.

19. S. Weinberg, *Phys. Rev. Letters* **29**, 1698 (1972).

20. The application of functional integrals to quantum mechanics is due to Feynman, and the standard physics text is R. Feynman and A. Hibbs, *Quantum Mechanics and Path Integrals* (McGraw-Hill, 1965). This follows Feynman's original definition of the functional integral. This is apparently different from

the definition I will give, but the two can be shown to be equivalent. A good mathematical reference is I. Gelfand and N. Vilenkin, *Generalized Functions*, Vol. 4 (Academic Press, 1964).

21. The clearest derivation I know of is that of K. S. Cheng, *J. Math. Phys.* 13, 1723 (1972). Cheng evaluates the functional integral *à la* Feynman,[20] and shows that it defines the same dynamics as the Schrödinger equation with an appropriate ordering of the *p*s and *q*s.

22. The Hamiltonian form first appears in Appendix B of R. P. Feynman, *Phys. Rev.* 84, 108 (1951). It was subsequently rediscovered many times by many authors, but I can not find any reference where its dangerous ambiguities are discussed. Indeed, one commonly finds in the literature the false statement that the Hamiltonian form is a method of quantization that is invariant under general classical canonical transformations. (A possible exception is the work of B. S. DeWitt, *Rev. Mod. Phys.* 29, 377 (1957). Those portions of this paper that I can understand I believe to be correct.) The remarks in the text are the product of conversations with D. Gross, C. Callan, and S. Treiman.

23. L. Faddeev and V. Popov, *Phys. Letters* 25B, 29 (1969) and 'Perturbation theory for gauge invariant fields', Kiev lecture available in English as NAL-Thy-57.

24. R. Arnowitt and S. Fickler, *Phys. Rev.* 127, 1821 (1962).

25. A good review of the anomalies is R. Jackiw, in *Lectures on Current Algebra and its Applications* (Princeton U. Press, 1970). Implications for gauge theories are discussed in D. Gross and R. Jackiw, *Phys. Rev.* D6, 477 (1972) and W. A. Bardeen, in *Proceedings of the XVI International Conference on High Energy Physics*, Vol. 2 (NAL, 1972).

26. K. Fujikawa, B. Lee, and A. Sanda, *Phys. Rev.* D6, 2923 (1972).

27. So many people have made important contributions to this standard lore that a fair set of references would be longer than Sec. 6.1. A good brief review is C. G. Callan in *Proceedings of the International School of Physics "E. Fermi" Course LIV* (Academic Press, 1972).

28. More details, and references, can be found in Chapter 3 in this volume.

29. For a detailed derivation of the solution, see Note 28.

30. D. Gross and F. Wilczek, *Phys. Rev. Letters* 30, 1343 (1973); H. D. Politzer, *ibid.*, 1346. For pure Yang–Mills theory, the result was known to G. 't Hooft in the summer of 1972, but not published by him.

31. For detailed computations see D. Gross and F. Wilczek, *Phys. Rev.* D8, 3633 (1973); D9, 980 (1974); and H. Georgi and H. D. Politzer, *Phys. Rev.* D9, 416 (1974).

32. Gross and Wilczek.[30,31]

33. S. Coleman and D. Gross, *Phys. Rev. Letters* 31, 851 (1973).

34. The computation can also be done by functional integration.[15]

6

Classical lumps and their quantum descendants
(1975)

1   Introduction

A stone thrown into a still body of water makes ripples that spread out and eventually die away. The stone disturbs the water, gives it energy, but, even if we ignore friction, this energy tends in the course of time to spread out over the water. If we imagine the water to be infinite in extent then, if we wait long enough, at no point is the water appreciably different from its state before the stone was cast. The disturbance dissipates.

This concept of dissipation can be generalized beyond the hydrodynamic example I have given. Consider a classical field theory, with energy density Θ_{00}, such that Θ_{00} is always greater than or equal to zero and is everywhere zero for the ground state(s) of the theory. Then we will say that a solution of the classical equations of motion is dissipative if

$$\lim_{t \to \infty} \max_{\mathbf{x}} \Theta_{00}(\mathbf{x}, t) = 0. \tag{1.1}$$

Most of the simple field theories with which we are familiar have the property that all of their non-singular solutions of finite total energy are dissipative. This is the case for Maxwell's equations, the Klein–Gordon equation, etc. However, there are some perfectly sensible classical field theories that possess non-singular non-dissipative solutions of finite energy. Among these are some (though not all) of the spontaneously broken gauge theories that have risen to such prominence in recent years.

In the most striking and simplest case, and the one that will occupy most of our attention, these solutions may be time-independent, lumps of energy holding themselves together by their own self-interaction. There are also more complicated kinds of non-dissipative solutions, solutions that are periodic in time or have even more exotic time dependences, oscillating or quivering lumps. Also, if our theory is Lorentz-invariant (or Galileo-invariant), if we can construct stationary lumps, we can also

185

construct steadily moving lumps, and then it is reasonable to imagine that we can construct solutions that correspond to several widely separated lumps, all moving away from each other.

(I should tell you that the concept of 'lump' which I have vaguely defined here is called 'soliton' in much of the recent literature. I avoid the word 'soliton' here because it has a very precise and narrow definition in applied mathematics, and most of the lumps we will talk about are not solitons in this narrow sense.)[1]

Once one knows that non-dissipative solutions exist, many obvious questions arise. Which classical theories possess non-dissipative solutions? Time-independent non-dissipative solutions? What are the properties of these solutions? What about the quantum theories obtained from these classical theories by canonical quantization? Do the classical lumps become particles in the quantum theory? If so, how do we compute the properties of these particles?

No one knows the complete answer to any of these questions. However, much is known, and much of what we know, especially about the quantum problem, has been learned within the last year. These lectures are an attempt to survey this knowledge.

Section 2 deals with a set of classical field theories so simple that we can find lumps by quadrature. These are theories of a single scalar field in a two-dimensional space-time. Here we shall meet such famous lumps as the kink of ϕ^4 theory and the soliton of the sine–Gordon equation.

Unfortunately, direct integration of the equations of motion is not a very attractive method for more than a single field or more than two space-time dimensions. Section 3 discusses the powerful method of topological conservation laws, which enables us to prove the existence of non-dissipative solutions without getting tangled up in the detailed structure of the equations of motion. In the course of this section, we shall meet some more famous lumps: the flux lines of superconductivity (lumps in two spatial dimensions), and the lumps in three spatial dimensions discovered independently last year by Polyakov and 't Hooft, and called hedgehogs by the first and monopoles by the second.

(A warning: when I delivered these lectures at Erice, many of the students found the material in Sect. 3 the most difficult part of the lectures. If you have trouble with Sect. 3, *skip it*; nothing in the remainder of these notes depends on it.)

In Sect. 4, we turn to the quantum problem. I discuss in some detail two approximation methods for treating time-independent classical solutions, a systematic weak-coupling expansion and a variational method. I also discuss much more briefly a method for treating periodic classical solutions, based on WKB ideas.

Section 5 deals with a very special system, the quantum sine–Gordon equation, for which it is possible to gain some information without recourse to approximations. Everything we shall learn about this system that falls within the expected domain of validity of the approximation methods of the previous section is consistent with them. However, we shall also discover some surprises.

Much of what I know about this subject has been learned from Ludwig Faddeev, Sheldon Glashow, Jeffrey Goldstone, Roman Jackiw, T. D. Lee, André Neveu, and Gian-Carlo Wick, all of whom I gratefully acknowledge.

2   Simple examples and their properties[2]

2.1   *Some time-independent lumps in one space dimension*

(Notation: The signature of the metric tensor is such that a timelike vector has positive norm. x always denotes a space-time point, with time and space components x^0 and x. An exception is two-dimensional space-time, where x is used indiscriminately for a space-time point and its space component; which is meant should always be clear from the context. Greek indices run over all space-time coordinates, latin indices from the middle of the alphabet over space coordinates only, latin indices from the beginning of the alphabet over internal-symmetry variables. Summation over repeated indices is always implied unless otherwise stated. $\hbar = c = 1$.)

Let us consider the simplest family of classical relativistic field theories: theories of a single scalar field in one space and one time dimension, with non-derivative interactions. The dynamics of such a theory are described by the Lagrange density,

$$\mathscr{L} = \tfrac{1}{2}\partial_\mu \phi \, \partial^\mu \phi - U(\phi), \tag{2.1}$$

where μ equals 0 and 1, and U is some arbitrary function of ϕ, but not of its derivatives. The energy of any field configuration is given by

$$E = \int dx [\tfrac{1}{2}(\partial_0 \phi)^2 + \tfrac{1}{2}(\partial_1 \phi)^2 + U]. \tag{2.2}$$

We will sometimes have occasion to write this as the sum of kinetic and potential energy,

$$E = T + V, \tag{2.3}$$

where T, as usual, is the term quadratic in time derivatives (the first term in Eq. (2.2)), and V, also as usual, is the term involving no time derivatives (the sum of the second two terms).

If the energy is to be bounded below, U must be bounded below. In this case, we can always add a constant to U such that the minimum value of U is zero; I will assume in what follows that this has always been done. It is evident from Eq. (2.2) that a state of minimum energy (a ground state) is one where ϕ is independent of space and time and equal to one of the zeros of U. Of course, if U has several zeros, the theory has several (degenerate) ground states. The energy of a ground state is zero.

There are two theories which I will use throughout these lectures as examples. One is familiar ϕ^4 theory, with the sign of the quadratic term in U such that there are two ground states. (This is the situation that, in the quantum theory, is called 'spontaneous breakdown of symmetry'.) In equations,

$$U = \frac{\lambda}{2}\,\phi^4 - \mu^2\phi^2 + \frac{\mu^4}{2\lambda},\tag{2.4a}$$

with λ and μ^2 positive parameters. This can also be written as

$$U = \frac{\lambda}{2}\,(\phi^2 - a^2)^2,$$

where a^2 is μ^2/λ. The two ground states are $\phi = \pm a$.

The other example is somewhat less familiar. It is the so-called[3] sine–Gordon equation:

$$U = \frac{\alpha}{\beta^2}\,(1 - \cos\beta\phi),\tag{2.4b}$$

where α and β are real parameters. Since we are free to redefine the sign of ϕ, we can with no loss of generality take β to be positive. Likewise, since we are free to shift ϕ by π/β (and add a constant to U), we can also take α to be positive. The ground states are those for which ϕ is an integral multiple of $2\pi/\beta$. If we study small oscillations about one of these ground states, say $\phi = 0$, we expand the cosine in power series:

$$U = \frac{\alpha}{2}\,\phi^2 - \frac{\alpha\beta^2}{4!}\,\phi^4 + \cdots.$$

Thus we see that, for small oscillations, α is something like the parameter μ^2 in the previous example, while $\alpha\beta^2$ is similar to λ.

Now let us turn to the problem of finding possible time-independent solutions of the equations of motion. For time-independent solutions, Hamilton's principle reduces to

$$\delta V = \delta \int dx [\tfrac{1}{2}(\partial_1\phi)^2 + U(\phi)] = 0.\tag{2.5}$$

This is mathematically identical to a familiar problem in particle mechanics, the motion of a particle of unit mass in a potential equal to *minus* $U(x)$,

$$\delta \int dtL = \delta \int dt[\tfrac{1}{2}(dx/dt)^2 + U(x)] = 0. \tag{2.6}$$

Fig. 1 shows the potential for this equivalent mechanical problem in the case of ϕ^4 theory.

Every motion of the particle in the potential corresponds to a time-independent solution of the field equations. However, most of these solutions are not of finite energy. For the energy integral (2.2) to converge, it is necessary for ϕ to go to a zero of U as x goes to either plus or minus infinity. In terms of the particle problem of Fig. 1, this means that the particle must go to the zeros of the potential as t goes to plus or minus infinity. Of course, the ground states (motions where the particle stays forever at the top of one of the two potential hills) satisfy this criterion, but there are also less trivial motions that will do the job, motions where the particle starts on top of one hill and moves to the top of the other. These are motions of zero particle energy. If the particle energy were negative, the particle would never reach the top of the hill; if it were positive, the particle would overshoot and travel on forever. This analysis is in no way special to the ϕ^4 potential sketched in Fig. 1. It only depends on the fact that U has two adjacent zeros between which the particle can move; the detailed form of U between the zeros is irrelevant, as is the behavior of U outside the region between the zeros.

Thus we find:

(1) If U has only one zero – that is to say, if the ground state of the theory is unique – there are no non-trivial time-independent solutions of finite energy.

(2) On the other hand, if U has more than one zero – that is to say, if the theory has more than one ground state – there always exist non-

Fig. 1

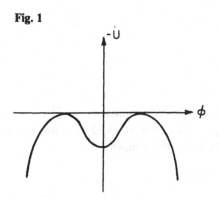

trivial time-independent solutions of finite energy. For such solutions, ϕ moves monotonically from one zero of U at x equals minus infinity to an adjacent zero of U at x equals plus infinity. By convention, we will call those solutions for which ϕ is monotone increasing 'lumps' and those for which it is monotone decreasing 'antilumps'.

(3) It is easy to reduce the problem of finding the explicit form of these solutions to quadratures. The particle motion is one of zero particle energy; hence, since the particle moves in the potential $-U(x)$,

$$\frac{1}{2}\left(\frac{dx}{dt}\right)^2 - U(x) = 0. \tag{2.7}$$

Translating this back into field language, we find

$$\tfrac{1}{2}(\partial_1\phi)^2 = U(\phi). \tag{2.8}$$

This has the solution

$$x = \pm \int_{\phi_0}^{\phi} d\phi\,[2U(\phi')]^{-\frac{1}{2}}. \tag{2.9}$$

Here the plus sign is for lumps, the minus sign for antilumps and the arbitrary parameter ϕ_0 is the value of ϕ at $x=0$, and can be any number between the two adjacent zeros.

Of course, the presence of the arbitrary parameter ϕ_0 is just an expression of spatial translation invariance. Another way of saying the same thing is to say that, if $f(x)$ is any solution of (2.9) with some fixed ϕ_0, then the general solution with arbitrary ϕ_0 is

$$\phi = f(x - b), \tag{2.10}$$

where b is arbitrary. In other words, the center of a lump can be anywhere.

(4) These formulae also enable us to simplify somewhat the expression for the energy of a lump. In particular, it follows from Eq. (2.8) that

$$E = \int dx [\tfrac{1}{2}(\partial_1\phi)^2 + U]$$

$$= \int dx (\partial_1\phi)^2$$

$$= \int d\phi\,[2U(\phi)]^{\frac{1}{2}}. \tag{2.11}$$

For both of our specific examples, these integrals are elementary. Just to have some concrete examples in front of us, I will write out the answers. For ϕ^4 theory, we find for the form of the lump

$$\phi = a\,\tanh(\mu x). \tag{2.12a}$$

The energy of the lump is given by

$$E = 4\mu^3/3\lambda. \tag{2.13a}$$

The lumps of ϕ^4 theory are frequently called 'kinks' in the literature.

For the sine–Gordon theory, we find

$$\phi = \frac{4}{\beta}\tan^{-1}\exp(\alpha^{\frac{1}{2}}x), \tag{2.12b}$$

where which branch of the inverse tangent we choose determines which two zeros we move between. The energy of the lump is

$$E = \frac{8\alpha^{\frac{1}{2}}}{\beta^2}. \tag{2.13b}$$

The lumps of sine–Gordon theory are frequently called 'solitons' in the literature. In both cases the antilumps are obtained simply by multiplying by minus one.

2.2   Small oscillations and stability

We have shown that any theory of a single scalar field in one space dimension with more than one ground state admits time-independent solutions of finite energy, but we have not yet shown that these solutions are stable under small perturbations. I now turn to this problem.

The equation of motion for our system is

$$\Box^2\phi + U'(\phi) = 0, \tag{2.14}$$

where the prime denotes differentiation with respect to ϕ. Let us consider a solution of the form

$$\phi(x, t) = f(x) + \delta(x, t), \tag{2.15}$$

where $f(x)$ is our time-independent solution and δ is the small perturbation. Inserting this expression in the equation of motion and only retaining terms of first order in the perturbation, we find

$$\Box^2\delta + U''(f)\delta = 0. \tag{2.16}$$

This equation is invariant under time translations, so we can express a general small perturbation as a superposition of normal modes. That is to say, the general solution is of the form

$$\delta(x, t) = \mathrm{Re}\sum_n a_n e^{i\omega_n t}\psi_n(x), \tag{2.17}$$

where the as are arbitrary complex coefficients, and the ψs and ωs obey the equation

$$-\frac{\mathrm{d}^2\psi_n}{\mathrm{d}x^2} + U''(f)\psi_n = \omega_n^2\psi_n. \tag{2.18}$$

Note that this is a one-dimensional Schrödinger equation, with potential $U''(f)$. Our solution is stable under small perturbations if and only if none of the energy eigenvalues of this Schrödinger equation are negative. I will now show that this is always the case.

Spatial translation invariance tells us that if $f(x)$ is a solution of the equation of motion, so is $f(x+a)$. Thus we already know one energy eigenfunction of our Schrödinger equation,

$$\psi_0(x) = df/dx. \tag{2.19}$$

We also know the associated eigenvalue; it is zero. In Sect. 2.1 I showed that f was always a monotone function of x; therefore ψ_0 has no nodes. It is a well-known theorem that for a one-dimensional Schrödinger equation with arbitrary potential the eigenfunction with no nodes is the eigenfunction of lowest energy.[4]   Q.E.D.

This takes care of the question of stability, but I should like to make some additional remarks about Eq. (2.18). If we could explicitly solve this equation, we could explicitly obtain (in the linearized approximation) a rich set of solutions to our equations of motion. These solutions are especially interesting if Eq. (2.18) possesses a discrete eigenvalue (other than the trivial zero eigenvalue). In this case, the associated eigenfunction yields a solution to the (linearized) equation of motion which is of finite energy and periodic in time. If we are daring enough to use quantum language before we get to the quantum theory, we can think of this situation as a meson bound to a lump. On the other hand, from continuum eigenfunctions, we can only form a solution of finite energy by forming a wave packet. In the same spirit, we can think of this wave packet as representing a meson scattering off a lump. (When we do get to the quantum theory, we shall see that these are not bad descriptions.)

2.3   *Lumps are like particles (almost)*

Although we are working with classical continuum theories, the objects we have found are much like classical particles (or, better yet, since the energy density is not concentrated at a point, classical extended bodies). They have a definite finite rest mass and a definite location, the location of the center-of-mass. Furthermore, from the stationary solutions we have found, we may construct, by Lorentz transformation, solutions corresponding to lumps moving with any velocity less than that of light. By Lorentz invariance, these moving solutions obey the traditional energy–momentum relation,

$$E = (P^2 + M^2)^{\frac{1}{2}}, \tag{2.20}$$

where M is the rest mass.

These are all properties which we would expect to hold for more traditional extended bodies (e.g. billiard balls) in relativistic classical mechanics.
However, there is one important sense in which lumps differ from classical
particles.

One classical particle at rest is always a solution of classical mechanics.
From this solution one can build an approximate many-particle solution
that consists of many particles at rest, all widely separated from each
other. I say 'approximate' because there may be interactions between the
particles (e.g. Yukawa forces); the solution becomes exact only as the
separations go to infinity. Can we construct the same sort of asymptotic
many-particle solutions for lumps?

To phrase the problem more precisely, let a_i be a set of points on the
line ($i = 1 \ldots n$), arranged in increasing order,

$$a_1 < a_2 \ldots < a_n. \tag{2.21}$$

Let $f_i(x)$ be a set of time-independent solutions of the field equation. Can
we find an approximate solution of the field equation, such that, in the
neighborhood of a_i,

$$\phi \approx f_i(x - a_i), \tag{2.22}$$

and such that the solution becomes exact as the distance between successive as goes to infinity?

The answer is obviously *no*, unless

$$f_i(\infty) = f_{i+1}(-\infty). \tag{2.23}$$

This simple condition for patching adjacent solutions together has far-
reaching consequences for any attempt to interpret lumps as particles.

To show this, let me begin with the sine–Gordon equation. For f_1, we
may choose any solution, soliton or antisoliton. For f_2, we may again
choose either soliton or antisoliton, but which soliton or antisoliton we
choose (i.e. which branch of the tangent) is determined by our previous
choice, and so on down the line. This strongly suggests that properly we
should not say that there are many solitons and many antisolitons,
corresponding to the many branches of the tangent, for this picture would
lead to a gross overcounting of the many-soliton states. We get the right
counting if we say instead that there is only one soliton and only one
antisoliton. If we adopt this position, we must also say that the only
observable functions of ϕ are those that are unchanged by the transformation

$$\phi \rightarrow \phi + 2\pi/\beta. \tag{2.24}$$

In ϕ^4 theory, the situation is even more peculiar. If we begin with a
kink, we do not have the option to follow it with either a kink or an anti-

kink; we must follow it with an antikink. Likewise, the next lump along the line must be a kink, etc. Just as above, this strongly suggests that we should not consider the kink and antikink as independent objects, but as the same object. (This observation is due to Goldstone and Jackiw.)[2] If we adopt this position, we must also say that the only observable functions of ϕ are those that are unchanged by the transformation

$$\phi \to -\phi. \tag{2.25}$$

I should emphasize that these are just questions of interpretation, of what words we use to describe the mathematically unambiguous properties of well-defined classical field theories. We could just as well say that the kink and the antikink are different, and that there are somewhat peculiar forces between two kinks that do not fall off with distance, but which are shielded by an intervening antikink. This is not so strange in in one spatial dimension as it would be in three; after all, in one dimension, even the familiar Coulomb force is independent of distance.

In Sect. 3 we will return to the question of patching together distant lumps, for gauge theories in two and three spatial dimensions. We will find the situation here more closely resembles the assembly of distant particles, but still there will be some novelties.

2.4   *More dimensions and a discouraging theorem*

The obvious next step is to attempt to find time-independent solutions for scalar field theories in more than one spatial dimension, and perhaps also with more than one scalar field. Unfortunately, this step leads us into a dead end, for such theories possess no such solutions.

Derrick's theorem. Let ϕ be a set of scalar fields (assembled into a big vector) in one time dimension and D space dimensions. Let the dynamics of these fields be defined by

$$\mathcal{L} = \tfrac{1}{2}\partial_\mu\phi \cdot \partial^\mu\phi - U(\phi), \tag{2.26}$$

and let U be non-negative and equal to zero for the ground state(s) of the theory. Then, for $D \geqslant 2$, the only non-singular time-independent solutions of finite energy are the ground states.[5]

Proof. Define

$$V_1 = \tfrac{1}{2}\int d^D x (\nabla\phi)^2, \tag{2.27}$$

$$V_2 = \int d^D x\, U(\phi). \tag{2.28}$$

V_1 and V_2 are both non-negative and are simultaneously equal to zero only for the ground states. Let $\phi(\mathbf{x})$ denote a time-independent solution.

Consider the one-parameter family of field configurations defined by

$$\phi(\mathbf{x}; \lambda) \equiv \phi(\lambda \mathbf{x}), \tag{2.29}$$

where λ is a positive number. For this family, the energy is given by

$$V(\lambda) = \lambda^{(2-D)} V_1 + \lambda^{-D} V_2. \tag{2.30}$$

By Hamilton's principle, this must be stationary at $\lambda = 1$. Thus,

$$(D-2)V_1 + DV_2 = 0. \tag{2.31}$$

For $D > 2$, this immediately implies that both V_1 and V_2 vanish, and the proof is complete. For $D = 2$, however, Eq. (2.31) only implies the vanishing of V_2, and a small amount of further argument is required. If V_2 vanishes it is stationary, since zero is its minimum value. Thus we may apply Hamilton's principle to V_1 alone, from which it trivially follows that V_1 also vanishes. Q.E.D.

I should emphasize that Derrick's theorem denies the existence of time-independent solutions *only*; it says nothing about time-dependent (but non-dissipative) solutions. Indeed, T. D. Lee[6] has constructed models involving scalar fields only which possess such solutions; I report on his construction in Appendix 1.

Nevertheless, the theorem is sufficiently discouraging to make us investigate theories of more than just scalar fields. The obvious next step is to look at theories of scalar fields and gauge fields.[7] This is not only the next step, it is almost as far as we can go, at least in three dimensions. Anything much more complicated would be non-renormalizable; this is no problem in classical physics, but it is a considerable problem if we want to go on to the quantum theory eventually.

Of course, we could consider fermions. However, I remind you that the classical limit of a Fermi field is not an ordinary c-number but an anticommuting c-number. Thus, if we were to consider fermions, the values of our classical fields would not be numbers but elements of a Grassman algebra, a situation I would just as soon avoid in these lectures.

There is, of course, another reason for looking at gauge field theories: there is reason to believe that they have something to do with the real world. At any rate, it is gauge theories that will occupy us in the next section.

3   Topological conservation laws[8]

3.1   *The basic idea and the main results*

In the previous section we demonstrated the existence of non-dissipative solutions for a family of simple field theories by explicitly displaying time-independent solutions. I would now like to establish exactly the same result by a more indirect line of argument. This would be a bootless exercise were it not that the indirect argument is generalizable to cases where explicit solutions are hard to come by, in particular to theories of gauge fields and scalar fields in three dimensions.

Instead of focusing on non-singular solutions of finite energy, let us focus on non-singular initial-value data (ϕ and $\partial_0\phi$ at some fixed time) of finite energy. Of course, these two sets of objects are in one-to-one correspondence; for every solution, there is initial-value data, and for every set of initial-value data, there is a solution. (For the class of theories under consideration, non-singular initial-value data can not develop a singularity in the course of time, for this would violate the conservation of energy.[9] Mathematical purists can take 'non-singular' to mean 'continuously differentiable'.)

For such initial-value data, just as for the time-independent solutions of Sect. 2.1, finiteness of the energy implies that

$$\lim_{x \to \pm\infty} \phi(x, t) \equiv \phi(\pm\infty, t),\tag{3.1}$$

must exist and must be zeros of U. Now, non-singular solutions are, in particular, continuous in time. Since, for our examples, the zeros of U form a discrete set, this implies that

$$\partial_0\phi(\pm\infty, t) = 0.\tag{3.2}$$

If U has more than one zero, this equation is non-trivial, and can be used to prove the existence of non-dissipative solutions.

To show how this works, let me take ϕ^4 theory as an example. I will show that any solution for which

$$\phi(\infty, t) = -\phi(-\infty, t),\tag{3.3}$$

is non-dissipative. Proof: By continuity in x, for any t, there must be some x for which $\phi = 0$. At this point,

$$\Theta_{00}(x, t) \geqslant U(0) = \lambda a^4/2.\tag{3.4}$$

Hence,

$$\max_x \Theta_{00}(x, t) \geqslant \lambda a^4/2,\tag{3.5}$$

which contradicts Eq. (1.1).

A slightly more abstract way of expressing Eq. (3.2) is to say that we have divided the space of non-singular finite-energy solutions at a fixed time into subspaces, labeled by $\phi(\pm\infty, t)$. (Thus, for ϕ^4 theory, we have four such subspaces.) These subspaces are disconnected components of the whole space, in the normal topological sense; it is not possible to continuously change a solution in one component into a solution in another component. Since time evolution is continuous, this implies that if a solution is in one component at any one time, it is in the same component at any other time.

This way of putting things emphasizes the difference between Eq. (3.2) and familiar conservation laws. Eq. (3.2) is a conservation law; just like the conservation of energy, it states that something is independent of time. However, unlike the conservation of energy, it is not a consequence of a symmetry of the theory. It is instead a consequence of a topological property of the space of non-singular finite-energy solutions, the fact that it is not connected. For this reason we shall call Eq. (3.2), and similar results we shall establish for more complicated theories, 'topological conservation laws'.

The remainder of this section will be devoted to the study of such topological conservation laws for spontaneously broken gauge field theories, in both two and three spatial dimensions. Since the analysis is lengthy, I will try to whet your appetite by stating some of the main results for three dimensions now.

(1) If the theory has no spontaneous symmetry breakdown, the space of non-singular finite-energy solutions has only one component, and there are no non-trivial topological conservation laws.

(2) The same disgusting situation obtains if the symmetry breakdown is total, that is to say, if no massless gauge mesons survive symmetry breakdown.

(3) We now come to the interesting case in which only one massless gauge meson (by convention, called the photon) survives symmetry breakdown. In this case there are two possibilities:

(3a) The gauge group when written as a product of simple Lie groups contains a U(1) factor, and the generator of this U(1) factor enters into the expression for the electric charge. (This is the case, for example, in the Weinberg–Salam model.) In this case, there are again no non-trivial topological conservation laws.

(3b) In all other cases (for example, in the model of Glashow and Georgi), the space of finite-energy non-singular solutions has an infinite number of components; there are non-trivial topological conservation laws. Just as above, these can be used to prove the existence of non-dissipative solutions. The components can be labeled by a quantity that can be interpreted physically as magnetic flux emerging from the solution. (This despite the fact that the theory contains no magnetic monopoles in the sense of Dirac, just ordinary charged and neutral scalar and vector fields!) This magnetic flux is quantized, always comes in integral multiples of a basic unit. (This despite the fact that these are classical, not quantum, theories!)

(4) Similar results hold if many massless gauge mesons survive symmetry breakdown, but I can not explain the results in detail at this early stage, because they involve the generalization of the concept of magnetic flux to non-Abelian theories. (This concept will be developed in full as our analysis proceeds, and the appropriate results stated precisely.)

Along with these appetizers, I should give you two warnings:

(1) Topological conservation laws enable us to establish the existence of non-dissipative solutions, *not* of time-independent solutions. For example, in ϕ^4 theory every set of initial conditions obeying Eq. (3.3) is non-dissipative, but, as we have seen in Sect. 2.1, only a small subset of these are time-independent. Another example: for the sine–Gordon equation it is possible to show, by arguments identical to those given above, that any solution for which

$$\phi(\infty, t) - \phi(-\infty, t) = 4\pi/\beta, \tag{3.6}$$

is non-dissipative, but we know that none of these are time-independent. (For all time-independent solutions the field changes by $2\pi/\beta$, half this amount.)

(2) Topological conservation laws give sufficient conditions for non-dissipative solutions, *not* necessary ones. It is quite possible for there to be non-dissipative solutions even when there are no non-trivial topological conservation laws. (See Appendix 1 for an example.)

These warnings emphasize that topological conservation laws do not tell us everything we want to know. Nevertheless, they tell us quite a lot, and at a very low cost in analytic effort, and thus are very much worth pursuing.

3.2   Gauge field theories revisited

Before proceeding to our main task, I would like to briefly[10] review gauge field theories, and to develop some of their properties which will be important to us later.

The theories we shall consider will involve a set of n scalar fields (not necessarily real), which we assemble into an n-vector, ϕ. We also have a compact connected Lie group, G, called the gauge group, and an n-dimensional unitary representation of G, $D(g)$. Given any function from space-time to G, $g(x)$, we define a transformation of the scalar fields, called a gauge transformation, by

$$g(x): \phi(x) \to D(g(x))\phi(x). \tag{3.7}$$

To avoid long equations with nested parentheses, we will identify the abstract group G with the representation $D(g)$, and write this as

$$g(x): \phi(x) \to g(x)\phi(x). \tag{3.8}$$

The generators of $D(g)$, a set of $n \times n$ orthonormal Hermitian matrices, are denoted by T^a. These obey the commutation relations

$$[T^a, T^b] = \mathrm{i} c^{abc} T^c, \tag{3.9}$$

where the cs are the structure constants of G, and the sum over repeated indices is implied. With these generators we associate a set of real vector fields, A_μ^a, called the gauge fields, with gauge transformation properties given by

$$A_\mu^a(x) T^a \to g(x) A_\mu^a(x) T^a g(x)^{-1} + \mathrm{i} e^{-1} \partial_\mu g(x) g(x)^{-1}, \tag{3.10}$$

where e is a real constant, called the gauge coupling constant. For an infinitesimal gauge transformation,

$$g(x) = 1 - \mathrm{i} T^b \omega^b(x), \tag{3.11}$$

this becomes

$$\delta A_\mu^a = c^{abc} A_\mu^b \delta \omega^c + e^{-1} \partial_\mu \delta \omega^a. \tag{3.12}$$

Equation (3.12) is more common in the literature than Eq. (3.10), but they are equivalent, and (3.10) is better suited for our purposes. (If G is not simple, there may be several gauge coupling constants, but for notational simplicity I will ignore this possibility here. None of our conclusions will be affected by the presence of several coupling constants.)

From these objects, we define the covariant derivative of ϕ,

$$\mathrm{D}_\mu \phi = \partial_\mu \phi + \mathrm{i} e A_\mu^a T^a \phi, \tag{3.13}$$

and the field strengths,

$$F_{\mu\nu}^a = \partial_\mu A_\nu^a - \partial_\nu A_\mu^a - e c^{abc} A_\mu^b A_\nu^c. \tag{3.14}$$

These have simple gauge-transformation properties:

$$\mathrm{D}_\mu \phi(x) \to g(x) \mathrm{D}_\mu \phi(x), \tag{3.15}$$

and

$$F_{\mu\nu}^a(x) T^a \to g(x) F_{\mu\nu}^a(x) T^a g(x)^{-1}. \tag{3.16}$$

From these objects, we construct the gauge-invariant Lagrange density that defines our theory,

$$\mathscr{L} = -\tfrac{1}{4} F_{\mu\nu}^a F^{a\mu\nu} + \mathrm{D}_\mu \phi^\dagger \cdot \mathrm{D}^\mu \phi - U(\phi), \tag{3.17}$$

where U is some invariant function of the scalar fields.[11] Just as before, we assume U is always greater than or equal to zero and equals zero for the ground state(s) of the theory. The simplest theory of this kind is the electrodynamics of charged scalar mesons (possibly with self-interactions). In this case, ϕ consists of a single complex field, and G is U(1), the group of complex numbers of unit modulus.

What about spontaneous symmetry breakdown? The ground states of the theory can easily be seen to all have energy zero; up to a gauge transformation, they are states where the gauge fields vanish and ϕ is a constant and a zero of U. For such a ground state, defined by $\phi = \phi_0$, we define H, a subgroup of G, called the unbroken group, by saying that h is an element of H if and only if $h\phi_0 = \phi_0$. If we study small oscillations about this ground state, we find that the gauge fields associated with the generators of H remain massless, while the other gauge fields acquire masses.

Because U is G-invariant, if ϕ_0 is a zero of U, so is $g\phi_0$, for any g in G. I will now make an important simplifying assumption: I will assume that *all* the zeros of U are of the form $g\phi_0$ for some g in G. This assumption excludes both the case of accidental degeneracy, in which U has zeros that are not forced on it by any symmetry, and the case in which the theory has an ordinary (non-gauge) internal symmetry group in addition to the gauge group, G. Accidental degeneracy is of no real interest to us, because we are interested in classical theories only as limits of quantum theories, and, in general, quantum corrections remove accidental degeneracies. The case of an internal symmetry group is more interesting, and I exclude it only to simplify the discussion. The generalization of all of our results to this case is sketched in Appendix 3.

The simplifying assumption is equivalent to the statement that the set of zeros of U can be identified with the coset space G/H, and therefore I will denote this set by G/H in the sequel. (If you do not remember what a coset space is, do not worry; just accept 'G/H' as my eccentric shorthand for 'the set of zeros of U' and you will have no difficulties.)

This completes the lightning review of standard lore. I now turn to less familiar material. In particular, I shall describe a method of using the gauge fields to associate an element of G with every path in space-time; these path-dependent group elements will be very useful later.

Let P be a path in space-time, going from some initial point x_0 to some final point x_1. Let us consider a field ϕ such that its covariant derivative vanishes along P:

$$\frac{dx^\mu}{ds} D_\mu \phi = 0 \qquad (3.18)$$

where s is some parameter along the path, chosen such that s goes from 0 to 1 as x goes from x_0 to x_1. Eq. (3.18) can be rewritten as

$$\frac{d\phi}{ds} = -ie \frac{dx^\mu}{ds} A_\mu^a T^a \phi. \qquad (3.19)$$

Aside from the fact that this involves finite-dimensional matrices rather than infinite-dimensional ones, this is structurally identical to the familiar time-evolution equation of interaction-picture perturbation theory, and can be solved in the same way, by Dyson's formula. In particular, ϕ at the end of the path is given in terms of ϕ at the beginning of the path by

$$\phi(x_1) = g(P)\phi(x_0),\qquad(3.20)$$

where

$$g(P) = S \exp\left(-ie \int_0^1 ds\, \frac{dx^\mu}{ds} A_\mu^a T^a\right),\qquad(3.21)$$

and S indicates s-ordering (defined in the same way as T-ordering in Dyson's formula.)

The $g(P)$s are very useful objects. If we know $g(P)$ for every path, we know the gauge fields, by differentiation. Also, they have simple gauge-transformation properties,

$$g(P) \to g(x_1)g(P)g(x_0)^{-1}.\qquad(3.22)$$

I shall now use the $g(P)$'s to prove the following:

Theorem. It is always possible to make a gauge transformation (or, as we say, choose a gauge) such that

$$A_0^a = 0.\qquad(3.23)$$

Proof. For any space-time point x, define P_x to be the straight-line path from $(\mathbf{x}, 0)$ to x. The desired gauge transformation is defined by

$$g(x) = g(P_x)^{-1},\qquad(3.24)$$

for, under this transformation,

$$g(P_x) \to g(P_x)^{-1}g(P_x)g(P_0) = 1.\qquad(3.25)$$

from which Eq. (3.23) follows by differentiation.

Note that if the original fields are non-singular, so are the transformed fields. In the future, we shall always assume that we have chosen our gauge such that Eq. (3.23) holds. (This still leaves us the freedom to make time-independent gauge transformations.) The advantage of working in such a gauge is that the time-derivative terms in the Lagrangian (3.17) simplify enormously:

$$D_0\phi = \partial_0\phi,\qquad(3.26a)$$

and

$$F_{0i}^a = \partial_0 A_i^a.\qquad(3.26b)$$

Thus, the structure of the initial-value problem is the same as that for a theory of scalar fields only; a complete and independent set of initial-value data is given by the fields and their first time derivatives at any fixed time. Also, the expression for the energy simplifies:

$$E = T + V,\qquad(3.27)$$

where

$$T = \int d^D x [\tfrac{1}{2}(\partial_0 A_i^a)^2 + \partial_0 \phi^\dagger \cdot \partial_0 \phi], \qquad (3.28)$$

$$V = \int d^D x [\tfrac{1}{4}(F_{ij}^a)^2 + \mathbf{D}\phi^\dagger \cdot \mathbf{D}\phi + U(\phi)], \qquad (3.29)$$

and D is the number of spatial dimensions. Finally, we have nothing to lose by working in such a gauge, since the definition of a dissipative solution, Eq. (1.1), is gauge-invariant.

3.3   Topological conservation laws, or homotopy classes

We are now ready to tackle the problem of finding topological conservation laws, that is to say, to divide the space of initial value data (fields and their first time derivatives at some fixed time) into disconnected components. Since we are always working at a fixed time, I will drop the explicit t-dependence from our formulae, to simplify notation. I will also use our freedom to make time-independent gauge transformations to insist that

$$A_r^a(\mathbf{x}) = 0, \ r \geqslant 1, \qquad (3.30)$$

where the subscript r indicates the radial component. (I have excluded a sphere around the origin because the radial component is ill-defined at the origin.) The existence of an appropriate gauge transformation follows from an argument identical with that at the end of Sect. 3.2, with t replaced by r.

I will carry out the necessary arguments in detail for two spatial dimensions; the generalization to three dimensions will be straightforward.

From the formula for the energy,

$$E \geqslant \int_1^\infty r \, dr \int_0^{2\pi} d\theta [\partial_r \phi^\dagger \cdot \partial_r \phi + U(\phi)]. \qquad (3.31)$$

Thus, if the energy is to be finite, by the same reasoning as in one dimension,

$$\lim_{r \to \infty} \phi(r, \theta) = \phi(\infty, \theta) \qquad (3.32)$$

must exist and must be a zero of U, that is to say, an element of G/H, for every θ.

Now we come to the key point: can $\phi(\infty, \theta)$ depend non-trivially on θ? At first glance, one would think not, for, if it did,

$$\mathbf{e}_\theta \cdot \nabla\phi \xrightarrow[r \to \infty]{} \frac{1}{r} \frac{d\phi(\infty, \theta)}{d\theta}, \qquad (3.33)$$

and this would make the energy integral diverge. However, the quantity that enters the energy is not the gradient, but the covariant gradient,

$$\mathbf{e}_\theta \cdot \mathbf{D}\phi = \frac{1}{r}\frac{d\phi}{d\theta} + ieA_0^a T^a \phi, \tag{3.44}$$

and for any given non-singular $\phi(\infty, \theta)$, we can choose the gauge fields so this goes to zero for large r as fast as we please. Note that our simplifying assumption – that the set of zeros of U is the coset space G/H – is critical for this argument, for it is only this assumption that allows us to assert that we can cancel an infinitesimal change in ϕ (the first term in the covariant derivative) with an infinitesimal group transformation (the second term). Observe that the finiteness of the energy links the large-r behavior of the gauge fields with that of the scalar fields:

$$\lim_{r\to\infty} rA_\theta^a T^a \phi(\infty, \theta) = -ie^{-1}\frac{d\phi(\infty, \theta)}{d\theta} \tag{3.35}$$

In particular, this implies that the gauge fields must go like $1/r$ at large r. This is perfectly consistent with the finiteness of the energy, for the field strengths then go like $1/r^2$, and their squares like $1/r^4$, which causes no large-r blowup in the energy integral.

The energy also has a contribution from the time derivatives of the fields, Eq. (3.28), but the time derivatives are independent initial-value data and can be independently adjusted to make only a finite contribution.

All of this may be trivially extended to three space dimensions; the only difference is that the angle θ is replaced by the two polar angles θ and ϕ. However, the process of extension stops at three dimensions; in four or more dimensions, a term in the energy density proportional to $1/r^4$ at large r does make the total energy blow up. (Fortunately, physical interest also stops at three dimensions.)

Let us summarize what we have found. In two dimensions, with every non-singular finite-energy set of initial value data we have associated the function $\phi(\infty, \theta)$. This function is a mapping of a circle, or, as mathematicians call it, S^1 (i.e. a one-dimensional sphere), into G/H. Likewise, in three dimensions, we have a mapping of S^2 (i.e. an ordinary two-dimensional sphere) into G/H.

If two sets of initial-value data have the same mapping associated with them, it is obvious that we can continuously turn one into the other without making the energy blow up; once we have tacked things down at spatial infinity, nothing continuous we do in the finite part of space can cause any trouble. On the other hand, if the two sets of initial-value data have different mappings associated with them, they can be continuously deformed into each other if and only if the associated mappings can be continuously deformed into each other.

Thus, we have greatly simplified our search for topological conservation laws. To determine whether two sets of initial-value data lie in the same connected component of the space of non-singular finite energy solutions, we need only determine whether the associated mappings from S^1 (in two space dimensions) or S^2 (in three) into G/H can be continuously deformed into each other.

Now, mathematicians[12] have a special word for 'continuously deformable'; they call it homotopic. For the benefit of purists, I will give the definition: let X and Y be two topological spaces, and let $f_0(x)$ and $f_1(x)$ be two continuous functions from X to Y. Let I denote the unit interval on the real line, $1 \geqslant t \geqslant 0$. We say the two functions f_0 and f_1 are homotopic if and only if there is a continuous function from the Cartesian product of X and I to Y, $F(x, t)$, such that $F(x, 0) = f_0(x)$ and $F(x, 1) = f_1(x)$. The function F is called a homotopy. As promised, this is just a fancy way of saying that f_0 is continuously deformable into f_1.

Homotopy is obviously an equivalence relation, and thus we can divide all functions from X to Y into homotopy classes, such that two functions are in the same class if they are homotopic and in different classes if they are not. Thus, with no loss in content and with a considerable elevation in tone, we can rephrase the result of two paragraphs back as the following:

Theorem. The connected components of the space of nonsingular finite-energy solutions are in one-to-one correspondence with the homotopy classes of mappings from S^{D-1} to G/H, where D is the number of space dimensions.

As a trivial corollary, if U has only one zero, so G/H consists of a single point, there is only one homotopy class (indeed, only one mapping), so there are no non-trivial topological conservation laws. This is result (1) of Sect. 3.1. On the other hand, if there is more than one homotopy class, there are non-trivial topological conservation laws.

You might worry that even if we do find several homotopy classes, the topological conservation laws might be physically uninteresting, because the solutions in different components might be gauge transforms of each other. This worry is eliminated by the following:

Theorem. Let $\phi_1(x)$ and $\phi_2(x)$ be two non-singular finite-energy field configurations such that

$$\phi_2(x) = g(x)\phi_1(x). \tag{3.36}$$

where $g(x)$ is non-singular. Then the associated mappings of S^{D-1} into G/H are homotopic.

Proof. For notational simplicity, I will give the proof only for $D=2$. The extension to $D=3$ is trivial. Since G, like all gauge groups, is connected, we can continuously distort $g(\mathbf{x})$ so that $g(0)=1$. This obviously defines a homotopy, so, with no loss of generality, we can restrict ourselves to the case $g(0)=1$. We now rewrite Eq. (3.36) in polar coordinates:

$$\phi_2(r, \theta)=g(r, \theta)\phi_1(r, \theta), \tag{3.37}$$

where, by assumption, $g(0, \theta)=1$. We now define a homotopy:

$$F(\theta, t)=g\left(\frac{t}{1-t}, \theta\right)\phi_1(\infty, \theta). \tag{3.38}$$

This obviously equals $\phi_1(\infty, \theta)$ when $t=0$ and $\phi_2(\infty, \theta)$ when $t=1$. Q.E.D.

3.4 Three examples in two spatial dimensions

I shall now try to put some flesh on these bare and abstract bones by giving some specific examples. I begin with three examples in two spatial dimensions. In each case, the existence of non-trivial topological conservation laws can be used to prove the existence of non-dissipative solutions, along the lines of the arguments in Sect. 3.1; I will not bother to give these arguments explicitly, though, and will restrict myself to finding the topological conservation laws.

Example 1. G is U(1), the multiplicative group of complex numbers of unit modulus, and the set of scalar fields consists of a single complex field, ϕ, on which G acts by multiplication. The self-interactions of ϕ are given by

$$U=\frac{\lambda}{2}(\phi^*\phi-a^2)^2, \tag{3.39}$$

where λ and a are positive numbers, just as in the ϕ^4 theory of Sect. 2.1. The set of zeros of U, G/H, consists of all ϕs of the form

$$\phi=ae^{i\sigma}, \tag{3.40}$$

where σ is a real number. That is to say, G/H is a circle, S^1. Thus we must find the homotopy classes of mappings from S^1 (in ordinary space) to S^1 (in field space).

This is a famous problem with a well-known and obvious solution. The homotopy classes are characterized by an integer, the 'winding number', which tells how many times you wind around the circle in field space when you go once around the circle in ordinary space. In equations,

$$n=\int_0^{2\pi}\frac{d\theta}{2\pi}\frac{d\sigma}{d\theta}. \tag{3.41}$$

Thus, the space of finite-energy non-singular solutions decomposes into an infinite number of disconnected components, each labeled by the winding number, n. An easy argument (construct it yourself) shows that any solution with $n \neq 0$ is non-dissipative. In fact, although topological conservation laws are not strong enough to show it, there are actually time-independent solutions with $n = \pm 1$. Solutions with higher values of n tend to decompose into widely separated solutions with $n = \pm 1$.

This theory, at least for time-independent solutions, is mathematically identical with the Landau–Ginzburg theory of Type II superconductors, and the lumps we have found are called 'flux lines' in superconductor theory.[13] They are called lines because superconductor theorists think of the two-space in which we are working as the x–y plane in a three-dimensional superconductor in which everything is independent of the z coordinate. The 'flux' requires more explanation:

By Stokes's theorem, the magnetic flux passing through our plane is given by

$$\Phi = \lim_{r \to \infty} r \int_0^{2\pi} d\theta A_\theta(r, \theta). \qquad (3.42)$$

By Eqs. (3.35) and (3.40),

$$\Phi = \int_0^{2\pi} d\theta e^{-1} \frac{d\sigma}{d\theta} = \frac{2\pi n}{e}, \qquad (3.43)$$

whence 'flux lines'.

It is a bit surprising that we have obtained a quantization condition, the quantization of flux, from a purely classical theory. (You should find it especially surprising if you have encountered more traditional treatments of flux quantization, which emphasize (quite properly) the essentially quantum-mechanical nature of the effect.) The reason for the apparent contradiction is that our e is the coupling constant for a field, ϕ. In classical physics, this quantity has no connection with the electric charge of a particle; the two quantities even have different dimensions. It is only the wave–particle duality (essentially quantum-mechanical) that enables us to connect the two, through

$$\hbar e_{\text{field}} = e_{\text{particle}}. \qquad (3.44)$$

Thus, when expressed in terms of the electron charge (or, more properly, the Cooper-pair charge), Eq. (3.43) does have Planck's constant in it, and is quantum-mechanical.

***Example* 2.** *G* is SO(3), and the scalar fields form an isovector, ϕ. *U* is as before, with ϕ^2 replacing $\phi^*\phi$. (This is the Georgi–Glashow model, without fermions, and in two space dimensions.) G/H is now a two-dimensional sphere, S^2.

Thus we must find the homotopy classes of mappings from S^1 to S^2. It is easy to show that every such mapping is homotopic to the trivial mapping (that which maps all of S^1 into a single point in S^2); this is the famous theorem that it is impossible to lasso a basketball. A more formal proof goes like this: S^1 is one-dimensional and S^2 is two-dimensional. Thus there is at least one point of S^2 into which no point in S^1 is mapped. Remove this point from S^2. S^2 with one point removed is topologically identical with an open disc. Define a homotopy by shrinking the disc to its central point. You cannot lasso a basketball.

Thus there is only one homotopy class, and no non-trivial topological conservation laws.

***Example* 3.** *G* is SO(3), as before. ϕ is an isotensor, which we represent, as usual, as a traceless real symmetric 3×3 matrix. I will assume that *U* is such that a general zero of *U* is of the form

$$\phi = 2aee^{T} - a(I - ee^{T}). \tag{3.45}$$

where a is some parameter, determined by the detailed form of *U*, and e is a real unit vector. (This is easy to arrange with a *U* that is a sum of terms proportional to $\mathrm{Tr}\phi^2$, $\mathrm{Tr}\phi^3$, and $\mathrm{Tr}\phi^4$.) Thus, there is one eigenvector of ϕ, e, with eigenvalue $2a$, and two orthogonal eigenvectors, both with eigenvalue $-a$. The infinitesimal structure of symmetry breakdown here is the same as in Example 2; there is one massless gauge meson corresponding to infinitesimal rotations about the axis e. However, G/H is very different in structure. For although the vectors e do indeed lie on a sphere (S^2), $-e$ defines the same ϕ as e. Thus G/H is a sphere with antipodal points identified.

Let us represent mappings from S^1 into G/H by drawing the closed path in G/H traced out as θ goes around S^1. Fig. 2 shows two such drawings. (Fig. 2(b) may not look like a closed path to you, but remember, antipodal points are identified.) Clearly Fig. 2 shows the only two possible closed paths, up to homotopies. Either a path is manifestly closed when drawn on a sphere (Fig. 2(a)) or it goes between antipodal points (Fig. 2(b)), and there is no way to continuously distort a path of the first type into one of the second.

Fig. 2

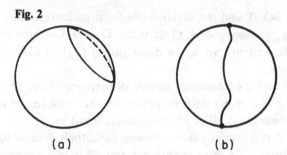

(a)   (b)

Thus there are two homotopy classes, and the space of non-singular finite-energy solutions has two connected components. I emphasize the contrast between this example and Example 1. There, the components were labeled by integers; in its algebraic structure, the topological conservation law was much like the conservation of electric charge in quantum field theory. Here, the topologically conserved quantity can only take on two values; the situation is much more like parity, for example.

3.5   Three examples in three dimensions
Example 4. The field theory is the same as in Example 2 (the Georgi–Glashow model), except that we are now in three space dimensions. Thus, we must study the homotopy classes of mappings from S^2 into S^2, rather than of S^1 into S^2.

Mappings of spheres into spheres are harder to visualize than mappings of circles into spheres, but I will try and convince you (by a very non-rigorous argument) that in this case there is at least one[14] non-trivial homotopy class. Argument: you cannot peel an orange without breaking the skin. The skin of the orange (idealized as being infinitely thin and infinitely flexible) is a sphere, and so is the surface of the flesh of the orange upon which it lies. When the skin lies on the surface it defines a mapping of a sphere (the skin) onto a sphere (the surface of the flesh). If this mapping were homotopic to the trivial mapping, we could distort the skin on the surface, without breaking it, until the whole skin lies over a single point. We could then lift the skin off the orange. This is obviously impossible.

Thus there is at least one mapping of a sphere into a sphere, to wit, the identity mapping, that is not in the same homotopy class as the trivial mapping. By arguments which should by now be standard, this implies that any set of non-singular finite-energy initial value data such that

$$\lim_{r \to \infty} \phi^b(\mathbf{x}) = ax^b/r, \tag{3.46}$$

where $b = 1, 2, 3$, cannot dissipate.

In fact, there are not only non-dissipative solutions obeying the boundary condition (3.46) but actual time-independent solutions, found independently by Polyakov[15] and 't Hooft.[16] (It is not so hard to find such solutions as one might think; see Appendix 4.) Polyakov called the solution a 'hedgehog', for obvious reasons; 't Hooft called it a 'magnetic monopole', for reasons that will become clear in Sect. 3.7.

Example 5. The group is SU(2), and the scalar fields transform like a complex isospinor, K. The zeros of U are all fields such that $\bar{K}K$ is some positive number, a^2. (This is the Weinberg–Salam model, without leptons and without the U(1) gauge field.)

If we write the two-component complex K field in terms of four real fields, we see that G/H is a hypersphere, S^3. All mappings from S^2 to S^3 are homotopic to the trivial mapping, by the argument given for Example 2, with 'disc' replaced by 'ball'. Thus, there are no non-trivial topological conservation laws.

Example 6. G is any compact connected Lie group, H is any discrete subgroup.

At this point, for the first and last time in this discussion, I will have to appeal to authority: it is a trivial corollary of a famous theorem of Elie Cartan[17] that, if H is discrete, all mappings of a sphere into G/H are homotopic to the trivial mapping.

Thus, there are no non-trivial topological conservation laws.[18] This is a result (2) of Sect. 3.1. (Note that the previous example is a special case of this one.)

3.6   *Patching together distant solutions, or, homotopy groups*
    I now turn to the question of patching together distant lumps, the question we analyzed for our simple models in Sect. 2.3. In one sense, the problem is much simpler for gauge theories than for the simple scalar models: we need not worry about patching together the gauge fields, for they vanish at large distances, and we need not worry about patching together the scalar fields, for we can always force them to match by an appropriate gauge transformation.

    Fig. 3 explains this statement in more detail for the case of two spatial dimensions. Fig. 3(a) shows the plane for some fixed time scalar field, $\phi_1(r, \theta)$. In order to fit the drawing onto a finite page, I have assumed that ϕ_1 attains its asymptotic value, $\phi_1(\infty, \theta)$, and $D\phi_1$ vanishes, at some finite radius, everywhere outside the shaded disc in the drawing. (The wedge protruding from the disc will be explained shortly.) Now, starting from some fixed angle, say $\theta = 0$, we can gauge transform $\phi_1(\infty, \theta)$ to some

Fig. 3

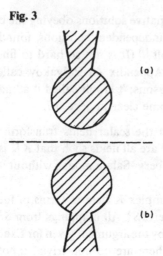

standard value, say $\phi_1 = \phi_0$, and continue gauge-transforming ϕ_1 to ϕ_0 as we move around the circle. Of course, we can not go all the way around the circle, for we would surely reach a discontinuity when we return to our starting point, if the solution is not in the trivial homotopy class. Thus we gauge transform ϕ_1 to ϕ_0 only in the unshaded region. Fig. 3(b) shows the plane for a second solution, ϕ_2. Here also we have gauge-transformed to make $\phi_2 = \phi_0$ in the entire unshaded portion of the plane. We can now path the two halves of the figure together, along the dashed line, with no discontinuity.

Fig. 3 also serves to illustrate the situation in three spatial dimensions, if we think of it as a cross-section of a surface of revolution. The discs are then balls and the wedges cones, and the argument is unchanged.

Thus, in both two and three dimensions, the assembly of distant lumps looks much more like the assembly of distant particles than it did in one dimension. Nevertheless, there is one tricky point: I said that we had to make a gauge transformation such that ϕ_1 became ϕ_0 everywhere in the unshaded region of Fig. 3(a). However, such a gauge transformation is not uniquely defined; once we have set ϕ_1 equal to ϕ_0, we do not change the situation if we make a gauge transformation, $g(\mathbf{x})$, such that g is an element of H everywhere in the unshaded region. If we consider two such gauge-equivalent versions of ϕ_1, and patch them on to the same ϕ_2, it is not clear that the two compound solutions obtained in this way will be gauge-equivalent. (Of course they are equivalent under a discontinuous (across the dashed line) gauge transformation, but this is not allowed; a discontinuous gauge transformation introduces terrible singularities in the gauge fields.)

It is easy to see that no ambiguity of this kind can arise in our patching prescription if H is connected, as it is for Examples 1 and 4, for in this case we can continuously distort $g(\mathbf{x})$ to one as we approach the dashed line, without changing ϕ_1 at all. Even if H is disconnected, as it is for Example 3, the same argument shows that we need check for ambiguity only in the case when $g(\mathbf{x})$ is a constant in the unshaded region; indeed, we need only check for some one constant g from each component of H. But check we must, for there is no general theorem that assures us there is no ambiguity in our patching prescription. (As we shall see shortly, there is no ambiguity even for Example 3. However, it is possible to construct somewhat more complicated models for which the patching prescription is ambiguous.)[19]

I have given a prescription for patching together two widely separated solutions of our field equations. Now, each of our two original solutions is associated with a homotopy class, as is the compound solution, and our patching prescription is obviously homotopically invariant. Thus the (possibly ambiguous) physical operation of patching together two solutions to make a third defines a (possibly ambiguous) mathematical law for composing two homotopy classes to make a third. That is to say, it gives a group structure to the set of homotopy classes. The groups we obtain in this way are objects well-known to mathematicians; they are called homotopy groups.[20]

I will now give the standard mathematical definitions of homotopy groups,[21] and show that they correspond to our patching prescription. As usual, I will begin in two dimensions, and later go on to three.

Consider a subset of the set of continuous mappings from S^1 into G/H, mappings such that some fixed point in S^1 (say $\theta = 0$) is mapped into some fixed point in G/H (say ϕ_0). Any such mapping can be thought of as a continuous function $\phi(\theta)$ from the interval $[0, 2\pi]$ into G/H, such that

$$\phi(0) = \phi(2\pi) = \phi_0. \tag{3.47}$$

Given two such mappings ϕ_1 and ϕ_2, we define the compound mapping $\phi_1 \cdot \phi_2$ simply by putting the two intervals together end to end (with the first on the left and the second on the right). This defines a continuous function on the interval $[0, 4\pi]$. We now rescale the coordinate by two to turn this into $[0, 2\pi]$, thus producing a new function of the standard form. In equations,

$$\begin{aligned} \phi_1 \cdot \phi_2(\theta) &= \phi_1(2\theta), && 0 \leqslant \theta \leqslant \pi \\ &= \phi_2(2\theta - 2\pi), && \pi \leqslant \theta \leqslant 2\pi. \end{aligned} \tag{3.48}$$

Another way of thinking of this is to think of ϕ_1 and ϕ_2 as describing two closed paths in G/H, each of which begins and ends at ϕ_0. $\phi_1 \cdot \phi_2$ then describes the path obtained by first going around ϕ_1 and then going around ϕ_2. This way of describing things makes it especially clear why we must impose Eq. (3.47) to get a definition of multiplication; there is no sense in the prescription to go first around one path and then around the other unless the second path begins where the first ends.

We define two such mappings to be homotopic just as in Sect. 3.3, except that we restrict the continuous deformation, $F(\theta, t)$ to obey Eq. (3.48) for each t. More loosely, we are allowed to continuously distort our closed paths as we will, except that the starting point is nailed down. Our definition of multiplication is obviously invariant under (restricted) homotopies, and thus defines a multiplication law among (restricted) homotopy classes. This multiplication obeys all the group axioms: associativity is trivial, the identity is the trivial homotopy class, and the inverse is obtained by going around the path in the opposite direction.

The group obtained in this way is called the first homotopy group, and is denoted by $\pi_1(G/H)$. We do not bother to specify ϕ_0 in this expression, because the group structure does not depend on ϕ_0. (Proof: gauge invariance.) From now on, we will reserve the term 'homotopy classes' for our old homotopy classes, and refer to our restricted homotopy classes (with ϕ_0 nailed down) as 'elements of the homotopy group'.

Given a general mapping from S^1 into G/H, we can always turn it into a mapping of the form (3.47) by multiplying it by an appropriate element of G. Thus we can always assign to every homotopy class an element of the homotopy group. It is easy to see that the trivial homotopy class is always assigned to the unit element of the homotopy group;[22] however, for other homotopy classes, there may be ambiguity in this prescription, because the 'appropriate element' of G is not unambiguously defined. Given one such element, we can always obtain another by left-multiplying by an element of H. If H is connected, there is no problem, because we can then continuously distort the element of H to 1, and this defines a homotopy. However, if H is disconnected, we have to check to see if there is or is not ambiguity.

You may have the feeling that you have read the preceding paragraph before, because it is exactly like the description of possible ambiguities in the patching procedure. This is as it should be, because the composition of homotopy group elements *is* our patching procedure. This can be seen from Fig. 3: if we draw a big circle around the figure, and traverse it in a positive sense, starting at $\theta = 0$, we begin at ϕ_0 in G/H, then circle the path associated with ϕ_1 (as we pass through the top shaded wedge), then that associated with ϕ_2 (the bottom wedge) and then return to ϕ_0.

It is an easy exercise to check that, for Example 1, π_1 is the additive group of the integers; this is the law of addition of winding number (or of magnetic flux). For Example 3, π_1 is the additive group of the integers modulo 2 (i.e. $1+1=0$). If, in Fig. 2, we take a path that goes from the north pole to the south pole, this is the same as a path that goes from the south pole to the north pole, since antipodal points are identified. Thus if we go around the path twice, we obtain a path that goes from the north pole to the north pole, that is to say, one that is homotopic to the trivial path. ($1+1=0$.)

The extension to three dimensions, that is to say, from S^1 to S^2, is straightforward. We consider mappings from S^2 into G/H such that some fixed point in S^2 (say the north pole) is mapped into some fixed point in G/H (say ϕ_0). Just as we can represent the circle as a line interval with boundary points identified, so we can represent the sphere as a square with boundary points identified. Fig. 4 shows such a representation; to make things clear, I have drawn three paths on S^2 and their images in the square. Note that the entire perimeter of the square represents a single point on the sphere, the north pole. (If, instead of a square, I had drawn a (topologically equivalent) closed disc, this representation would be that used in polar projection maps of the earth, like the United Nations flag, where the entire boundary of the disc represents the south pole.)

We can now define the multiplication of two mappings by putting two squares side by side and rescaling the horizontal coordinate to produce a new square. The remainder of the construction is step-by-step the same as in the two-dimensional case. The group we arrive at in this way is called the second homotopy group, and is denoted by $\pi_2(G/H)$.

There is one important new feature in three dimensions: the second homotopy group is always Abelian. (Although the first homotopy group is Abelian in our examples, this is not always the case.) Fig. 5 is a sequence of frames from a motion picture of the homotopy which proves this assertion. We first map the two squares with their sides in common into two discs with a portion of their perimeters in common; we then con-

Fig. 4

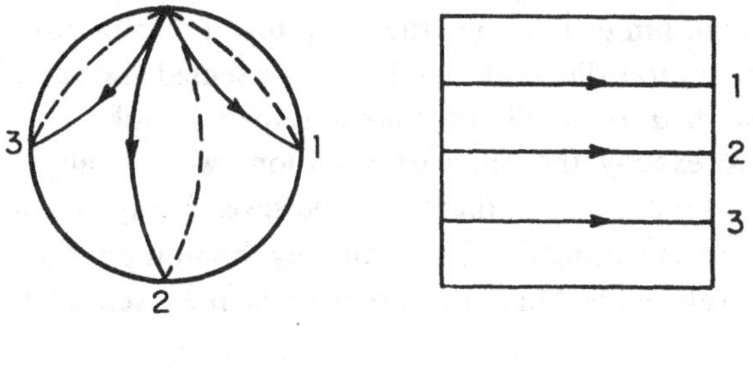

Fig. 5

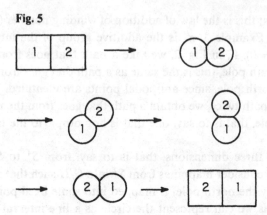

tinuously change the common portion of the perimeter until it has moved by π; we then map the two discs back into two squares. The essential reason we can perform this construction is that the boundary of a square is connected; we can not perform a similar construction for π_1 because the boundary of a line interval is not connected.

Of course, we can also see the same thing in terms of patching together solutions, as in Fig. 3. In two dimensions, we can not, by a continuous deformation, exchange the top and the bottom of Fig. 3, without either moving a shaded wedge through the ray $\theta = 0$ (destroying Eq. (3.47)) or moving the wedges through each other (destroying our ability to make sense of the drawing). In contrast, in three dimensions, the shaded wedges become cones, and there is plenty of room in three-space to move one cone around another.

This takes care of all of the physics and all of the mathematics of patching together distant solutions, but I would like to waste a few minutes worrying about problems of interpretation, that is to say, about the best words to use to describe the situation we have uncovered.

Let us suppose, for simplicity, that instead of patching together general solutions, we are patching together time-independent solutions, like the lumps of Sect. 2. In Sect. 2.3 we found that we could not always patch together widely separated lumps. This meant that if we were to make the assembly of distant lumps look like the assembly of distant particles, we had to say that several different solutions represented the same object. (In ϕ^4 theory, we had to identify the kink and the antikink.)

We now have exactly the opposite situation: we can always patch together distant solutions, but there may be several ways of doing the patching, if there is ambiguity in identifying homotopy classes with homotopy group elements. Thus, if we are to make the assembly of distant lumps look like the assembly of distant particles, we must adopt the opposite stratagem, and say that a single solution represents several different objects, one for each way of assigning a homotopy group element to the solution. (As Rudolf Haag has pointed out,[23] this is closely parallel to the situation in quantum parastatistics. Here, if you restrict yourself to the one-particle subspace of Fock space, there is no way of telling that there is more than a single species of particle. It is only by doing experiments in the many-particle subspace that you can detect that there are in fact several species.)

Even this stratagem does not work in the special case of two-dimensional theories for which $\pi_1(G/H)$ is non-Abelian. When you assemble distant particles in two dimensions, the two-particle state you obtain does not depend on the angular order of imaginary wedges radiating from the particles. In this special case, I know of no way of interpreting things such that the assembly of distant lumps is like the assembly of distant particles.

3.7   *Abelian and non-Abelian magnetic monopoles, or, $\pi_2(G/H)$ as a subgroup of $\pi_1(H)$*

From now on we will restrict ourselves to three spatial dimensions. Although some of the phenomena we shall encounter have two-dimensional analogues, these are essentially trivial; the heavy artillery is needed only in three dimensions.

We shall use heavily the representation of a sphere by a square of Fig. 4. In particular, we shall introduce Cartesian coordinates on the square, t_1 and t_2, $1 \geqslant t_{1,2} \geqslant 0$, and represent the asymptotic values of our scalar fields by a function $\phi(t_1, t_2)$, which equals ϕ_0 on the boundary of the square (the north pole of the sphere).

Although we have not been paying much attention to them lately, we of course have gauge fields as well as scalar fields in our theory. These go to zero like $1/r$ for large r; thus we can associate a group element with every path on the sphere (equivalently, on the square), defined by Eq. (3.21). For every point in the square, we will choose a standard path, defined to be a horizontal line starting from the t_2 axis and terminating at the given point. We thus define a function from the square into G, $g(t_1, t_2)$. Note that $g(t_1, t_2)$ does *not* define a continuous function on the sphere. It is equal to 1 on the left-hand side of the square, because this is the starting point of all our paths, and it is equal to 1 at the top and bottom of the square, because these are trivial paths that never leave the north pole, but it is not necessarily equal to 1 on the right-hand side of the square. This situation is shown in Fig. 6.

Fig. 6

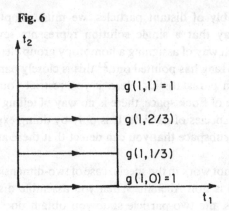

$\phi(t_1, t_2)$ and $g(t_1, t_2)$ are not independent functions; they are connected by the three-dimensional analogue of Eq. (3.35), the statement that $D\phi$ goes to zero for large r more rapidly than $1/r$ (indeed, in three dimensions, more rapidly than $1/r^{\frac{3}{2}}$). Thus,

$$\phi(t_1, t_2) = g(t_1, t_2)\phi_0, \tag{3.49}$$

which implies that

$$\phi_0 = g(1, t_2)\phi_0. \tag{3.50}$$

Thus, $g(1, t_2)$ defines a path in H, which starts at 1 and ends at 1. A continuous distortion of our field configuration (restricted so ϕ at the north pole stays fixed at ϕ_0) produces a continuous distortion of the path in H. That is to say, Fig. 6 defines a mapping from $\pi_2(G/H)$ into $\pi_1(H)$.

From Fig. 6, we can read off some interesting properties of this mapping:

(1) The arguments associated with Fig. 5 show that we can define group multiplication in π_2 by putting our two squares together any which way. If we put them together one on top of the other, it is clear that the product of two elements in $\pi_2(G/H)$ is mapped into the product of the corresponding two elements in $\pi_1(H)$. That is to say, our mapping is a homomorphism of one group into the other.

(2) Every line of fixed t_1 on the square defines a path in G (not necessarily in H), which begins at 1 and ends at 1. As t_1 goes from 0 to 1, this path changes continuously from the trivial path to $g(1, t_2)$. Thus, $g(1, t_2)$ can not be an arbitrary path in H, but must be one that is homotopic to the trivial path when H is embedded in G.

This result can be rephrased in group-theoretical language: every closed path in H is also a closed path in G; thus there is a natural homomorphism of $\pi_1(H)$ into $\pi_1(G)$. What we have found is that our homomorphism maps $\pi_2(G/H)$ into the kernel of this natural homomorphism. (I remind you that the kernel of a homomorphism is the set of all elements that are mapped into 1.)

(3) Conversely, given a closed path in H that is homotopic to the trivial closed path when H is embedded in G, we can run through our construction backwards and find a $\phi(t_1, t_2)$ consistent with this path being $g(1, t_2)$. That is to say, we can use the homotopy to define a function on the square, $g(t_1, t_2)$, and then use Eq. (3.49) to define ϕ. In group-theoretical language, our homomorphism maps $\pi_2(G/H)$ *onto* the kernel of this natural homomorphism.

Thus we have established that our homomorphism is onto. I shall now show that it is also one-to-one, and thus we have an invertible homomorphism, i.e. an isomorphism.

Theorem. $\pi_2(G/H)$ is isomorphic to the kernel of the natural homomorphism of $\pi_1(H)$ into $\pi_1(G)$.

Proof. As stated, we have to establish the one-to-one nature of our homomorphism. Since we are dealing with a homomorphism, a mapping that preserves group multiplication, all we need to show is that only the identity element of $\pi_2(G/H)$ is mapped into the identity element of $\pi_1(H)$. In other words, given that $g(1, t_2)$ is homotopic to the trivial path $[g(t) = 1]$ in H, we wish to show that $\phi(t_1, t_2)$ is continuously deformable to the constant function, $\phi = \phi_0$. We can obviously continuously deform g (and therefore ϕ) so that it runs through all its changes in the left-hand half of the square, and is independent of t_1 in the right-hand half, $t_1 \geqslant \frac{1}{2}$. Now we can replace g for $t_1 \geqslant \frac{1}{2}$ by the homotopy that turns $g(1, t_2)$ into 1. This is an abrupt change, not a continuous deformation, but it does not affect ϕ at all, since the homotopy stays in H. We now have arranged matters such that $g(t_1, t_2) = 1$ on the entire boundary of the square. That is to say, $g(t_1, t_2)$ defines a continuous mapping from S^2 into G. By the theorem of Cartan, quoted in Example 6, such a mapping can always be continuously deformed into the trivial mapping. Through Eq. (3.49), this defines a continuous deformation of $\phi(t_1, t_2)$ into the trivial mapping. Q.E.D.[24]

As a specific instance, let us consider the field theory of Example 4. Here G is SO(3) and H is U(1). Thus $\pi_1(H)$ is the additive group of the integers. If we embed U(1) in SO(3), only paths with even winding numbers, rotations by even multiples of 2π, are deformable to the trivial path. (This is why SO(3) has double-valued representations.) Thus, the kernel of the natural homomorphism is the subgroup of the even integers.

We have finally found the structure of the topological conservation laws for Example 4; each connected component of the space of non-singular finite-energy solutions is labeled by an even integer. This is reminiscent of Example 1, where the components were labeled by integers, and where these integers were identified with quantized magnetic flux. I shall now show that the same identification holds here.

To do this, I will first have to find the definition of magnetic flux. We have three field-strength tensors, and thus three candidates for the magnetic field vector:

$$H_i^b = \varepsilon_{ijk} F^{bjk}. \tag{3.51}$$

What linear combination of these is the magnetic field depends on what linear combination of our three gauge fields is massless, that is to say, is the photon field. Thus, for a ground state, a constant ϕ which is a zero of U, the answer is unambiguous:

$$H_i = \phi^b H_i^b / a, \tag{3.52}$$

where a is the magnitude of ϕ at a zero of U. This is certainly the quantity that would be measured by a magnetometer in this case. It is also the quantity that would be measured by a magnetometer if $U(\phi)$ and $\mathbf{D}\phi$ were approximately zero over the region of space occupied by the magnetometer. ('Approximately zero' means small compared to the characteristic length and energy scales of the theory.) However, if this condition is not satisfied, then you need the detailed theory of the mechanism of the magnetometer to know what the magnetometer measures. The mathematical reflection of this is that we can add all sorts of terms to Eq. (3.52) (proportional to $\mathbf{D}\phi$, etc.) that vanish for a ground state. Thus we have an infinite variety of possible definitions of the magnetic field, and any one of them is as good as any other; they just describe different models of idealized magnetometers. Fortunately, we do not wish to compute the magnetic field point by point, but only to compute the net magnetic flux emerging from our solution. For this we only need the field for infinite r, where all definitions reduce to Eq. (3.52). Thus,

$$\Phi = \lim_{r \to \infty} r^2 \int d\Omega H_r. \tag{3.53}$$

To evaluate this quantity, let us choose ϕ_0 to point in the 3-direction in isospin space, and let us gauge transform ϕ so that it equals ϕ_0 over the whole of the sphere at infinity except for a small region near the north pole. (It is easy to show that such a gauge transformation always exists: the sphere with the small region excised is topologically equivalent to a disc with the north pole at its center. On such a disc, we can define the gauge transformation at every point by transforming continuously outward from the center along radial lines.) Fig. 7 shows this situation using the by-now-familiar representation of the sphere as a square; the excised

Fig. 7

region is indicated by the black dot. Since $\mathbf{D}\phi$ vanishes on the sphere at infinity, outside the black dot the only non-vanishing component of A_i^b is A_i^3; thus,

$$\mathbf{H} = \nabla \times \mathbf{A}^3, \tag{3.54}$$

outside the black dot. Furthermore, since H is a continuous gauge-invariant function, if we make the black dot sufficiently small, we can evaluate Eq. (3.53) with negligible error by integrating only over the portion of the sphere that lies above the path P in Fig. 7. Thus

$$\Phi = \oint_P \mathbf{A}^3 \cdot \mathbf{dx}, \tag{3.55}$$

by Stokes's theorem.

Now, in the same gauge, let us evaluate the element of $\pi_1(H)$ associated with this field configuration. For any path across the square, P', that avoids the black dot, the formula for $g(P')$, Eq. (3.21), is trivial to evaluate, because only $\mathbf{A}^3$ is non-vanishing:

$$g(P') = \exp\left[-ieT_3 \oint_{P'} \mathbf{A}^3 \cdot \mathbf{dx} \right]. \tag{3.56}$$

Just as before, by gauge invariance and continuity, we can evaluate the winding number with negligible error by just going down to the path P. Thus,

$$2\pi n = e \oint_P \mathbf{A}^3 \cdot \mathbf{dx}. \tag{3.57}$$

(Of course, I am assuming that T^3 has been normalized in the standard way, such that the gauge fields carry charges e, $-e$, and 0.) I remind you that we showed above that the winding number must be an even integer.

Thus we find

$$\Phi = 2\pi n/e, \tag{3.58}$$

where n is an even integer. This is a remarkable result: all solutions to this field theory that do not belong to the trivial homotopy class are magnetic monopoles, carrying quantized magnetic charge, and all our topological conservation laws are equivalent to the conservation of magnetic charge.

Nearly all of this analysis extends trivially to a general field theory for which the connected part of H is isomorphic to $U(1)$. (Only the connected part of H is relevant to $\pi_1(H)$.) Here also all topological conservation laws are equivalent to the conservation of magnetic charge. The only part of the analysis that depends on the details of the theory, on how H is embedded in G, is the condition that determines the allowed values of magnetic charge, the generalization of Eq. (3.58). (We shall shortly develop a simple algorithm for finding this generalization in any given theory.) The extension to general Abelian H is also trivial. In this case we may have several massless vector mesons, several photons, and for each photon we have a magnetic charge. On the other hand, if H is not Abelian, there is no way of extending our analysis. Of course, we can always state flatly that the non-Abelian generalization of magnetic charge is an element of $\pi_1(H)$, but there is no general way of representing such a quantity as the integral of a numerical valued field over a large sphere. For example, $\pi_1(SO(3))$ is the additive group of the integers modulo 2, and there is simply no way of obtaining this structure by the addition of ordinary numbers.

I should emphasize that the objects we have found are very different from the old magnetic monopoles of Dirac. The field of a Dirac monopole is simply the magnetostatic analog of the usual electrostatic Coulomb field. Like the Coulomb field, the Dirac monopole field is a singular solution of free electrodynamics carrying infinite electromagnetic energy. To make the Coulomb field part of a dynamical theory, it is necessary to introduce non-electromagnetic dynamical degrees of freedom, charged particles that constitute the source of the Coulomb field; the masses and spins of these charged particles are free parameters. Likewise, to make the Dirac monopole field part of a dynamical theory, it is necessary to introduce non-electromagnetic dynamical degrees of freedom, the magnetic monopoles; the masses and spins of these monopoles are free parameters.

Our monopoles could not be more different. They have nothing to do with ordinary electrodynamics, and only appear in theories of non-Abelian gauge fields interacting with scalar fields. They are totally nonsingular and carry finite energy. They necessitate no new dynamical degrees of freedom, and all of their properties are determined in terms of the same parameters that govern the dynamics of the ordinary scalar and vector mesons.

There is one point of similarity: Dirac monopoles also carry quantized magnetic charge. We can see how this arises using our homotopy methods.

Outside the Dirac monopole, all that exists is electromagnetic field: $G = H = U(1)$. In addition to the gauge-independent singularity at the origin, the gauge field A has a gauge-dependent line of singularities ('the string') going from the origin to infinity. The string can be moved about by gauge transformations, but can never be gauged away altogether.[25] (The relation between the string and the monopole is rather like that between a branch line (movable) and a branch point (immovable) in the theory of a multi-valued analytic function.) We can now reinterpret Fig. 7 as representing a sphere about the monopole (not necessarily of infinite radius), with the black dot representing the location of the string. Just as before, we can associate with this field configuration an element of $\pi_1(H) = \pi_1(G)$, and identify the winding number with the magnetic flux. Since there is no way of getting rid of the string (the black dot), there is no constraint on this element; thus we obtain Eq. (3.58) for arbitrary (integer) n. We can also see that if n is not zero, there must be a real singularity, not removable by a gauge transformation, somewhere within the sphere. For if there is no real singularity as we shrink the sphere to zero, n must change continuously. Since n is an integer, this means it must be independent of the radius of the sphere. But unless there is a singularity, n must vanish for a sphere of zero radius. We can also see the generalization of a Dirac monopole field to a gauge theory with a non-Abelian gauge group, G. It is a field configuration, necessarily singular, associated with an element of $\pi_1(G)$ other than the identity.[26]

I would now like to pick up a question I mentioned a few paragraphs back, and find the allowed values of the magnetic charge for a general theory in which the connected part of H is isomorphic to U(1). Every connected Lie group, G, has a simply-connected covering group, $\bar{G}$. All multiple-valued representations of G are ordinary single-valued representations of $\bar{G}$. (For example, the covering group of SO(3) is SU(2), and the covering group of U(1) is the additive group of the real numbers.) Let us change our conventions and declare that the gauge group of our theory is not G but $\bar{G}$. Of course, our scalar fields may not form a faithful representation of $\bar{G}$ – several elements of $\bar{G}$ may effect the same transformation of the fields – but faithfulness was never used in our arguments. Also of course, when we enlarge G we must also enlarge H, but let it be so enlarged. Once we have done this enlargement, the kernel condition of our previous theorem is trivial, for $\bar{G}$ is simply connected, and every closed path in $\bar{G}$ can be shrunk to the trivial path. Thus the only thing we have to check is that when we run down the right-hand side of Fig. 6, we really get a closed path in the enlargement of H, that is to say a closed path in $\bar{G}$. To check this, it is sufficient to check that the path is closed in every representation of $\bar{G}$. In terms of magnetic flux, this means that for every non-zero electric charge Q occurring in any representation of $\bar{G}$, the magnetic flux must be an integral multiple of $2\pi/Q$.

There are now two possibilities: (1) the representations of $\bar{G}$ may contain arbitrarily small non-zero charges. In this case, only zero magnetic flux is possible, and there are no non-trivial topological conservation laws; (2) there is a minimum positive charge, Q_{min}. Now, let Q be some other charge. By forming the direct product of the representation that contains Q with a string of representations that contain Q_{min} (or their complex conjugates), we can form a representation that contains an object of charge $Q + nQ_{min}$, for any positive (or negative) integer n. Thus Q must be an integral multiple of Q_{min}, if we are not to obtain a contradiction. Hence the only constraint on the magnetic flux is that it is an integral multiple of $2\pi/Q_{min}$.

We summarize all this in the following:

Theorem. For a general theory for which the connected part of H is isomorphic to U(1), all topological conservation laws are equivalent to the conservation of magnetic flux. Magnetic flux comes in integral multiples of a minimum unit, given by

$$\Phi_{min} = 2\pi/Q_{min}, \tag{3.59}$$

where Q_{min} is the smallest positive electric charge occurring in any single-valued or multiple-valued representation of G. If there is no smallest positive charge, there are no non-trivial topological conservation laws.

Corollary. If G contains a U(1) factor, and if the generator for this U(1) factor enters into the expression for the electric charge, there are no non-trivial topological conservation laws.

Proof. For, in this case, the U(1) factor has infinitely-multiple-valued representations, and thus we can make the electric charge as small as we please.

In all other cases, the standard representation theory of Lie groups tells us that there is a smallest non-zero charge, and thus there are non-trivial topological conservation laws.

This theorem and its corollary constitute result (3) of Sect. 3.1; result (4) is distributed throughout this subsection.

It might be helpful to see these principles at work in some examples:

Example 7. G is SU(3). H is U(2), embedded in SU(3) in the same way the isospin–hypercharge U(2) group is embedded in strong-interaction SU(3). Although this is not strong-interaction SU(3), I will save on definitions by using strong-interaction language and referring to the generators of U(2) as 'isospin' and 'hypercharge'. For orientation, let me begin by considering the case in which the symmetry breaks down all the way to hypercharge U(1). This is topologically the same as electric charge, treated above. Thus, in this case, all topological conservation laws are equivalent to the conservation of 'hypermagnetic flux' and the minimum value of this flux is 2π divided by the minimum hypercharge, that of the non-strange quarks. If U(2) were isomorphic to SU(2)⊗U(1), the additional factor of SU(2) would have no effect; SU(2) is simply connected and makes no contribution to the homotopy structure. However, the mapping of SU(2)⊗U(1) into U(2) is two-to-one. (If we represent an element of the first group by (U, λ), where U is an SU(2) matrix and λ is a complex number of unit modulus, then (U, λ) is mapped into the same element of U(2), $U\lambda$, as $(-U, -\lambda)$.) As a consequence of this, paths that go only half-way around SU(2)⊗U(1), from $(1, 1)$ to $(-1, -1)$, are mapped into closed paths in U(2), and the minimum unit of magnetic flux is half what it would be if H were U(1).

Example 8. G is SU(3) and H is SO(3), the subgroup of real unimodular unitary matrices. Since SU(3) is simply connected, the kernel condition is trivial. Thus, $\pi_2(G/H)$ is $\pi_1(H)$ is the additive group of the integers, modulo two. This is the same algebraic structure we found in two spatial dimensions back in Example 3.

Example 9. This is the same field theory as in Example 3, but in three spatial dimensions. G is SO(3), and the connected part of H is U(1), the subgroup of z-rotations. Thus, as far as the second homotopy group goes, this is the same as Example 4, and the same conclusions hold: the magnetic flux is given by Eq. (3.58), etc. However, there is one novelty; H has two components, and the component that does not contain the identity contains a rotation by π about the x-axis. This changes the sign of a z-rotation, and thus changes the sign of electric charge and magnetic flux. Thus we finally have a concrete example of the patching ambiguity we worried so much about in Sect. 3.6. In this theory, a monopole and an antimonopole are represented by gauge-equivalent solutions of the field equations; it is only when we try and put them together to make many-monopole states that we find that they are in fact different objects.

4   Quantum lumps

4.1   *The nature of the classical limit*

For example, consider the sine–Gordon equation:

$$\mathcal{L} = \tfrac{1}{2}(\partial_\mu \phi)^2 + \frac{\alpha}{\beta^2}(\cos \beta\phi - 1). \tag{4.1}$$

If we define

$$\phi' = \beta\phi, \tag{4.2}$$

then

$$\mathscr{L} = \frac{1}{\beta^2} \left[\tfrac{1}{2}(\partial_\mu \phi')^2 + \alpha(\cos \phi' - 1) \right]. \tag{4.3}$$

We see that, in classical physics, β is an irrelevant parameter; if we can solve the sine–Gordon equation for any non-zero β, we can solve it for any other β. The only effect of changing β is the trivial one of changing the energy and momentum assigned to a given solution of the equation.

This is not true in quantum physics, because the relevant object for quantum physics is not $\mathscr{L}$ but

$$\frac{\mathscr{L}}{\hbar} = \frac{1}{\beta^2\hbar} \left[\tfrac{1}{2}(\partial_\mu \phi')^2 + \alpha(\cos \phi' - 1) \right]. \tag{4.4}$$

Another way of saying the same thing is to say that in quantum physics we have one more dimensional constant of nature, Planck's constant, than in classical physics. Thus the quantum version of the sine–Gordon theory involves one more dimensionless parameter, $\beta^2\hbar$, than the classical theory.

Because the only way either β or $\hbar$ enters Eq. (4.4) is through the combination $\beta^2\hbar$, the classical limit, vanishing $\hbar$, is exactly the same as the small-coupling limit, vanishing β. Once we have this knowledge, there is no need to keep cluttering up our equations with $\hbar$s, so from now on I will adopt standard quantum units, and set $\hbar$ equal to one.

Nothing in this analysis is special to the sine–Gordon equation; every step goes through in exactly the same way for ϕ^4 theory, with λ replacing β^2. Here again the classical limit is the small-coupling limit. Likewise, for two-coupling-constant theories, like the gauge theory defined by Eq. (3.39), the classical limit is vanishing e, with λ/e^2 fixed.

These manipulations are trivial, but they teach us something important: if there are particles in the quantum theory that correspond to classical lumps, they are most likely to resemble their classical ancestors for weakly coupled theories. (Conversely, there is no more reason to trust classical analysis for strongly coupled theories than there is to trust the Born approximation.)[27] This suggests that the most direct way to construct quantum lumps is by an expansion in powers of the coupling constant. The leading term in such an expansion should give the classical results, appropriately reinterpreted in quantum language, and the higher terms should give quantum corrections.

I will now derive the first few terms in such an expansion for the case of a time-independent lump.[28]

4.2   *Time-independent lumps: power-series expansion*

For simplicity, I will do the analysis for a single scalar field in one spatial dimension, the sort of theory we discussed in Sect. 2. The generalization to more complicated theories, like those of Sect. 3, will be straightforward.

I will describe the theory in terms of a rescaled field, as in Eq. (4.2), and, for uniformity of notation, I will in all cases denote the coupling constant by β^2. Thus,

$$\beta^2 \mathcal{L} = \tfrac{1}{2}(\partial_\mu \phi')^2 - U(\phi'). \tag{4.5}$$

The canonical momentum density is given by

$$\pi' = \frac{\partial \mathcal{L}}{\partial(\partial_0 \phi')} = \frac{\partial_0 \phi'}{\beta^2}, \tag{4.6}$$

and the Hamiltonian by

$$\beta^2 H = \int \left[\beta^4 \pi'^2/2 + (\nabla \phi')^2/2 + U(\phi') \right] dx$$

$$= \int (\beta^4 \pi'^2/2) \, dx + V[\phi']. \tag{4.7}$$

I will assume that U has at least two zeros, so there exist finite-energy non-singular stable time-independent classical solutions. I denote these by

$$\phi'(x) = f(x - b). \tag{4.8}$$

I also define

$$E_0 \equiv V[f]. \tag{4.9}$$

I emphasize that these are rescaled quantities, and thus independent of β^2.

Equation (4.7) defines a rather peculiar Hamiltonian from the viewpoint of ordinary perturbation theory. Firstly, there is an explicit β^2 on the left-hand side of the equation. Of course, this is a trivial peculiarity; if we can find a power-series expansion for the energy eigenfunctions and eigenvalues of $\beta^2 H$, we can find one for those of H. Secondly, the small parameter multiplies the kinetic energy rather than the potential energy. This is very strange indeed; have we ever encountered such a system before?

Yes, we have. For this is precisely the situation for a diatomic molecule:

$$H = \frac{\mathbf{P}^2}{2M} + V(r), \tag{4.10}$$

where M is the nuclear reduced mass. The standard expansion procedure in the study of the spectra of diatomic molecules uses as the small parameter $1/M$, the coefficient of the kinetic energy. Of course, our system is not exactly a diatomic molecule. What it is exactly is a polyatomic molecule,

$$\sum_{i=1}^{N} \frac{\mathbf{P}^{(i)2}}{M_i} + V[\mathbf{x}^{(1)} \ldots \mathbf{x}^{(N)}].$$ (4.11)

To be precise, it is an infinitely polyatomic molecule, where all the nuclei have mass $1/\beta^4$. Thus the problem of constructing quantum lumps is one that was solved completely more than forty years ago.

I will now explain this solution, first by reminding you of the familiar results for a diatomic molecule,[29] then by telling you the trivial extension to a polyatomic molecule, and, finally, by making the even more trivial transcription of this extension into the language of field theory.

For the diatomic molecule, we assume the interatomic potential is as shown in Fig. 8. The minimum of V is at $r = r_0$, and $V(r_0) = E_0$. The first

Fig. 8

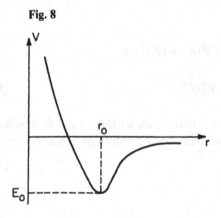

three approximations to the low-lying energy eigenstates and their eigenvalues are shown in Table 1.

Table 1. *The diatomic molecule*

Order of approximation	Energy eigenstate	Energy eigenvalue
0	$|r_0, \Omega\rangle$	E_0
1	$|n, \Omega\rangle$	$+(n+\tfrac{1}{2})[V''(r_0)/M]^{\tfrac{1}{2}}$
2	$|n, l, m\rangle$	$+\dfrac{l(l+1)}{2Mr_0^2} + \cdots$

As we see from the table, the proper expansion parameter for energy eigenvalues is $1/M^{\tfrac{1}{2}}$; this will become β^2 in the field-theory problem.

(The right-hand column of the table is additive; that is to say, the energy in first order is the sum of the first two entries, etc.) I will now explain the origin of the table.

In zeroth order, we neglect the kinetic energy altogether. The particle sits at the bottom of the potential, in an eigenstate of the position operator, x. The magnitude of the particle position is fixed at r_0, but its angular position is arbitrary. (As usual, Ω is shorthand for the pair of spherical coordinates θ and ϕ.) This is not much like the real spectrum of low-lying states; in particular, there is a totally spurious degeneracy in Ω, which, as we shall see, is removed only in second order.

To first order, we begin to see the effects of the vibration of the particle about its equilibrium position. Since M is very large, the particle does not vibrate very far, and, to first order, we can replace the potential near equilibrium by a harmonic potential

$$V(r) = E_0 + \tfrac{1}{2}V''(r_0)(r - r_0)^2. \qquad (4.12)$$

The energy eigenfunctions are now harmonic-oscillator wave functions in r, but still delta-functions in Ω. They are labeled by the usual harmonic-oscillator excitation number, n, and have the usual harmonic-oscillator energies. (These are the famous vibrational levels.) They are no longer eigenstates of the position operator, but it is still easy to compute the expectation value of any component of x. For example, if we normalize the eigenstates such that

$$\langle n', \Omega' | n, \Omega \rangle = \delta_{nn'}\delta(\Omega - \Omega'), \qquad (4.13)$$

then, for the z-component of the position operator, in the ground state,

$$\langle 0, \Omega' | z | 0, \Omega \rangle = r_0 \cos \theta \, \delta(\Omega - \Omega'). \qquad (4.14)$$

It is only in second order that we begin to see the effects of rotation; this is because the zeroth-order moment of inertia of the molecule is Mr_0^2. The degeneracy in Ω is removed; angle eigenstates are replaced by angular-momentum eigenstates

$$|n, l, m\rangle = \int d\Omega \, Y_{lm}(\Omega) |n, \Omega\rangle, \qquad (4.15)$$

and a rigid-rotator term is added to the energy. (These are the famous rotational levels.) Note that the rotational structure is determined purely by group theory; it involves no properties of V that have not entered earlier approximations. In addition, we begin to see the effects of departures from the harmonic approximation, Eq. (4.12). I have indicated these terms (vibrational–vibrational coupling) in the table by triple dots. They depend on the detailed form of V (in particular, on $V'''(r_0)$ and $V^{iv}(r_0)$).

Unlike the rotational term, they do not affect the qualitative features of the problem, nor do the higher terms in the expansion.

The extension of all this to a polyatomic molecule is trivial. Unless the equilibrium configuration of the molecule is one in which all the nuclei are aligned, the equilibrium configurations are labeled, like the positions of a rigid body, by three Euler angles rather than two polar angles. As a consequence of this, the rotational spectrum, once it appears in second order, will be that of a rigid body, rather than that of a rigid rotator. Also, there are many ways to vibrate about equilibrium, and the single integer n is replaced by a string of integers n_i, one for each normal mode. Likewise, in the first-order eigenvalue,

$$\left(\frac{V''}{M}\right)^{\frac{1}{2}} (n+\tfrac{1}{2}) \rightarrow \sum_i \left(\frac{K_i}{M}\right)^{\frac{1}{2}} (n_i+\tfrac{1}{2}), \tag{4.16}$$

where the Ks are the spring constants for the normal modes. (There is a possible error here into which one is less likely to fall in the diatomic case. The count of normal modes should not include the three zero-frequency normal modes that correspond to infinitesimal rotations; these degrees of freedom are already taken care of by the Euler angles, and it would be double counting to include them with the ns.)

Now, as promised, to treat the Hamiltonian (4.7) requires hardly more than a change of notation. This produces Table 2, which I will now explain.

To zeroth order, all we have to do is replace $1/M^{\frac{1}{2}}$ by β^2, and divide the β^2 out of the left-hand side of Eq. (4.7). Instead of a degenerate family of eigenstates of the position operator, connected by rotations, we have a degenerate family of eigenstates of the field operator, connected by space translations. I have labeled these by b, the position of the center of the lump.

Table 2. *The quantum lump*

Order of approximation	Energy eigenstate	Energy eigenvalue
1	$\|b\rangle$ $\|n_1, n_2 \ldots ; b\rangle$	E_0/β^2 $+\sum_i (n_i+\tfrac{1}{2})\omega_i$
2	$\|n_1, n_2 \ldots ; p\rangle$	$+\dfrac{\beta^2 p^2}{2E_0}+\ldots$

To first order, we merely have to replace the normal-mode spring constants of Eq. (4.16) by the eigenfrequencies of Eq. (2.18). Just as for the molecule, the count of normal modes does not include the zero-frequency mode, Eq. (2.19). (Of course, the eigenfrequencies form a continuous set, so the sum should be an integral. For the moment, let me imagine we have put the system in a box, so the eigenfrequencies are discrete; I will take care of the continuum problem shortly.) As usual when passing from particle mechanics to field theory, we reinterpret harmonic-oscillator excitation numbers as meson normal-mode occupation numbers. This justifies the last paragraph of Sect. 2.2; the small perturbations about the classical solution do indeed describe mesons interacting with a lump. The state where all the ns vanish, which we denote by $|0; b\rangle$, is an isolated lump; states with non-vanishing n correspond to one or more mesons bound to or passing by the lump. To this order, there is no sign of meson–meson interactions in the energy because meson–meson interactions are of order β^2; their effects are analogous to the vibrational–vibrational coupling in the molecule, and, like it, they are lurking behind the three dots in the second-order energy. In analogy to Eqs. (4.13) and (4.14), if we normalize the first-order ground states, for example, such that

$$\langle 0; b'|0; b\rangle = \delta(b' - b), \tag{4.17}$$

then

$$\langle 0; b'|\phi'(x)|0; b\rangle = \delta(b' - b)f(x - b). \tag{4.18}$$

In second order, the spurious degeneracy is lifted, and the translation eigenstates are replaced by momentum eigenstates. Just as for the molecule, the proper eigenstates are determined purely by group theory, by the symmetries of the problem,

$$|n_1 \ldots ; p\rangle = \int \frac{db}{(2\pi)^{\frac{1}{2}}} e^{ipb}|n_1 \ldots ; b\rangle, \tag{4.19}$$

as is the degeneracy-lifting part of the energy,

$$(p^2 + M^2)^{\frac{1}{2}} = M + p^2/2M + O(M^{-3})$$
$$= \frac{E_0}{\beta^2} + \frac{\beta^2 p^2}{2E_0} + O(\beta^6). \tag{4.20}$$

As a consequence of Eqs. (4.18) and (4.19), we can give an interpretation of the classical solution, $f(x)$, in terms of the conventional concepts of particle physics:

$$\langle 0; p'|\phi'(0)|0; p\rangle = \int \frac{dx}{(2\pi)} e^{-i(p - p')x}f(x). \tag{4.21}$$

In words, $f(x)$ is the leading approximation to the Fourier transform of the ϕ' form factor of the quantum lump.

Now let me return to the problems associated with the fact that the sum over normal modes is actually an integral. Obviously, we can take care of everything if we can take care of the isolated lump, so let us concentrate on this.

I remind you that Eq. (2.18) tells us that the *squares* of the eigenfrequencies are the eigenvalues of the positive operator

$$K = -\frac{d^2}{dx^2} + U''(f). \tag{4.22}$$

If we put the system in a box, so the eigenfrequencies are discrete, the sum of the eigenfrequencies is the trace of the operator $K^{\frac{1}{2}}$, the positive square root of K. Of course, what we are really interested in is not the energy of the lump, but the difference between the energy of the vacuum and that of the lump. Thus we define

$$K_0 = -\frac{d^2}{dx^2} + U''(\phi_0), \tag{4.23}$$

where ϕ_0 is a zero of U. The desired energy difference is

$$\tfrac{1}{2}\left(\sum_i \omega_i\big|_{\text{lump}} - \sum_i \omega_i\big|_{\text{vacuum}}\right) = \tfrac{1}{2}\mathrm{Tr}(K^{\frac{1}{2}} - K_0^{\frac{1}{2}}). \tag{4.24}$$

As we send the box to infinity, the left-hand side of this equation becomes difficult to interpret, but the right-hand side remains unambiguous; it is what should properly appear in Table 2.

Although I will not attempt to explicitly compute Eq. (4.24) here for any of our models, I should tell you that the trace is ultraviolet divergent, and also that this is as it should be. There are no counterterms in Eq. (4.7), so the parameters occurring in it (like α for the sine–Gordon equation, or μ^2 for ϕ^4 theory) should be interpreted as bare parameters. As we shall see explicitly in Sect. 4.3, these bare parameters are finite quantities plus divergent terms proportional to β^2 (and sometimes also to higher powers of β). Thus the zeroth order expression for the energy, E_0/β^2, generates a divergence of order β^0, which must cancel against a corresponding divergence in Eq. (4.24) if everything is to remain finite to this order.

I do not have time in these lectures to show explicitly these divergence cancellations, and must refer you to the literature if you want a demonstration. However, I should stress that it would be a genuine shock if the divergences did not cancel. After all, for all renormalizable theories, our experience has been that the usual counterterms are sufficient to get rid of all the divergences in all observable processes – not just scattering amplitudes but also bound-state energies, etc. A quantum lump is just another bound state (I will expand on this remark later); its energy is an observable quantity, and we would expect it to be purged of divergences by renormalization just like every other observable.

The extension to three spatial dimensions is straightforward. Of course, since three-dimensional theories with time-independent solutions necessarily involve much more than just a single scalar field, the actual computations are considerably more onerous. Indeed, so far as I know, no one has even made an approximate computation of the first-order correction to the lump energy, Eq. (4.24), for any three-dimensional theory.

The only qualitative difference occurs in the second-order degeneracy-removing term. This is because there is a larger group of geometrical symmetries in three dimensions (the six-parameter Euclidean group), and thus the manifold of minima of V may have as many as six dimensions. There are three cases. (1) The classical lump is spherically symmetric. In this case, the classical solutions are labeled by a three-vector, $\mathbf{a}$, the position of the center of the lump, and the degeneracy-removing term is simply $\beta^2 \mathbf{p}^2/2E_0$, just as in one dimension. (2) The classical lump is not spherically symmetric, but it has an axis of symmetry. In this case, the classical solutions are labeled by $\mathbf{a}$ and by the pair of angles θ and ϕ, that give the orientation of the symmetry axis. We then obtain an additional degeneracy-removing term, $\beta^2 s(s+1)/2I_0$, where s is the intrinsic angular momentum of the lump and I_0 is its classical moment of inertia. This is exactly like the situation for the diatomic molecule; the quantum lump has rotational excitations. (3) The classical lump has no axis of symmetry. In this case, in addition to $\mathbf{a}$, we need three Euler angles. This is exactly like the polyatomic molecule; the quantum lump has rotational excitations, and their spectrum is given by the quantum rigid-body formula.

Of course, in any number of dimensions, we can also have degeneracy if our classical lump is not invariant under some unbroken internal symmetry group of the theory. In this case the degeneracy would also be removed in second order, leading to what we might call 'internal rotational excitations'.[30]

We have learned a lot about the structure of those quantum lumps that correspond to time-independent classical solutions. They are really quite like more familiar composite systems, ordinary bound states. There is one main difference. Ordinary bound states are, for weak coupling, approximate eigenstates of particle number. (Thus, in lowest order, positronium is made up of one electron and one positron, but, as we go to higher orders, we find contributions from two electrons and two positrons, etc.) As a result of this, as the coupling vanishes, the mass of an ordinary bound state goes to some multi-particle threshold. In contrast, for weak coupling, the quantum lump is an approximate eigenstate of the field operator, ϕ. This does not commute with particle number, and thus there is no reason for the mass of the lump to go to a threshold as the coupling vanishes. Indeed, as we have seen, it goes to infinity.

4.3   Time-independent lumps: coherent-state variational method[31]

In the previous subsection, we tackled the quantum lump by a method from non-relativistic quantum mechanics. This is not the only non-relativistic method that can be used on this problem; the Rayleigh–Ritz variational method can also be used. Of course, to say 'the variational method' is to say nothing until one specifies the trial states; for a large enough set of trial states the variational method is guaranteed to be arbitrarily accurate and arbitrarily impractical.

I will restrict myself to theories of scalar fields in one spatial dimension with non-derivative interactions, like the theories of Sect. 2. This time, the restriction is not made for reasons of simplicity, but because I believe that these are the only theories for which I have a chance of making a reasonable guess for the trial states. The reason is that these theories are locally Fock; this means that if we restrict the theories to a box, their exact energy eigenstates are in ordinary Fock space. This is connected to the mild ultraviolet divergences of these theories; it is definitely not true for theories in more spatial dimensions, where the divergences are worse. Since all trial states that have been proposed in the literature are in Fock space (when restricted to a box), they are guaranteed to be grotesquely bad guesses for the true energy eigenstates in more than one spatial dimension. This does not mean that the variational method is *a priori* inapplicable; it just means that better trial states than any yet proposed are needed.

(In fairness, I should tell you that most workers in this field disagree completely with the preceding paragraph and think that I am a fussbudget.)

Another advantage of sticking to one spatial dimension is that the divergence structures of the theories are so simple that we can write the Hamiltonian in terms of finite parameters in closed form. Thus we preserve the great advantage of the variational method, that we always obtain an upper bound on the energy. Of course, even in three spatial dimensions we would obtain an upper bound, but it would be useless because it would be expressed in terms of the bare parameters of the theory, known in terms of finite parameters only as infinite power series. There is no advantage in obtaining exact upper bounds on the eigenvalues of a Hamiltonian that is only known approximately.

Thus, before I describe variational computations, I will discuss how to sum up the divergences which arise in all orders of perturbation theory for the theory defined by

$$\mathcal{H} = \tfrac{1}{2}\pi^2 + \tfrac{1}{2}(\partial_1\phi)^2 + U(\phi). \tag{4.25}$$

(Note that we are back with canonical fields, not rescaled fields as in Sect. 4.2.)

For this theory, the only ultraviolet divergences that occur in any order of perturbation theory come from graphs that contain a closed loop consisting of a single internal line, that is to say, graphs in which two fields at the same vertex are contracted with each other. For example, the graph of Fig. 9(a) is ultraviolet divergent, while that of Fig. 9(b) is not. Thus, all ultraviolet divergences can be removed by normal-ordering the interaction Hamiltonian in the interaction picture.

However, variational computations are not carried out in the interaction picture, but in the Schrödinger picture, where the Hamiltonian is given as a function of fixed-time coordinates and momenta, as in Eq. (4.25). It is easy to see what interaction-picture normal-ordering becomes in the Schrödinger picture: we simply define the 'creation' and 'annihilation' parts of ϕ and π as if they were free fields.

$$\phi^{(\pm)}(x, m) = \frac{1}{2}\left[\phi(x) \mp \frac{i}{(-\nabla^2 + m^2)^{\frac{1}{2}}} \pi(x) \right], \qquad (4.26a)$$

and

$$\pi^{(\pm)}(x, m) = \mp i(-\nabla^2 + m^2)^{\frac{1}{2}}\phi^{(\pm)}(x, m), \qquad (4.26b)$$

where $+$ indicates creation and $-$ annihilation. We then define the normal-ordered version of any function of these operators as the function rearranged with all the creation operators on the left and all the annihilation operators on the right.

Note that this prescription is ambiguous, because m is not specified. Of course, if we were really doing interaction picture perturbation theory, it would be senseless to choose m to be other than that mass which occurs in the free Hamiltonian. However, we are doing variational computations, and we should avoid premature decisions about the best way to divide H into a free and an interacting part. Thus, for the moment, I will keep m free, and denote by N_m the normal-ordering operation defined by the mass m.

(This is off the main line of argument, but it is amusing to consider what happens in interaction-picture perturbation theory, if m is chosen

Fig. 9

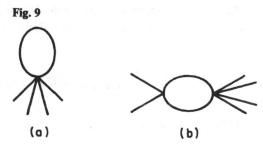

(a)   (b)

to be different from μ, the mass in the free Hamiltonian. In this case, it is easy to see that the divergent loop integral of Fig. 9(a) is replaced by a convergent integral according to the prescription:

$$\int \frac{d^2 k}{k^2 - \mu^2} \rightarrow \int d^2 k \left(\frac{1}{k^2 - \mu^2} - \frac{1}{k^2 - m^2} \right). \tag{4.27}$$

Note that if $m = \mu$ (the conventional rule) the graph is canceled completely (the conventional result). However, even if $m \neq \mu$, the divergent integral is converted into a convergent one.)

Whatever m we choose, we can normal-order the Hamiltonian by Wick's theorem. We begin by computing the contraction of two fields,

$$\phi^2(x) = N_m \phi^2(x) + \int \frac{dk}{4\pi} \frac{1}{(k^2 + m^2)^{\frac{1}{2}}}. \tag{4.28}$$

The integral is divergent; as usual, we introduce a high-momentum cutoff and find

$$\int_{-\Lambda}^{\Lambda} \frac{dk}{4\pi} \frac{1}{(k^2 + m^2)^{\frac{1}{2}}} = \frac{1}{4\pi} \ln \frac{4\Lambda^2}{m^2}, \tag{4.29}$$

for $\Lambda \gg m$. Whence, for any function of ϕ,

$$U(\phi) = N_m \left\{ \exp \left[\frac{1}{8\pi} \left(\ln \frac{4\Lambda^2}{m^2} \right) \frac{d^2}{d\phi^2} \right] U(\phi) \right\}. \tag{4.30}$$

This is just Wick's theorem; the successive terms in the expansion of the exponential give us the terms with no contractions, one contraction, etc.

We can also use Wick's theorem to pass from one normal-ordering prescription to another. Since, from Eq. (4.28),

$$N_m \phi^2 = N_\mu \phi^2 + \frac{1}{4\pi} \ln \frac{m^2}{\mu^2}, \tag{4.31}$$

it follows that

$$N_m U(\phi) = N_\mu \left\{ \exp \left[\frac{1}{8\pi} \left(\ln \frac{m^2}{\mu^2} \right) \frac{d^2}{d\phi^2} \right] U(\phi) \right\}. \tag{4.32}$$

Note that this is free of all reference to the cutoff, as it should be.

Equations (4.30) and (4.32) take care of the last term in the Hamiltonian density, Eq. (4.25). The first two terms can be treated similarly. For brevity of notation, I define

$$\mathscr{H}_0 = \tfrac{1}{2} \pi^2 + \tfrac{1}{2} (\partial_1 \phi)^2. \tag{4.33}$$

Normal-ordering this is trivial, since it is a mere quadratic function of ϕ and π; thus

$$\mathscr{H}_0 = N_m \mathscr{H}_0 + \text{divergent constant.} \tag{4.34}$$

An easy integration reveals that

$$N_m \mathcal{H}_0 = N_\mu \mathcal{H} + \frac{1}{8\pi} (\mu^2 - m^2). \tag{4.35}$$

To have a definite example, let me apply these formulae to the sine–Gordon equation,

$$\mathcal{H} = \mathcal{H}_0 - \frac{\alpha_0}{\beta^2} \cos \beta\phi + \gamma_0. \tag{4.36}$$

Here I have put a nought on α, in anticipation of its impending renormalization. I have also added a constant to the Hamiltonian density, to acknowledge the fact that we do not know the ground-state energy of the quantum theory. γ_0 is not a new free parameter; it is to be chosen at the end of our computations so the ground-state energy is zero. As we shall see immediately, there is no need to renormalize β.

Applying Eqs. (4.30) and (4.34), we see that, for any m,

$$\mathcal{H} = N_m \left[\mathcal{H}_0 - \frac{\alpha_0}{\beta^2} \left(\frac{m^2}{4\Lambda^2} \right)^{\beta^2/8\pi} \cos \beta\phi + \gamma_0 \right.$$

$$\left. + \text{divergent constant} \right]. \tag{4.37}$$

Thus we can absorb all divergences into a multiplicative renormalization of α and an additive renormalization of γ. That is to say, we define

$$\alpha = \alpha_0 \left(\frac{m^2}{4\Lambda^2} \right)^{\beta^2/8\pi}, \tag{4.38}$$

and

$$\gamma = \gamma_0 + \text{divergent constant}, \tag{4.39}$$

and obtain

$$\mathcal{H} = N_m \left[\mathcal{H}_0 - \frac{\alpha}{\beta^2} \cos \beta\phi + \gamma \right]. \tag{4.40}$$

We can now let the cutoff go to infinity, and we shall. From now on we will work only with the finite form of $\mathcal{H}$, Eq. (4.40). Of course, if at some future time we find we are unhappy with our original choice of m, we can change it in a moment, using Eqs. (4.32) and (4.36),

$$\mathcal{H} = N_\mu \left[\mathcal{H}_0 - \frac{\alpha}{\beta^2} \left(\frac{\mu^2}{m^2} \right)^{\beta^2/8\pi} + \gamma + \frac{1}{8\pi} (\mu^2 - m^2) \right]. \tag{4.41}$$

Note that the result of changing m can be absorbed in a redefinition of α and γ. In this sense the arbitrary normal-ordering mass m is like the arbitrary normalization mass that appears in the theory of the renormalization group.

We are now in a position to do a simple sample variational computation. Let us consider

$$\mathcal{H} = N_m[\mathcal{H}_0 + \tfrac{1}{2}M^2\phi^2]. \tag{4.42}$$

This is a free-field Hamiltonian density, perversely normal-ordered; that is to say, m is not necessarily equal to M. Of course, any fool can solve this problem exactly, and there is no real need to use the variational method. Nevertheless, the computation is instructive.

As our trial states, let us choose the vacuum states appropriate to a free field of mass μ. These are defined by

$$\phi^{(-)}(x, \mu)|0, \mu\rangle = \pi^{(-)}(x, \mu)|0, \mu\rangle = 0. \tag{4.43}$$

Since these states are translationally invariant, we need only compute the expectation value of the Hamiltonian density. This is made trivial by our re-normal-ordering equations,

$$\mathcal{H} = N_\mu\left[\mathcal{H}_0 + \tfrac{1}{2}M^2\phi^2 + \frac{1}{8\pi}\left(\mu^2 - m^2 - M^2\ln\frac{\mu^2}{m^2}\right)\right]. \tag{4.44}$$

Whence,

$$\langle 0, \mu|\mathcal{H}|0, \mu\rangle = \frac{1}{8\pi}\left(\mu^2 - m^2 - M^2\ln\frac{\mu^2}{m^2}\right). \tag{4.45}$$

This completes the first step of the Rayleigh–Ritz method, the computation of the expectation value of the Hamiltonian in the trial state. The next step is to differentiate this with respect to the variational parameter, μ,

$$1 - \frac{M^2}{\mu^2} = 0. \tag{4.46}$$

Hence the best trial state is $|0, M\rangle$. Of course, this is also the exact ground state. This is reassuring.

Now let us try exactly the same procedure, with exactly the same trial states, for the sine–Gordon equation. From Eq. (4.41),

$$\langle 0, \mu|\mathcal{H}|0, \mu\rangle = -\frac{\alpha}{\beta^2}\left(\frac{\mu^2}{m^2}\right)^{\beta^2/8\pi} + \gamma + \frac{1}{8\pi}(\mu^2 - m^2). \tag{4.47}$$

Now comes the surprise: this expression is unbounded below, as μ goes to infinity, if β^2 exceeds 8π. The variational method always gives a rigorous upper bound on the ground-state energy; in this case the upper bound is minus infinity. We have rigorously proved that the energy of the sine–Gordon theory is unbounded below if β^2 exceeds 8π; the theory has no ground state, even if we put it in a box, and is physically nonsensical. If β^2 is less than 8π, Eq. (4.47) does have a minimum; of course, this does not prove that the theory is physically sensible in this case.

Although we have gone surprisingly far with the states $|0, \mu\rangle$, they are obviously not a rich enough set to be good trial states for the vacuum in the general case, let alone for a lump. For one thing, the expectation value of ϕ vanishes in any such state. I will now describe a family of states that do not have this deficiency; they are called coherent states.

Coherent states are best explained for a very simple system, a harmonic oscillator,

$$H = \tfrac{1}{2}(p^2 + q^2). \tag{4.48}$$

As usual, we define the raising and lowering operators,

$$a = (q + \mathrm{i}p)/\sqrt{2}, \tag{4.49a}$$

$$a^\dagger = (q - \mathrm{i}p)/\sqrt{2}. \tag{4.49b}$$

A coherent state is labeled by an arbitrary complex number, z, and is defined by

$$|z\rangle = \mathrm{e}^{-\frac{1}{2}z^*z}\, \mathrm{e}^{za^\dagger}|0\rangle, \tag{4.50}$$

where $|0\rangle$ is the oscillator ground state. It is easy to check that

$$\langle z|z\rangle = 1, \tag{4.51}$$

and

$$a|z\rangle = z|z\rangle, \tag{4.52a}$$

or, equivalently,

$$\langle z|a^\dagger = \langle z|z^*. \tag{4.52b}$$

From these equations, we see that the complex number z can be given in terms of the expectation values of p and q,

$$z = \frac{1}{\sqrt{2}}(\langle q\rangle + \mathrm{i}\langle p\rangle). \tag{4.53}$$

For any normal-ordered function of p and q,

$$\langle :F(p, q): \rangle = F(\langle p\rangle, \langle q\rangle). \tag{4.54}$$

This is because normal ordering puts all the lowering operators on the right, where we can use Eq. (4.52a), and all the raising operators on the left, where we can use Eq. (4.52b).

All of this extends trivially to an infinite assembly of harmonic oscillators, a free field of mass μ. Here, starting from the state $|0, \mu\rangle$, we can construct a family of states labeled by two arbitrary functions of space $\langle \phi(x)\rangle$ and $\langle \pi(x)\rangle$. We will use these as trial states to find the energy eigenstates of the theory defined by

$$\mathcal{H} = N_m[\mathcal{H}_0 + U(\phi)]. \tag{4.55}$$

Just as before, the computation is simplified by re-normal-ordering,

$$\mathcal{H} = N_\mu[\mathcal{H}_0 + U(\phi, \mu)], \tag{4.56}$$

where, by our re-normal-ordering equations,

$$U(\phi, \mu) = \exp \left[\frac{1}{8\pi} \left(\ln \frac{m^2}{\mu^2} \right) \frac{d^2}{d\phi} \right] U(\phi) + \frac{1}{8\pi} (\mu^2 - m^2).$$ (4.57)

Thus,

$$\langle \mathcal{H}(x) \rangle = \tfrac{1}{2} \langle \pi(x) \rangle^2 + \tfrac{1}{2} (\partial_1 \langle \phi(x) \rangle)^2 + U(\langle \phi(x) \rangle, \mu).$$ (4.58)

The easiest thing to compute is the best trial state(s) for the ground state(s) of the theory. The first two terms in (4.58) are obviously minimized by choosing $\langle \pi(x) \rangle$ to be zero and $\langle \phi(x) \rangle$ to be a constant. The ground state(s) are then determined by the absolute minimum (minima) of $U(\langle \phi \rangle, \mu)$. This is almost as easy as the corresponding computation in the classical case; the only difference is that we have to search for minima in a two-parameter space rather than a one-parameter space. Once the best values of $\langle \phi \rangle$ and μ have been found, it is convenient to add a constant to Eq. (4.58) so the ground-state energy is zero.

Let us denote the value of μ found from the ground-state computation by μ^*. Any of our coherent states that do not have μ equal to μ^* will have an infinite energy expectation value, for the space integral of the last term in Eq. (4.58) will diverge. Thus the problem of finding other variational energy eigenstates is *identical* to the problem of Sect. 2.1, the problem of finding classical time-independent lumps, with the classical U equal to $U(\phi, \mu^*)$. Hence if the ground-state variational computation yields more than one ground state, we will always obtain a variational lump, the energy of the lump will be given by the classical formula, etc.

In atomic physics, variational computations cut across the perturbation expansion, picking up a part of every order of perturbation theory. This is also the case here. The weak-coupling expansion of Sect. 4.2 encountered infinities, which had to be removed by renormalization. For the sine–Gordon equation, as we can see from Eq. (4.38), these infinities occur in all orders of β^2. The variational computation is phrased in terms of finite parameters; it is 'already renormalized', and therefore contains at least a piece of all orders of the β^2 expansion. On the other hand, it does not contain all of each order. In particular, it does not contain the most significant feature of the second-order computation, the removal of the classical translational degeneracy. Our coherent states are not momentum eigenstates (except for the ground-state trial states), and we have an infinite degenerate family of variational energy eigenstates, labeled by the position of the center of the lump.

There is another way to look at this. Again, let me use the sine–Gordon equation as an example. The method of Sect. 4.2 gives us a power series in β^2, with α_0 held fixed. We can turn this into a renormalized expansion by expressing α_0 as a finite parameter, α, times a power series in β^2, with divergent coefficients, and regrouping terms in the usual way. There are, of course, many such renormalized expansions, because there are many ways of defining α, many possible renormalization conventions. If we choose α to be the coefficient which occurs in $U(\phi, \mu^*)$, then, for this definition (and for no other), the variational computation is identical with the zeroth order of the renormalized series expansion. In this way of looking at things, the variational computation of the ground state serves to determine the 'best choice' of renormalized parameters for the series expansion.

Of course, it is always possible to improve a variational computation by enlarging the set of trial states. For example, it is possible to enforce translational invariance by building momentum eigenstates out of super-positions of coherent states. Also, it is possible to enlarge the set of trial states by generalizing the concept of normal-ordering, replacing $(-\nabla^2 + m^2)$ in Eq. (4.26) by a more general positive operator. Such enlargements have been considered in the literature, but I will not discuss them here. The calculations become considerably more difficult, and the qualitative features of the output are unchanged.

As I said at the beginning, I do not believe there are any plausible variational computations using coherent states or simple generalizations of them in more than one spatial dimension. It would be nice to have reasonable variational states in this case. A good place to begin exploring would be a super-renormalizable theory in two spatial dimensions. Here, although all divergences are not removed by normal ordering, it is still possible to sum them up in closed form. Thus we have at least a minimal criterion for a reasonable trial state: the expectation value of the Hamiltonian density should not be infinite. Coherent states do not meet this test.

4.4   *Periodic lumps: the old quantum theory and the DHN formula*

Time-independent solutions are not the only non-dissipative solutions of classical field theories, although they are certainly the easiest to study. Among others, there are periodic solutions.

Once again, the sine–Gordon equation provides a useful example, for an exact periodic solution is known:

$$\phi(x, t) = \frac{4}{\beta} \tan^{-1} \left[\frac{\eta \sin \omega t}{\cosh \eta \omega x} \right], \tag{4.59}$$

where

$$\eta = (\alpha - \omega^2)^{\frac{1}{2}}/\omega, \tag{4.60}$$

and ω ranges from 0 to α. This solution has a simple physical interpretation, which can most easily be seen if $\omega \ll \alpha$. In this case,

$$\phi(x, t) \approx \frac{4}{\beta} \tan^{-1} \left[\frac{\alpha^{\frac{1}{2}} \sin \omega t}{\omega \cosh \alpha^{\frac{1}{2}} x} \right]. \tag{4.61}$$

This simplifies further if we restrict ourselves to portions of the period such that

$$\omega \ll \alpha^{\frac{1}{2}} |\sin \omega t|. \tag{4.62}$$

Let us also assume $\sin \omega t$ is positive. In this case, the argument of the tangent is very large near $x = 0$, and ϕ is approximately $2\pi/\beta$. As we go to large negative x, $\cosh \alpha^{\frac{1}{2}} x$ eventually grows large and ϕ departs from $2\pi/\beta$. In this region, we may approximate $\cosh \alpha^{\frac{1}{2}} x$ by an exponential, and obtain,

$$\phi(x, t) \approx \frac{4}{\beta} \tan^{-1} \exp[\alpha^{\frac{1}{2}} x + \ln(2\alpha^{\frac{1}{2}} \sin \omega t/\omega)]. \tag{4.63}$$

That is to say, there is a soliton far to the left. Likewise, there is an anti-soliton far to the right. As $\sin \omega t$ increases to one, the soliton and anti-soliton move farther apart from each other. When $\sin \omega t$ passes through one, they turn around and begin to approach each other. As $\sin \omega t$ comes down to zero, Eq. (4.62) no longer applies, and we can no longer describe ϕ in terms of a widely separated soliton–antisoliton pair. This is reasonable, because the soliton and antisoliton are on top of each other. However, we can pick up the description again when $\sin \omega t$ becomes negative, and the soliton and antisoliton have passed each other.

Thus, Eq. (4.59) can be thought of as a soliton and an antisoliton oscillating about their common center-of-mass. For this reason, it is called 'the doublet solution'.

For a later computation, we will need the energy of the doublet. The calculation is done most quickly at $t = 0$:

$$
\begin{aligned}
E &= \frac{1}{2} \int dx (\partial_0 \phi)^2 \\
&= \frac{8\eta^2 \omega^2}{\beta^2} \int dx \frac{1}{(\cosh \eta \omega x)^2} \\
&= \frac{16\eta \omega}{\beta^2} \\
&= 2M \left(1 - \frac{\omega^2}{\alpha} \right)^{\frac{1}{2}},
\end{aligned}
\tag{4.64}
$$

where

$$M = 8\alpha^{\frac{1}{2}}/\beta^2, \tag{4.65}$$

is the soliton mass. Note that the mass of the doublet is always less than twice the soliton mass, as we would expect for bound motion of a soliton–antisoliton pair.

We would expect such bound classical motions to become quantum bound states, energy eigenstates, in the quantum version of the theory. Our problem is to compute, in some plausible approximation, the energy eigenvalues. Fortunately for us, a crude but serviceable approximation is at hand, for this is the fundamental problem of the old quantum theory, solved by Bohr and Sommerfeld. Their quantization formula says that if we have a one-parameter family of periodic motions, labeled by the period, T, then an energy eigenstate occurs whenever

$$\int_0^T \mathrm{d}t p\dot{q} = 2\pi n, \tag{4.66}$$

where n is an integer. Note that this is *not* the WKB formula, which has $n + \frac{1}{2}$ rather than n on the right. The WKB formula is a refinement of Eq. (4.66), but it is a refinement only for special circumstances, a particle moving in a potential. (For example, there is no $\frac{1}{2}$ in the corresponding quantization condition for angular motion.) Eq. (4.66) is cruder than the WKB formula, but it is much more general; it is always the leading approximation for any dynamical system.

For the benefit of those of you who have forgotten the stratagems of the old quantum theory, I will briefly recapitulate the arguments that lead to Eq. (4.66). For any classical system, the energy is given by

$$E = p\dot{q} - L. \tag{4.67}$$

Integrating over a period, we find

$$ET = 2\pi n - \int_0^T \mathrm{d}t L \tag{4.68}$$

where n is defined by Eq. (4.66). Differentiating this with respect to T, with $q(0)$ fixed, we find

$$E\mathrm{d}T + T\mathrm{d}E = 2\pi \mathrm{d}n - L(T)\mathrm{d}T + p(T)\dot{q}(T)\mathrm{d}T. \tag{4.69}$$

(The last term comes from differentiating the integrand, integrating by parts, and using the Euler–Lagrange equations.) Thus,

$$\mathrm{d}E = \omega \mathrm{d}n. \tag{4.70}$$

Thus, if the motion is quantized at integer ns, the frequency of the quantum emitted when the system makes a transition from one quantum state to the next is the same as the frequency of the classical motion. This is Bohr's correspondence principle. (Of course, the WKB formula also obeys Eq.

(4.70). In this way of looking at things, the only new thing produced by WKB analysis is the integration constant for Eq. (4.70).)

There is nothing in this argument that makes any reference to the details of the dynamics, and it was only notational laziness that made me carry it through only for a system with a single degree of freedom. It is just as valid for a system with an infinite number of degrees of freedom, that is to say, a field theory:

$$\int_0^T dt \int dx \pi(x, t) \partial_0 \phi(x, t) = 2\pi n. \tag{4.71}$$

I will now apply this formula to the sine–Gordon doublet. The fastest way to do the integral is to integrate Eq. (4.70),

$$\frac{dn}{dE} = \omega^{-1} = \alpha^{-\frac{1}{2}} \left(1 - \frac{E^2}{4M^2}\right)^{-\frac{1}{2}}, \tag{4.72}$$

and use as boundary condition the obvious fact that the left-hand side of Eq. (4.71) vanishes when $E = 0$. Thus we find a sequence of quantum states,

$$E_n = 2M \sin(\beta^2 n/16) \tag{4.73}$$

where $n = 0, 1, 2, \ldots < 8\pi/\beta^2$.

We can check this formula for small β. In this case,

$$E_n = n\alpha^{\frac{1}{2}} + O(\alpha^{\frac{1}{2}}\beta^2). \tag{4.74}$$

Now, $\alpha^{\frac{1}{2}}$ is the mass of the fundamental meson, for small β. It is not surprising that the fundamental meson should emerge as the lowest quantum state of the doublet, other than the vacuum. For small E, the soliton and antisoliton never separate much, and Eq. (4.59) looks much more like a small perturbation of the ground state than an oscillating pair. The other quantum states also have a simple interpretation. For small coupling, the fundamental mesons have only weak interactions among themselves; if we expand the interaction in powers of β, the leading term is an attractive quartic interaction proportional to β^2. A weak attractive interaction makes only weakly bound states, for which non-relativistic reasoning is sufficient. In the non-relativistic limit, the quartic interaction becomes an attractive two-body delta-function potential. It is known that for a system of identical bosons in one dimension, such an interaction produces exactly one two-body bound state, exactly one three-body bound state, etc. These are the other quantum states in Eq. (4.74). (You might wonder if all of these states are really stable, since there is no obvious conservation law that keeps a four-meson bound state from decaying into two mesons. It turns out that, in tree approximation, all of the bound states *are* stable. The secret reason for this is that the classical sine–Gordon equation possesses an infinite number of non-obvious (but also non-trivial) conservation laws. Whether the higher states are stable when loops are taken into account is at the moment unknown.)

It is worth pointing out that the spacing of energy levels predicted by Eq. (4.74) is independent of β for small β. This is a general feature of Bohr–Sommerfeld quantization applied to any classical field theory with periodic solutions; it is simply a consequence of Eq. (4.70) and the fact that ω is not affected when we scale out β, as in Sect. 4.1. On the other hand, the lowest energy eigenvalue is independent of β for small β only because the energies of our classical periodic motions go all the way down to zero. If this were not the case, the lowest eigenvalue would be proportional to β^{-2}.

In the theory of a particle in a potential, there is a more accurate formula than the Bohr–Sommerfeld quantization formula, the WKB formula. Likewise, there is a more accurate formula for periodic lumps, the analog of the WKB formula. This formula was derived and applied to several cases in a brilliant sequence of papers by Dashen, Hasslacher, and Neveu;[32] I will call it the DHN formula. The derivation of the DHN formula is extremely lengthy, and I do not have time even to sketch it in these lectures. However, the final formula is simple enough to state, once some preliminary definitions are made. As usual, for notational simplicity, I will restrict myself to the case of a single scalar field in one space dimension; the generalization to more complicated theories is trivial.

Let us consider a one-parameter family of periodic solutions of our theory, parametrized by the period T. I denote these solutions by $\phi_T(x, t)$. Let us study small perturbations about one of these solutions,

$$\phi(x, t) = \phi_T(x, t) + \delta(x, t). \tag{4.75}$$

Inserting this in the equation of motion, and retaining only terms linear in δ, we obtain

$$\Box^2 \delta + U''(\phi_T)\delta = 0. \tag{4.76}$$

This equation is invariant under time translations by T. Thus, following closely the procedures of Sect. 2.2 for a time-independent lump, we look for solutions that are eigenfunctions of such a time translation, that is to say, solutions obeying

$$\delta_i(x, t + T) = e^{-iv_i}\delta_i(x, t).$$

The vs are called stability angles. The reason for this terminology is that ϕ_T is infinitesimally stable if and only if all the vs are real. From now on I will assume that this is the case. Because of space-time translational invariance, there are two eigenfunctions, $\partial_\mu \phi_T$, with zero stability angles. Because Eq. (4.76) is real, the eigenfunctions with non-zero stability angles occur in complex-conjugate pairs; the stability angles are of the same magnitude but opposite sign.

The interpretation of the δs is much the same as in Sect. 2.2. Discrete eigenfunctions can be thought of as mesons bound to the oscillating lump, continuum eigenfunctions as mesons scattering off the lump. For notational simplicity I will treat all eigenfunctions as discrete. (We can always put the system in a box, and let the box go to infinity at the end of our computations.)

I am now in a position to state the DHN formula.[33] Let n_i be a sequence of non-negative integers, one for each positive stability angle. For each such sequence, we select a set of periodic classical motions by the quantization condition

$$
\int_0^T dt \int dx \pi_T(x, t) \partial_0 \Phi_T(x, t)
$$
$$
+ \sum_{v_i > 0} (n_i + \tfrac{1}{2}) \left(T \frac{dv_i}{dT} - v_i \right) = 2n\pi, \tag{4.77}
$$

where n is an integer. Let us denote the energy of one of these selected classical motions by E_{cl}. Then there is a quantum energy eigenstate with eigenvalue

$$
E = E_{cl} + \sum_{v_i > 0} (n_i + \tfrac{1}{2}) \frac{dv_i}{dT}. \tag{4.78}
$$

Some remarks:

(1) There is an obvious similarity between Eq. (4.78) and the first-order term in the weak-coupling expansion of Sect. 4.2. Actually, the weak-coupling expansion can be thought of as a degenerate case of the DHN formula. A time-independent solution is *ipso facto* a periodic solution, of arbitrary period T, and thus can be thought of as constituting a one-parameter family of periodic solutions all by itself. In the notation of Sect. 4.2, E_{cl} is E_0/β^2, and v_i is $\omega_i T$. Eq. (4.77) is trivially satisfied for $n = 0$, and Eq. (4.78) becomes the first-order approximation of Sect. 4.2.

(2) Just like the sum of frequencies in Sect. 4.2, the sum over stability angles in the DHN formula is divergent. Just as in Sect. 4.2, the divergence is canceled once the bare parameters in the classical theory are re-expressed in terms of renormalized parameters.

(3) The DHN formula obeys the correspondence principle. The differential of Eq. (4.77), with fixed n_i, is

$$
T dE_{cl} + \sum (n_i + \tfrac{1}{2}) T \frac{d^2 v_i}{dT^2} \, dT = 2\pi dn. \tag{4.79}
$$

The differential of Eq. (4.78) is

$$dE = dE_{cl} + \sum (n_i + \tfrac{1}{2}) \frac{d^2 v_i}{dT^2} dT. \tag{4.80}$$

Thus,

$$T dE = 2\pi dn. \tag{4.81}$$

This is Eq. (4.70).

(4) Equation (4.77) is ambiguous. A stability angle is defined only modulo 2π, but if we add 2π to one of the stability angles in Eq. (4.77), we change the equation. On the other hand, if we add 4π to a stability angle (or 2π to each of two angles), we make no change; the result can be absorbed in n. Thus we are missing one binary bit of information. It is possible to give rules for eliminating this ambiguity, but I will not do so here. In the example at hand, the sine–Gordon doublet, the stability angles are unambiguously known for $T = 2\pi/\alpha^{\frac{1}{2}}$, when the doublet becomes the ground state, and are known for other T by continuity.

In a spectacular analysis, Dashen, Hasslacher, and Neveu[34] computed the stability angles for the sine–Gordon doublet and evaluated the sum over stability angles. Their results are: (1) There are no discrete stability angles. Thus there is only a single series of bound states, labeled by the integer n. (2) The energies of these bound states are given by

$$E_n = 2M \sin(\beta'^2 n/16), \tag{4.82}$$

where $n = 0, 1, 2 \ldots < 8\pi/\beta'^2$,

$$\beta'^2 = \frac{\beta^2}{1 - \beta^2/8\pi}, \tag{4.83}$$

and M is the soliton mass. M is not given by Eq. (4.65), but is the soliton mass corrected by the DHN formula, or, equivalently, by the first-order weak coupling expansion. (See remark (1) above.) I have written the equation in this form, rather than giving separate expressions for E_n and M, to eliminate α, and thus avoid worries about renormalization conventions. Note that the DHN formula is identical to the Bohr–Sommerfeld formula, except that β is replaced by β'.

Dashen *et al.* speculate that Eq. (4.82) is exact. They have three reasons for this. (1) Equation (4.83) blows up at 8π. This is precisely the place where we found the bottom dropping out of the energy spectrum in Sect. 4.3. (2) Since the first quantum doublet state is just the ordinary meson, and the second is just a two-meson bound state, the ratio E_2/E_1 can be computed in ordinary perturbation theory through the Bethe–Salpeter equation. Dashen *et al.* have done this computation up to and including terms of order β^8. The result agrees with Eq. (4.82) to this order. (3) We shall get to the third reason in Sect. 5.

I emphasize that even if the DHN formula does turn out to be exact for the sine–Gordon doublet, this is pure fortuity, like the exactness of the WKB formula for the hydrogen atom. If a ϕ^6 term is added to the sine–Gordon Lagrangian, the DHN formula and the perturbation expansion no longer agree. From our viewpoint, the possible exactness of the DHN formula for this special system is unfortunate. As we shall see in Sect. 5, the sine–Gordon equation is a marvelous theoretical laboratory for checking various approximation schemes, and the exactness of the DHN formula makes the sine–Gordon laboratory as useless for studying the errors in the DHN approximation as the hydrogen atom would be for studying the errors in the WKB approximation.

5   A very special system

5.1   A curious equivalence[35]

Several times in these lectures I have alluded to a peculiar property of the sine–Gordon equation. It is now time to reveal the secret: the sine–Gordon equation is equivalent, in a sense I shall make precise, to the massive Thirring model. This is a surprise, because the massive Thirring model is a canonical field theory whose Hamiltonian is expressed in terms of fundamental Fermi fields only. Even more surprising, when $\beta^2 = 4\pi$, the sine–Gordon equation is equivalent to a free massive Dirac theory, in one spatial dimension.

To establish these results, I will have to devote a little time to defining the massive Thirring model.

The (massless) Thirring model[36] is a theory of a single Dirac field in one spatial dimension, with

$$\mathcal{L} = \bar{\psi} i \gamma_\mu \partial^\mu \psi - \tfrac{1}{2} g j_\mu j^\mu, \tag{5.1}$$

where

$$j^\mu = \bar{\psi} \gamma^\mu \psi, \tag{5.2}$$

and g is a free parameter, the coupling constant. The formal definition of j_μ as a product of two fields at the same point is plagued with ambiguities, just as in higher-dimensional theories, but, again just as in higher-dimensional theories, these ambiguities can be removed by demanding that

$$[j^0(x, t), \psi(y, t)] = -\delta(x - y)\psi(y, t). \tag{5.3}$$

Once this has been done, the Hamiltonian derived from Eq. (5.1) requires no further renormalizations, aside from a zero-point energy subtraction. The model is exactly soluble and physically sensible for g greater than minus π.

Within the massless model, it is possible to define renormalized chiral eigendensities,

$$\sigma_\pm = Z\bar\psi(1 \pm \gamma_5)\psi, \tag{5.4}$$

where Z is a cutoff-dependent constant. (Note that Z is determined only up to a finite multiplicative constant.) The massive Thirring model is formally defined by adding a mass term to the Hamiltonian density of the massless model:

$$\mathscr{H} \to \mathscr{H} + \mathscr{M} \tag{5.5}$$

where

$$\mathscr{M} = \tfrac{1}{2}m'(\sigma_+ + \sigma_-). \tag{5.6}$$

Here m' is merely a real parameter; it is not to be identified with the mass of any presumed one-particle state. The massive model is not exactly soluble; indeed, it is a difficult and still not totally solved problem even to show that these equations define a physically sensible theory.[37] However, it is certainly true that the matrix elements of $\mathscr{M}$ between any two states of the massless model is a well-defined object, free of ultraviolet divergences. If the massive model does exist, these matrix elements should be sufficient to define it.

In fact, I will not study matrix elements between general states, but only states obtained by applying space-time smeared polynomials in $\sigma_\pm$ to the vacuum state. This family of states is invariant under both $\mathscr{M}$ and the massless Hamiltonian; therefore, a state in this subspace should stay in this subspace under the time evolution defined by the massive Hamiltonian. For this subspace, the massive model is defined by the vacuum expectation value of a string of $\sigma_\pm$s, computed in the massless model. Some of these will be smeared to make the state on the right, some smeared to make the state on the left, and one left unsmeared to make $\mathscr{M}$.

These vacuum expectation values can be found in the literature on the massless model.[38] I will write down the explicit form only for the case when all separations are space-like; the answers for other configurations are uniquely determined from this expression by analytic continuation.

$$\left\langle 0 \left| \prod_{i=1}^{n} \sigma_+(x_n)\sigma_-(y_n) \right| 0 \right\rangle = A^{2n} \frac{\prod\limits_{i>j} [(x_i - x_j)^2(y_i - y_j)^2]^\delta}{\prod\limits_{i,j} [(x_i - y_j)^2]^\delta}, \tag{5.7}$$

where

$$\delta = \frac{1}{1 + g/\pi}, \tag{5.8}$$

and A is an arbitrary constant, undetermined until we fix the renormalization convention that determines the finite part of Z. For this section only, I have defined

$$x^2 = -x_\mu x^\mu, \tag{5.9}$$

so the square of a spacelike vector is positive. Because of the chiral invariance of the massless model, expectation values of strings with unequal numbers of σ_+s and σ_-s vanish.

We now have all the information about the Thirring model we need to prove the equivalence.

Now let us consider the theory of a free massless scalar field

$$\mathcal{H} = \tfrac{1}{2} N_m [\pi^2 + (\nabla \phi)^2] = N_m \mathcal{H}_0, \tag{5.10}$$

where m is an arbitrary normal-ordering mass. In this theory, let us define

$$S_\pm = N_m e^{\pm i \beta \phi}, \tag{5.11}$$

where β is a real number.

I will shortly demonstrate that the vacuum expectation of a string of Ss is given by an expression of exactly the form of Eq. (5.7) with

$$\delta = \beta^2 / 4\pi. \tag{5.12}$$

Thus, we can identify the space of states made by applying strings of Ss to the vacuum in the scalar theory with the set of states made by applying strings of σs to the vacuum in the Thirring model. Furthermore, we can identify the mass term in the Thirring model with the sine–Gordon interaction,

$$\mathcal{M} = -\frac{\alpha}{\beta^2} N_m \cos \beta \phi. \tag{5.13}$$

Of course, in order to do this consistently, we must match Eqs. (5.8) and (5.12), that is, we must say

$$\frac{\beta^2}{4\pi} = \frac{1}{1 + g/\pi}. \tag{5.14}$$

Equations (5.13) and (5.14) define the correspondence between the massive Thirring model and the sine–Gordon equation.[39] Note that if $\beta^2 = 4\pi, g = 0$, and the sine–Gordon equation is the theory of a free massive Dirac field.

Now for the computation of the expectation value of a string of Ss. The computation is complicated by the extreme infrared divergences of a massless scalar field in two dimensions. For example, for a free scalar field of mass μ,

$$\triangle_+(x; \mu) = -\frac{1}{4\pi} \ln(c\mu^2 x^2) + O(\mu^2 x^2), \tag{5.15}$$

where c is a numerical constant, related to Euler's constant. This blows up badly when μ goes to zero. I will circumvent this problem by adding a small mass term to Eq. (5.10),

$$N_m \mathcal{H}_0 \to N_m[\mathcal{H}_0 + \tfrac{1}{2}\mu^2\phi^2] = N_\mu[\mathcal{H}_0 + \tfrac{1}{2}\mu^2\phi^2], \qquad (5.16)$$

plus an irrelevant constant. At the end of the computation I will let μ go to zero.

The computation is simplified by re-normal-ordering Eq. (5.11),

$$N_m e^{\pm i\beta\phi} = \left(\frac{\mu^2}{m^2}\right)^{\beta^2/8\pi} N_\mu e^{\pm i\beta\phi}. \qquad (5.17)$$

Thus, what we wish to compute is

$$\left(\frac{\mu^2}{m^2}\right)^{\Sigma\beta_i^2/8\pi} \left\langle 0, \mu \left| \prod_i N_\mu e^{i\beta_i\phi(x_i)} \right| 0, \mu \right\rangle, \qquad (5.18)$$

where the β_is are constants, equal in the case at hand to either plus or minus β. The computation is a trivial application of Wick's theorem. Because the operators are normal-ordered with respect to the free-field mass, μ, there are no contractions of an operator with itself, and thus we obtain

$$\left(\frac{\mu^2}{m^2}\right)^{\Sigma\beta_i^2/8\pi} \prod_{i>j} e^{-\beta_i\beta_j\Delta_+(x_i-x_j;\,\mu)}. \qquad (5.19)$$

We now let μ go to zero, and use Eq. (5.15). We find, in the small μ limit,

$$\left(\frac{\mu^2}{m^2}\right)^{\Sigma\beta_i^2/8\pi} \prod_{i>j} [c\mu^2(x_i-x_j)^2]^{\beta_i\beta_j/4\pi}. \qquad (5.20)$$

The most important thing about this expression is that it is proportional to

$$(\mu^2)^{(\Sigma\beta_i)^2/8\pi}. \qquad (5.21)$$

Thus, if the βs do not sum to zero, that is to say, if there are not equal numbers of S_+s and S_-s, the limit is zero. This is the recovery in the scalar theory of the selection rule we obtained from chirality conservation in the Thirring model. On the other hand, if the βs do sum to zero, (5.20) is independent of μ. Indeed it is identical to Eq. (5.7), aside from terms that can be absorbed in the arbitrary constant A.

This completes the argument.

Although I will not show it explicitly here, very similar arguments can be used to express the Thirring model current in sine–Gordon language.[38] The result is

$$j^\mu = \frac{\beta}{2\pi} \varepsilon^{\mu\nu}\partial_\nu\phi, \qquad (5.22)$$

where $\varepsilon^{01} = -\varepsilon^{10} = 1$.

5.2   *The secret of the soliton*

One of the most intriguing features of the equivalence we have found is the relation between the coupling constants, Eq. (5.14). Large g is small β. Thus we have a rare animal indeed, a non-trivial relativistic field theory for which it is possible to make sensible strong-coupling approximations. The theory is the massive Thirring model, and the approximations are the small-β approximations for the sine–Gordon equation discussed in Sect. 4.

The central topic of Sect. 4 was the appearance of a coherent bound state, the quantum soliton. At first glance, one might think that our equivalence was proved only for states made from the vacuum by the application of strings of local field operators, the Ss. Any such state has the property that the expectation value of ϕ has the same value at plus infinity as at minus infinity. Thus we can not investigate a one-soliton state, where the expectation value of ϕ changes by $2\pi/\beta$.

However, this objection does not apply to a state consisting of a widely separated soliton–antisoliton pair. For such a state, for small β,

$$\langle \phi(x) \rangle = f(x - b) - f(x + b), \tag{5.23}$$

where f is the classical soliton solution, and the separation is $2b$. If we apply Eq. (5.22) to the widely separated pair, we find

$$\langle j_0 \rangle = \frac{\beta}{2\pi} \langle \partial_1 \phi \rangle$$

$$= \frac{\beta}{2\pi} \frac{\mathrm{d}}{\mathrm{d}x} [f(x - b) - f(x + b)]. \tag{5.24}$$

The right-hand side of this equation is peaked about the points $-b$ and b, the locations of the antisoliton and soliton. The integrated charge over the soliton peak is $+1$; that over the antisoliton peak is -1. Now, the charge in question is fermion number, normalized such that the ψ field carries charge -1, as we see from Eq. (5.3). Thus the soliton is a particle of fermion number 1, and the antisoliton is a particle of fermion number -1. Also, unless we have missed something in Sect. 4, the soliton and the antisoliton are the lightest particles in the theory with non-zero fermion number. Thus the soliton of the sine–Gordon equation is nothing other than the fundamental fermion of the massive Thirring model.

It is a bit surprising to see a fermion appearing as a coherent state of a Bose field. Certainly this could not happen in three dimensions, where it would be forbidden by the spin-statistics theorem. However, there is no spin-statistics theorem in one dimension, for the excellent reason that there is no spin.

In fact, the appearance of fermions occurs even in a much simpler one-dimensional theory, the theory of a free massless scalar boson. The two-particle subspace of this theory contains states where each particle is in a normalizable state with spatial momentum support restricted to the positive axis. Even though this is a normalizable state, it is still an eigenstate of $P_\mu P^\mu$, with eigenvalue zero; all the two-momenta in the individual wave functions are aligned null vectors. Thus, if we adopt the usual definition of a particle, a normalizable eigenstate of P^2, we must say that the two-particle sector of Fock space contains single massless particles, as does the three-particle sector, the four-particle sector, and the coherent-state sector. As Streater and Wilde[40] discovered some time ago, some of these coherent states are fermions; they can be created from the vacuum by fields which anticommute with each other for space-like separations. (These fields are non-local functions of ϕ, but that is neither here nor there.)

Thus, the theory of a massless scalar field is chock-full of massless particles. (This is one way of understanding its terrible infrared divergences.) Typically, when we add an interaction, most of these particles dissolve into the continuum; which particles survive depends on the detailed form of the interaction. For example, if we add an interaction proportional to ϕ^2, only the particle in the one-particle sector of Fock space survives, and we have the theory of a free massive boson. On the other hand, as we have seen, if we add an interaction proportional to $\cos(2\pi^{\frac{1}{2}}\phi)$, only a fermion survives, and we have a theory of a free massive fermion.

So much for the soliton. What is the fundamental meson of sine–Gordon theory in Thirring-model language? In the Thirring model, positive g corresponds to an attractive force between fermions and anti-fermions. In the non-relativistic limit, this is a delta-function potential, and produces a bound state for arbitrarily small g. As we increase g, we would expect the bound state mass to decrease, perhaps becoming as small as possible (zero) as g goes to infinity. That is to say, we expect the fermion/bound-state mass ratio to go to infinity for vanishing β. But this is just where the soliton/meson mass ratio goes to infinity. Thus the lowest fermion–antifermion bound state of the massive Thirring model is an obvious candidate for the fundamental meson of sine–Gordon theory.

A check on this picture: The bound state has a symmetric wavefunction. Since the particles being bound are fermion and antifermion, this means the bound state is odd under both parity and charge conjugation. Equation (5.22) tells us how to transport the definitions of parity and charge conjugation from the Thirring model to the sine–Gordon equation. We see that the ϕ field is pseudoscalar and charge-conjugation odd.

This is consistent with the computations of Sect. 4.4, where the fundamental meson emerged as the lowest bound state of a soliton–antisoliton doublet. At this point I can state the postponed third reason for believing the DHN formula is exact: equation (4.82) predicts that all the doublet bound states disappear when β^2 exceeds 4π. This is precisely the point where the Thirring model interaction switches from attractive to repulsive.

The picture we have developed for this special system has some of the features of nuclear democracy in the sense of Chew. There is no way of deciding whether the fermion is fundamental and the boson a bound state or the boson is fundamental and the fermion a quantum lump. One is the natural way of putting things if one is describing the massive Thirring model and the other is the natural way of putting things if one is describing the sine–Gordon equation, but these two theories define identical physics. Which you choose to use is purely a matter of taste.

5.3   *Qualitative and quantitative knowledge*

We can learn both qualitative and quantitative information from the sine–Gordon equation.

Qualitatively, the sine–Gordon equation demonstrates explicitly that 'a lump' is a classical concept, like 'a wave' or 'a particle'. To ask whether the sine–Gordon soliton is a lump or a fundamental particle is as silly as asking whether light is a particle or a wave. Light acts very much like a wave in a certain approximation and very much like a particle in a different approximation. Whether it is better to think of it as a classical wave with quantum corrections or a classical particle with quantum corrections depends upon which approximation is appropriate. Likewise, for small β, it is certainly useful to think of the quantum soliton as a classical lump with quantum corrections. However, when β^2 increases to 4π, the quantum soliton becomes the most fundamental particle imaginable, the quantum of a free field theory.

Quantitatively, we can use the sine–Gordon equation as a theoretical laboratory in which we can test the approximation methods of Sect. 4. In Table 3 I have computed the predictions of these methods for the ratio of the soliton mass to the meson mass for three values of β^2 : 4π (where the qualitative picture of the soliton as a lump totally breaks down), 2π, and π. At 4π we know the exact answer, because this is the point where the meson (in its guise as a fermion–antifermion bound state) is just on the verge of binding. (I happen to know the exact answer for 2π, so I have included this in the table.)[41]

Some features of the table could have been anticipated: The variational method gives the same answer as zeroth-order weak-coupling, the DHN

Table 3. *Soliton/meson mass ratio in sine–Gordon theory*

Method	$\beta^2 = \pi$	$\beta^2 = 2\pi$	$\beta^2 = 4\pi$
Zeroth-order weak-coupling expansion[a]	2.55	1.27	0.64
Coherent-state variation[b]	2.55	1.27	0.64
First-order weak-coupling expansion[c]	2.23	0.95	0.32
Bohr–Sommerfeld quantization[d]	2.56	1.31	0.71
DHN formula[e]	2.25	1.00	0.50
Exact	?	1.00	0.50

[a] Equation (2.13b).
[b] This always gives the same result as the previous method.
[c] The explicit formula is not given in the text; it is $(8/\beta^2) - (1/\pi)$.
[d] Equation (4.64).
[e] Equation (4.82).

formula is exact, wherever it can be checked, and the stronger the coupling, the more the various approximation methods differ from each other and from the exact result.

However, the smallness of the errors, even for large coupling, is surprising. $\beta^2 = 4\pi$ is very strong coupling; the fundamental meson decays into a soliton–antisoliton pair, and the whole classical picture of the system is nonsense. Nevertheless, all of our approximations are still good to within 40% or so, not at all bad for a strong-coupling problem. $\beta^2 = 2\pi$ is still fairly strong coupling: the soliton and meson masses are identical; this is certainly very far from the small β limit, in which the soliton is very much heavier than the meson, and we would expect the weak-coupling expansion to be very bad. Nevertheless, the first-order weak-coupling approximation is good to within 5%.

Before I computed this table, I did not anticipate that the picture of the soliton as a lump would give such accuracy for such strong coupling. Unfortunately, since I know of no system other than the sine–Gordon equation soluble in the strong-coupling regime, I have no idea of whether this accuracy is a general feature of these approximations or one peculiar to this very special system.

5.4   Some opinions

This has been a long series of physics lectures with no reference whatsoever to experiment. This is embarrassing.

Certainly objects much like the lumps we have discussed play an important role in many fields of physics. There are lumps in superconductivity theory, in plasma physics, in solid-state physics. However, this is a school on subnuclear physics, and all of my lectures have been oriented

to relativistic quantum field theory. Is there any chance that the lump
will be more than a theoretical toy in our field?

I can think of two possibilities.

One is that there will appear a theory of strong-interaction dynamics
in which hadrons are thought of as lumps, or, more probably, as systems
of quarks bound to lumps. (The SLAC bag[42] involves something like this
idea, as, in a different way, does the Nielsen–Olesen string.[43]) At the
moment, I am pessimistic about the success of such a theory. Classical
ideas seem to me to be linked to weak couplings, as we saw in Sect. 4.1,
and therefore unlikely to be useful in the study of the strong interactions.
However, this intuition can be wrong, as we saw in Sect. 5.3, and I stand
ready to be converted in a moment by a convincing computation.

The other possibility is that a lump will appear in a realistic theory with
weak coupling constants, a theory of weak and electromagnetic interactions. If such a lump were to be kept from dissipating by the topological
conservation laws of Sect. 3, the theory would have to imbed the
$U(1) \otimes SU(2)$ group of the Weinberg–Salam model in a larger group
without $U(1)$ factors, but there is no shortage of such models. In this case,
the lump, once discovered, would be unmistakable. It would be much
heavier than the other particles in the theory (including the intermediate
vector bosons) and it would be a magnetic monopole. Such monopoles
are theoretically much more attractive than Dirac monopoles; for example, the renormalization difficulties that plague a field theory of Dirac
monopoles are totally absent for lumps. Thus, if a monopole is found, it
will be a very attractive possibility that it is a lump, and the properties of
the monopole will tell us a great deal about the structure of the weak and
electromagnetic interactions. Unfortunately, all monopole searches con-
ducted to date have been failures.

Appendix 1: A three-dimensional scalar theory with non-dissipative solutions

Consider, in three spatial dimensions, the theory of a real scalar
field ϕ and a complex scalar field ψ, with

$$U = (a\phi - m)^2 \psi^* \psi + b\phi^2 + c\phi^4, \tag{A.1}$$

where $a, b, c,$ and m are positive numbers. T. D. Lee showed that this model
has non-dissipative solutions.[6]

The unique ground state is $\psi = \phi = 0$. For any dissipative solution, the
fields become arbitrarily small for sufficiently large times, and thus we can
compute the energy with arbitrary accuracy by only retaining the quadratic
terms in U. In equations,

$$E = \lim_{t \to \infty} \left[\int d^3x(\partial_0\psi^*\partial_0\psi + \nabla\psi^* \cdot \nabla\psi + \tfrac{1}{2}\partial_0\phi\partial_0\phi + \tfrac{1}{2}\nabla\phi \cdot \phi \right.$$

$$\left. + m^2\psi^*\psi + b\phi^2) \right], \tag{A.2}$$

for a dissipative solution.

The theory is invariant under ψ phase transformations and thus possesses a conserved charge,

$$Q = \text{Im} \int d^3x\psi^*\partial_0\psi. \tag{A.3}$$

For any dissipative solution,

$$Q^2 \leqslant \left| \int d^3x\psi^*\partial_0\psi \right|^2$$

$$\leqslant \left[\int d^3x\psi^*\psi \right] \left[\int d^3x\partial_0\psi^*\partial_0\psi \right]$$

$$\leqslant E^2/m^2. \tag{A.4}$$

Here I have used the Schwarz inequality on the second line, and Eq. (A.2) on the third. I will now show that there exist non-dissipative solutions by displaying initial-value data that violate this inequality.

The initial-value data are:

$$\phi = m/a, \qquad r \leqslant L,$$
$$\quad = me^{-(r-L)}/a, \quad r \geqslant L, \tag{A.5}$$
$$\partial_0\phi = 0, \tag{A.6}$$
$$\psi = \frac{A}{r} \sin(\pi r/L), \quad r \leqslant L,$$
$$\quad = 0, \qquad r \geqslant L, \tag{A.7}$$

and

$$\partial_0\psi = i\pi\psi/L, \tag{A.8}$$

where A and L are positive numbers.

ϕ and ψ have been chosen to make the first term in U vanish everywhere. Also, ψ has been chosen such that

$$\int d^3x\nabla\psi^* \cdot \nabla\psi = \pi^2 L^{-2} \int d^3x\psi^*\psi. \tag{A.9}$$

Thus it is trivial to compute

$$\lim_{A \to \infty} (Q^2/E^2) = L^2/4\pi^2. \tag{A.10}$$

This violates the inequality for $L > 2\pi/m$.

Appendix 2: A theorem on gauge fields

Theorem. In the standard theory of gauge fields only,

$$\mathscr{L} = -\tfrac{1}{4}F^a_{\mu\nu}F^{a\mu\nu}, \tag{A.11}$$

in D spatial dimensions, the only non-singular time-independent finite-energy solutions are gauge transforms of $A^a_\mu = 0$, for $D \neq 4$.

Note that this theorem does not depend on any special choice of gauge. If it did, it would be uninteresting, for a solution that is time-independent in one gauge can be horribly time-dependent in another gauge. The theorem asserts that there are no time-independent solutions in any gauge.

Proof. Evaluating the Lagrangian for a time-independent solution, we find

$$L = L_1 - L_2, \tag{A.12}$$

where

$$L_1 = \tfrac{1}{2}\int d^D x (F^a_{0i})^2 = \tfrac{1}{2}\int d^D x (\partial_i A^a_0 + ec^{abc}A^b_0 A^c_i)^2, \tag{A.13}$$

and

$$L_2 = \tfrac{1}{4}\int d^D x (F^a_{ij})^2. \tag{A.14}$$

Note that this is not the same as the total energy. The energy is the sum of these two terms, not the difference. However, finiteness of the energy does imply that each of the two terms must be finite.

Hamilton's principle for a time-independent solution implies that L must be stationary under time-independent variations. Consider the two-parameter family of field configurations defined by

$$A^a_0(\mathbf{x}; \sigma, \lambda) = \sigma\lambda A^a_0(\lambda\mathbf{x}), \tag{A.15}$$

$$A^a_i(\mathbf{x}; \sigma, \lambda) = \lambda A^a_i(\lambda\mathbf{x}). \tag{A.16}$$

An easy computation shows that

$$L(\sigma, \lambda) = \sigma^2\lambda^{(4-D)}L_1 - \lambda^{(4-D)}L_2. \tag{A.17}$$

This must be stationary at $\sigma = \lambda = 1$. Thus,

$$L_1 = L_2 = 0, \tag{A.18}$$

for $D \neq 4$. This, in turn, implies

$$F^a_{\mu\nu} = 0, \tag{A.19}$$

from which the stated result follows by standard arguments. (If you are not familiar with the standard arguments, there is a proof at the end of Appendix 5.)

Appendix 3: A trivial extension

In this appendix I sketch the extension of the analysis of Sect. 3 to the case where our theory of gauge fields and scalar fields possesses ordinary internal symmetries as well as gauge symmetries.

Let the full symmetry group of the theory be some compact Lie group G', not necessarily connected. The gauge group, G, is a closed connected invariant subgroup of G'. Let the full group of unbroken symmetries (for some choice of ϕ_0) be H'. H, the group of unbroken gauge symmetries, is the intersection of G and H'.

So far, this is just definitions. I will now make the assumption announced in Sect. 3.2. I will assume there is no accidental degeneracy; thus the set of zeros of U can be identified with the coset space G'/H'. This is much weaker than the simplifying assumption of Sect. 3.2; G'/H' can be much bigger than G/H.

G'/H' is the union of the orbits of G, the minimal sets invariant under the action of G. One of these orbits is G/H. All the orbits are topologically identical to G/H, since any orbit can be obtained from any other by the action of some element of G'.

The scalar fields at infinity must lie on a single orbit, for otherwise we could not cancel their space derivatives with gauge fields. (See the discussion after Eq. (3.34).) Thus, at least at first glance, it seems as if we have a richer classification of solutions than in Sect. 3. Our solutions are labeled not just by an element of $\pi_{D-1}(G/H)$, but also by an orbit of G in G'/H'.

However, this richness is illusory. Solutions associated with different orbits can not exist in the same world. For, if we try to patch together two distant solutions associated with different orbits, as in Fig. 3, we can not make a gauge transformation such that the solutions are identical in the unshaded region. (If you find this argument unconvincing, consider a theory with spontaneous symmetry breakdown but no gauge fields, and instead of a lump, consider an ordinary meson, like the sigma meson in the sigma model. Does the fact that there are many vacuum states mean that an experimenter exploring the universe of the sigma model would find many kinds of sigma mesons?)

Thus, the existence of the larger group G' is totally irrelevant to the phenomena discussed in Sect. 3. The only relevant objects are the gauge group G and its unbroken subgroup H.

Appendix 4: Looking for solutions

When studying a linear partial differential equation, it is frequently useful to write the field variables as sums over a set of functions that transform according to the irreducible representations of the symmetry group of the theory. In this way, for example, the Schrödinger equation for a particle in a central potential is converted into a system of uncoupled ordinary differential equations, one for each angular-momentum eigenstate. Of course, this stratagem does not work if we deal with nonlinear equations, as we must if we look for a time-independent lump. However, there is a stratagem based on the same ideas which yields less far-reaching, but still useful, results.

Let G^* be some subgroup of the full symmetry group of the theory (now to be thought of as including space-time symmetries). Every field configuration can be written as the sum of two terms, a symmetric part, invariant under G^*, and an asymmetric part, a sum of terms each of which transforms according to some non-trivial irreducible representation of G^*. If we plug this decomposition into the expression for the energy, we find that the energy contains no terms linear in the asymmetric part, by Schur's Lemma. Thus the energy, evaluated at a symmetric field configuration, is automatically stationary under asymmetric variations. If we wish to search for symmetric time-independent solutions, we need only insert the most general symmetric field configuration into the energy, and stationarize this restricted problem. There is no need to check for stationarity under asymmetric variations. Similar remarks apply to time-dependent solutions, if 'energy' is replaced by 'action'.

I will give two examples.

(1) The field theory of Example 4, the 't Hooft–Polyakov model. The boundary conditions for the simplest kind of non-dissipative solution, Eq. (3.46), are invariant under simultaneous space–isospin rotations. They are also invariant under parity, if we define the ϕ fields to be pseudoscalar. We choose G^* to be this group of combined rotations plus parity. The most general field configuration invariant under G^* is of the form

$$\phi^b = r^b f(r),$$
$$A^{ib} = \varepsilon^{ibc} r^c g(r), \tag{A.20}$$

where f and g are arbitrary functions of r. Thus the problem has been reduced from one that involves six functions of three variables to one that involves only two functions of one variable. The reduced problem is not beyond the reach of numerical analysis, even if one is armed only with a desk calculator.

(2) The model of Appendix 1. Inspired by the non-dissipative initial-value data, we look for solutions that are invariant under (a) rotations, (b) simultaneous time translations by T and phase transformations by ωT, where ω is some real number, and (c) the transformation

$$\Psi(\mathbf{x}, t) \rightarrow \psi^*(\mathbf{x}, -t)$$
$$\phi(\mathbf{x}, t) \rightarrow \phi(\mathbf{x}, -t). \tag{A.21}$$

The most general set of fields symmetric under this choice of G^* is

$$\psi = e^{i\omega t} f(r),$$
$$\phi = g(r) \tag{A.22}$$

where f and g are arbitrary real functions of r. Once again the problem has been reduced to one that involves only two unknown functions of r.

Some remarks:

(1) This method does not help in finding general solutions, only in finding symmetric solutions. Still, knowing only a few symmetric solutions is better than knowing nothing at all.

(2) Schur's Lemma guarantees stationarity, but not stability, for stability is a question of second variations. After a symmetric solution is found, stability under asymmetric variations has to be checked. To the best of my knowledge, this has not yet been done even for the solution of Example 4 obeying Eq. (A.20) found numerically by 't Hooft.

Appendix 5: Singular and non-singular gauge fields

This is an appendix to the discussion of Dirac monopoles in Sect. 3.7. A Dirac monopole has a genuine gauge-invariant singularity at the origin of coordinates, where the magnetic field blows up; it also has a gauge-variant singularity, the string. The string can be moved around by gauge transformations, but not eliminated altogether. We want to understand why the singularity at the origin necessitates the string. As a by-product, we shall gain a deeper understanding of non-singular gauge fields, and prove the theorem mentioned at the end of Appendix 2. The arguments have nothing to do with the metric structure of Minkowski space, and I will therefore work in a general real n-space, with points $x = (x^1 \ldots x^n)$.

Definition. Let R be a region in n-space. We say that there is a locally non-singular field defined in R if R can be covered with a family of open sets such that (1) in each of these open sets there is defined a non-singular gauge field, $A_i^a(x)$, (2) whenever two open sets overlap, there is a non-singular gauge transformation defined in the overlap region that turns the gauge field in one set into the gauge field in the other.

The first condition is the statement that for every point in R there is some neighborhood of the point that contains no gauge-invariant singularities; the second condition is the statement that when two open sets overlap, all gauge-invariant quantities are uniquely defined in the region of overlap.

An ordinary non-singular gauge field is *a fortiori* locally non-singular. The Dirac monopole is locally non-singular for all of three-space with the origin excised. Our fundamental problem is, given a locally non-singular gauge field, when can we make a gauge transformation in each of the open sets such that the end result is a non-singular gauge field defined in all of R? Phrased more briefly, when are locally non-singular gauge fields gauge-equivalent to non-singular gauge fields?

Theorem. If R is a closed n-cube of side L, then any locally non-singular gauge field is gauge-equivalent to a non-singular gauge field.

Proof. Choose the origin of coordinates to lie within R and construct a regular cubic lattice of points based at the origin with lattice spacing ε. Let each point of this lattice be the center of an open cube with side 2ε. We thus obtain a covering of R by a regular lattice of open cubes. Consider the family of such coverings with $\varepsilon = (\frac{1}{2})^m$, m an integer. Every point of R is contained in some open cube from one of these coverings that is, in turn, contained in one of the open sets in the definition of local non-singularity. Thus we can replace the covering of the definition by a covering by these open cubes. R is compact, and thus we can replace this covering, in turn, by a finite subcovering. A finite covering must contain a smallest cube. Thus we can replace the covering in the definition by a covering by a regular lattice of open cubes, that which contains this smallest cube. We are now ready to construct the required gauge transformation.

The construction is exactly the same as that which we used to construct the gauge $A_0^a = 0$ in Sect. 3.2. Starting from the hyperplane $x^1 = 0$, we construct the group elements $g(P)$ associated with paths parallel to the x^1 axis. There is no problem in changing gauge as we pass from cube to cube, because the gauge-transformation properties of $g(P)$ depend only on the value of the gauge transformation at the end point of the path. (See Eq. (3.22).) We then use $g(P)$ to gauge transform the fields in each cube such that $A_1^a = 0$ everywhere. This implies that the gauge transformation connecting two overlapping cubes is independent of x^1. (See Eq. (3.10).) Thus, we can move along a column of cubes parallel to the x^1 axis, and make this gauge transformation in each successive cube. We have thus transformed the open cubes into open columns parallel to the x^1 axis, such that in each column a single-valued non-singular gauge field is defined. We may still have to gauge transform between adjacent columns, but the gauge transformation is independent of x^1.

We now iterate this procedure. On the hyperplane $x^1 = 0$, we gauge transforms such that $A_2^a = 0$, and apply the same (x^1-independent) gauge transformation throughout the covering. We thus turn the columns into slabs parallel to the x^1-x^2 plane. After $n-1$ such iterations, we have gauge transformed such that

$$A_r^a = 0, \ x^1 = x^2 = \cdots = x^{r-1} = 0, \tag{A.23}$$

and have arrived at a single-valued non-singular gauge field defined in the entire cube. Q.E.D.

Theorem. If R is all n-space, any locally non-singular gauge field is gauge-equivalent to a non-singular gauge field.

Proof. Consider two n-cubes, both centered at the origin, with one larger than the other. If we do the construction of the previous theorem in the larger cube, and then restrict the answer to the smaller cube, it is easy to see that we get the same result we would have obtained if we had originally done the construction only in the smaller cube. Thus we can fill n-space with an ever-increasing set of cubes, such that when we increase the size of the cube, we do not have to make further gauge transformations in already conquered territory. Thus, this procedure trivially has a limit, which defines the required gauge transformation in all of n-space.

Theorem. If R is all of n-space, excluding a single point, then any locally non-singular gauge field is gauge-equivalent to a non-singular gauge field defined in all of R except for a line going from the excluded point to infinity, parallel to the x^1 axis. (This is the Dirac string.)

Proof. Choose coordinates such that the excluded point is $(1, 0, 0\ldots)$. Exclude from R a small open sphere about this point. The intersection of R with a small n-cube is now a compact set, and we can do the construction of the first theorem. However, as we merrily gauge-transform along, we can not reach that portion of the columns of the covering that lies above the excluded sphere; this remains unknown territory. As the excluded sphere becomes arbitrarily small, this unknown territory shrinks down to the string.

Comments. (1) Since we have never used the metric structure of n-space, we can do all of our constructions in any non-singular coordinate system. This means that the string can be any non-self-intersecting non-singular curve going from the excluded point to infinity. (2) The generalization to any finite set of excluded points is trivial. There is one string for each point.

Theorem. Any non-singular gauge field in all n-space such that $F_{ij}^a = 0$ everywhere is gauge equivalent to a vanishing gauge field. (This is the theorem referred to at the end of Appendix 2.)

Proof. We go to the gauge of Eq. (A.23). Since A_1^a vanishes everywhere,

$$F_{1i}^a = \partial_1 A_i^a = 0. \tag{A.24}$$

Thus A_i^a is independent of x^1. On the surface $x^1 = 0$, A_2^a vanishes. But it is independent of x^1; therefore it vanishes everywhere. Thus,

$$F_{2i}^a = \partial_2 A_i^a = 0. \tag{A.25}$$

Etc.

Notes and references

1. For the narrow definition, and also for much information about a system we shall encounter shortly, the sine–Gordon equation (and about many similar systems), see A. Scott, F. Chu, and D. McLaughlin, *Proc. IEEE* **61**, 1443 (1973).
2. Much of the material in this section is plagiarized from two long papers on the quantum problem: J. Goldstone and R. Jackiw, *Phys. Rev.* **D11**, 1486 (1975); N. Christ and T. D. Lee, *Phys. Rev.* **D12**, 1606 (1975).
3. This equation has an interesting history. It was studied extensively by differential geometers in the last quarter of the nineteenth century, because of its role in the theory of surfaces of constant negative curvature. Bianchi called it, 'l'equazione fondamentale di tutta la teoria delle superficie pseudosferiche'. (*Lezioni di geometria differenziale*, 3rd ed., I, 658.) The equation enters geometry in the following way. On any two-dimensional Riemannian manifold, it is possible to choose coordinates in some neighborhood of any point such that

    $$ds^2 = du^2 + dv^2 + 2du \, dv \cos \theta(u, v).$$

    In terms of such coordinates, a simple computation shows that the statement that the manifold has constant negative curvature is equivalent to

    $$\partial^2 \theta / \partial u \, \partial v = \alpha \sin \theta,$$

    where α is a constant related to the magnitude of the curvature. This is the sine–Gordon equation, in light-cone coordinates.
    The equation entered particle physics through the work of Skyrme (*Proc. Roy. Soc.* **A247**, 260 (1958); **A262**, 237 (1961)) who studied it as a simple model of a nonlinear field theory. Skyrme's work prefigured many of the results we shall obtain in Sect. 5.
    The silly name is of recent origin. From a letter from David Finkelstein: 'I am sorry that I ever called it the sine–Gordon equation. It was a private joke between me and Julio Rubinstein, and I never used it in print. By the time he used it as the title of a paper he had earned his Ph.D. and was beyond the reach of justice.'
4. P. M. Morse and H. Feshbach, *Methods of Theoretical Physics*, (McGraw-Hill, New York, 1953), I, 721.
5. G. H. Derrick, *J. Math. Phys.* **5**, 1252 (1964).
6. R. Friedberg, T. D. Lee and A. Sirlin, *Phys. Rev.* **D13**, 2739 (1976).
7. An extension of Derrick's theorem can be proved for theories of gauge fields alone. See Appendix 2 (but read Sec. 3.2 first).
8. I have reorganized the material in this section so much that it is difficult to give specific credits at appropriate places in the text, but I would be astonished if any of its contents comes as a surprise to the authors of the following papers: D. Finkelstein, *J. Math. Phys.* **7**, 1218 (1966); J. Arafune, P. Freund, and C.

Goebel, *J. Math. Phys.* **16**, 433 (1975); Yu. S. Tyupkin, V. A. Fateev, and A. S. Shvarts, *JETP Lett.* **21**, 42 (1975); M. I. Monastyrskiĭ and A. M. Perelemov, *ibid.*, 43 (1975); A. Patrascioiu, *Phys. Rev.* **D11**, 523 (1975). The contents of several long conversations with J. Goldstone and L. Faddeev are also distributed uniformly throughout this section.

9. The initial-value theorem established by C. Parenti, F. Strocchi, and G. Velo, (*Phys. Lett.* **59B**, 157 (1975), *Comm. Math. Phys.* **53**, 65 (1977)) should be sufficient to demonstrate this rigorously for all of the models we shall consider.

10. For a lengthier discussion, see Chapter 5 in this volume.

11. If the scalar fields are real, a factor of $\frac{1}{2}$ is conventionally inserted in front of the second term in Eq. (3.17).

12. Two standard mathematical expositions of homotopy theory are: P. J. Hilton, *An Introduction to Homotopy Theory*, No. 43, Cambridge Tracts in Mathematics and Mathematical Physics; N. E. Steenrod, *The Topology of Fibre Bundles*, No. 14, Princeton Mathematics Series.

13. See, for example, A. Fetter and P. Hohenberg, Chapter 14 of *Superconductivity*, ed. by R. Parks.

14. We shall find them all, by and by.

15. A. M. Polyakov, *JETP Lett.* **20**, 194 (1974).

16. G. 't Hooft, *Nucl. Phys.* **B79**, 276 (1974).

17. E. Cartan, *Œuvres complètes*, I, 2, 1307.

18. This is a result of Tyupkin *et al.*[8] It was also found independently, and by a very different method, by Patrascioiu.[8]

19. We shall construct one at the end of Sec. 3.7.

20. To my knowledge, the first use of homotopy groups to derive topological conservation laws was by D. Finkelstein and C. Misner, *Ann. Phys.* **6**, 230 (1959).

21. For slightly more details and much more rigor, see the texts of Note 12.

22. If it is not obvious to you, here is the proof. We want to show that given $\phi(\theta)$ such that $\phi(0)=\phi_0$, and given a homotopy $\phi(\theta, t)$ such that $\phi(\theta, 0)=\phi(\theta)$ and $\phi(\theta, 1)=$ a constant, then there exists a second homotopy, $\phi'(\theta, t)$, obeying all the previous conditions and in addition obeying the condition $\phi'(0, t)=\phi_0$ for all t. Proof. Since $\phi(0, t)$ is continuous, there exists continuous $g(t)$ such that $g(t)\phi(0, t)=\phi_0$. Define $\phi'(\theta, t)=g(t)\phi(\theta, t)$.

23. I thank Professor Haag for clarifying this situation for me, and also for paying for the coffee.

24. Homotopy experts will recognize the preceding eight paragraphs as the world's most ham-handed exposition of the exact homotopy sequence,
$$\pi_2(G)\to\pi_2(G/H)\to\pi_1(H)\to\pi_1(G).$$

25. For more on strings, and singular gauge fields in general, see Appendix 5.

26. This is a discovery of E. Lubkin, *Ann. Phys.* **23**, 233 (1963). Lubkin may have been the first to realize the utility of homotopy theory in describing the global properties of gauge fields.

27. This casts doubt on the utility of the analysis of Sec. 2 when H is non-Abelian. The effective coupling constant (in the sense of the renormalization group) grows with distance for non-Abelian gauge fields, so there is not much point in using classical reasoning to study the fields at large distances. The same thing phrased another way: if quarks are indeed confined because they carry non-Abelian charges, then the same mechanism should confine non-Abelian monopoles.

28. This expansion has been derived by many authors using many different methods. In addition to the papers of Note 2, there are: J.-L. Gervais and B. Sakita, *Phys. Rev.* **D11**, 2943 (1975); **D12**, 1038 (1975); V. E. Korepin,

P. P. Kulish, and L. Faddeev, *JETP Lett.* **21**, 139 (1975); C. Callan and D. Gross, *Nucl. Phys.* **B93**, 29 (1975); E. Tomboulis, *Phys. Rev.* **D12**, 1678 (1975); A. Klein and F. Krejs, *Phys. Rev.* **D13**, 3295 (1981).

29. For more details, see any graduate-level text on quantum mechanics.
30. R. Rajaraman and E. J. Weinberg, *Phys. Rev.* **D11**, 2950 (1975).
31. Two early papers on lumps as coherent states are: P. Vinciarelli, *Lett. Nuovo Cimento* **2**, 4, 905 (1972); K. Cahill, *Phys. Letters* **53B**, 174 (1974).
32. R. C. Dashen, B. Hasslacher, and A. Neveu, *Phys. Rev.* **D10**, 4114; 4130; 4138 (1974). A pedagogical review of these methods has been written by R. Rajaraman (*Phys. Reports* **21**, 227 (1975)).
33. The form given is essentially that of Rajaraman's review.[32]
34. R. Dashen, B. Hasslacher, and A. Neveu, *Phys. Rev.* **D11**, 3424 (1975). A lot of the early material in this section is plagiarized from this paper.
35. Most of this section is based on S. Coleman, *Phys. Rev.* **D11**, 2088 (1975), although some of the arguments have been rearranged.
36. The Thirring model was proposed by W. Thirring, *Ann. Phys.* (N.Y.) **3**, 91 (1958). A formal operator solution was found by V. Glaser, *Nuovo Cimento* **9**, 990 (1958), and a rigorous solution by K. Johnson, *Nuovo Cimento* **20**, 773 (1961). More references are in Note 35.
37. J. Fröhlich has been able to show that the model exists and obeys the usual axioms for a range of coupling constants. His work is summarized in his lectures at the 1975 Erice School on mathematical physics. See J. Fröhlich, *Comm. Math. Phys.* **47**, 233 (1976), J. Fröhlich and E. Seiler, *Helv. Phys. Acta* **49**, 889 (1976).
38. See Note 35 for more details.
39. S. Mandelstam has gone further than this and actually constructed the Fermi fields as (non-local) functions of the Bose field (*Phys. Rev.* **D11**, 3026 (1975)).
40. R. Streater and I. Wilde, *Nucl. Phys.* **B24**, 561 (1970).
41. S. Coleman, *Ann. Phys.* **101**, 239 (1976).
42. See, for example, the lectures of S. Drell at this school.
43. H. Nielsen and P. Olesen, *Nucl. Phys.* **B61**, 45 (1973).

7

The uses of instantons
(1977)

1 Introduction

In the last two years there have been astonishing developments
in quantum field theory. We have obtained control over problems pre-
viously believed to be of insuperable difficulty and we have obtained deep
and surprising (at least to me) insights into the structure of the leading
candidate for the field theory of the strong interactions, quantum chromodynamics. These goodies have come from a family of computational
methods that are the subject of these lectures.

These methods are all based on semiclassical approximations, and,
before I can go further, I must tell you what this means in the context of
quantum field theory.

To be definite, let us consider the theory of a single scalar field in four-
dimensional Minkowski space, with dynamics defined by the Lagrangian
density

$$\mathcal{L} = \tfrac{1}{2}\partial_\mu \phi \partial^\mu \phi - \tfrac{1}{2}m^2\phi^2 - g^2\phi^4. \tag{1.1}$$

For classical physics, g is an irrelevant parameter. The easiest way to see
this is to define

$$\phi' = g\phi. \tag{1.2}$$

In terms of ϕ',

$$\mathcal{L} = \frac{1}{g^2}\left(\tfrac{1}{2}\partial_\mu \phi' \partial^\mu \phi' - \tfrac{1}{2}m^2\phi'^2 - \phi'^4\right). \tag{1.3}$$

Thus, g does not appear in the field equations; if one can solve the theory
for any positive g, one can solve it for any other positive g; g is irrelevant.
Another way of seeing the same thing is to observe that, in classical physics,
g is a dimensionful parameter and can always be scaled to one.

Of course, g *is* relevant in quantum physics. The reason is that quantum

265

physics contain a new constant, $\hbar$, and the important object (for example, in Feynman's path-integral formula) is

$$\frac{\mathscr{L}}{\hbar} = \frac{1}{g^2\hbar}(\tfrac{1}{2}\partial_\mu\phi'\partial^\mu\phi' + \ldots). \tag{1.4}$$

As we see from this expression, the relevant (dimensionless) parameter is $g^2\hbar$, and thus semiclassical approximations, small-$\hbar$ approximations, are tantamount to weak-coupling approximations, small-g approximations.

At this point you must be puzzled by the trumpets and banners of my opening paragraph. Do we not have a perfectly adequate small-coupling approximation in perturbation theory? No, we do not; there is a host of interesting phenomena which occur for small coupling constant and for which perturbation theory is inadequate.

The easiest way to see this is to descend from field theory to particle mechanics. Consider the theory of a particle of unit mass moving in a one-dimensional potential,

$$L = \tfrac{1}{2}\dot{x}^2 - V(x; g), \tag{1.5}$$

where

$$V(x; g) = \frac{1}{g^2}F(gx), \tag{1.6}$$

and F is some function whose Taylor expansion begins with terms of order x^2. Everything I have said about the field theory defined by Eq. (1.1) goes through for this theory. However, let us consider the phenomenon of transmission through a potential barrier (Fig. 1). Every child knows that the amplitude for transmission obeys the WKB formula,

$$|T(E)| = \exp\left\{-\frac{1}{\hbar}\int_{x_1}^{x_2} dx[2(V-E)]^{\frac{1}{2}}\right\}[1 + O(\hbar)], \tag{1.7}$$

where x_1 and x_2 are the classical turning points at energy E. This is a semiclassical approximation. Nevertheless, transmission, barrier penetra-

Fig. 1

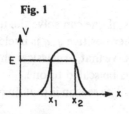

tion, is not seen in any order of perturbation theory, because Eq. (1.7) vanishes more rapidly than any power of $\hbar$, and therefore of g.

I can now make my first paragraph more explicit. There are phenomena in quantum field theory, and in particular in quantum chromodynamics, analogous to barrier penetration in quantum particle mechanics. In the last two years a method has been developed for handling these phenomena. This method is the subject of these lectures.

The organization of these lectures is as follows. In Sect. 2 I describe the new method in the context of particle mechanics, where we already know the answer by an old method (the WKB approximation). Here the instantons which play a central role in the new method and which have given these lectures their title first appear. In Sect. 3 I derive some interesting properties of gauge field theories. In Sect. 4 I discuss a two-dimensional model in which instantons lead to something like quark confinement and explain why a similar mechanism has (unfortunately) no chance of working in four dimensions. In Sect. 5 I explain 't Hooft's resolution of the U(1) problem. In Sect. 6 I apply instanton methods to vacuum decay. Only this last section reports on my own research; all the rest is the work of other hands.[1]

I thank C. Callan, R. Dashen, D. Gross, R. Jackiw, M. Peskin, C. Rebbi, G. 't Hooft, and E. Witten for patiently explaining large portions of this subject to me. Although I have never met A. M. Polyakov, his influence pervades these lectures, as it does the whole subject.[2]

A note on notation. In these lectures we will work in both Minkowski space and in four-dimensional Euclidean space. A point in Minkowski space is labeled x^μ, where $\mu = 0, 1, 2, 3$, and x^0 is the time coordinate. In Minkowski space I will distinguish between covariant and contravariant vectors, $x_\mu = g_{\mu\nu} x^\nu$, where the metric tensor has signature $(+ - - -)$. Euclidean space is obtained from Minkowski space by formal analytic continuation in the time coordinate, $x^4 = -ix^0$. A point in Euclidean space is labeled x^μ, where $\mu = 1, 2, 3, 4$. The signature of the metric tensor is $(+ + + +)$. Thus covariant and contravariant vectors are component-by-component identical, and I will not bother to distinguish between them. Note that $x \cdot y$ in Minkowski space continues to $-x \cdot y$ in Euclidean space. The Euclidean action is defined as $-i$ times the continuation of the Minkowskian action. When discussing particle problems, I will use t for both Euclidean and Minkowskian time; which is meant will always be clear from the context. In Sect. 2 explicit factors of $\hbar$ are retained; elsewhere, $\hbar$ is set equal to one.

2   Instantons and bounces in particle mechanics

2.1   *Euclidean functional integrals*

In this section we will deal exclusively with the theory of a spinless particle of unit mass moving in a potential in one dimension:

$$H = \frac{p^2}{2} + V(x). \tag{2.1}$$

We will rederive some familiar properties of this much-studied system by unfamiliar methods. For the problem at hand, these methods are *much* more awkward than the standard methods of one-dimensional quantum mechanics; however, they have the great advantage of being immediately generalizable to quantum field theory.

Our fundamental tool will be the Euclidean (imaginary time) version of Feynman's[3] sum over histories:

$$\langle x_f | e^{-HT/\hbar} | x_i \rangle = N \int [dx] e^{-S/\hbar}. \tag{2.2}$$

Both sides of this equation require explanation:

On the left-hand side, $|x_i\rangle$ and $|x_f\rangle$ are position eigenstates, H is the Hamiltonian, and T is a positive number. The left-hand side of Eq. (2.2) is of interest because, if we expand in a complete set of energy eigenstates,

$$H|n\rangle = E_n|n\rangle, \tag{2.3}$$

then

$$\langle x_f | e^{-HT/\hbar} | x_i \rangle = \sum_n e^{-E_n T/\hbar} \langle x_f | n \rangle \langle n | x_i \rangle. \tag{2.4}$$

Thus, the leading term in this expression for large T tells us the energy and wave-function of the lowest-lying energy eigenstate.

On the right-hand side, N is a normalization factor, S is the Euclidean action[4]

$$S = \int_{-T/2}^{T/2} dt \left[\frac{1}{2} \left(\frac{dx}{dt} \right)^2 + V \right], \tag{2.5}$$

and $[dx]$ denotes integration over all functions $x(t)$, obeying the boundary conditions, $x(-T/2) = x_i$ and $x(T/2) = x_f$. To be more specific, if $\bar{x}$ is any function obeying the boundary condition, then a general function obeying the boundary conditions can be written as

$$x(t) = \bar{x}(t) + \sum_n c_n x_n(t), \tag{2.6}$$

where the x_ns are a complete set of real orthonormal functions vanishing at the boundaries,

$$\int_{-T/2}^{T/2} dt\, x_n(t) x_m(t) = \delta_{nm},$$ (2.7a)

$$x_n(\pm T/2) = 0.$$ (2.7b)

Then, the measure $[dx]$ is defined by

$$[dx] = \prod_n (2\pi\hbar)^{-\frac{1}{2}}\, dc_n.$$ (2.8)

(This measure differs in normalization from the measure defined by Feynman;[3] this is why we need the normalization constant N. However, as we shall see, we shall never need an explicit formula for N.)

The right-hand side of Eq. (2.2) is of interest because it can readily be evaluated in the semiclassical (small $\hbar$) limit. In this case the functional integral is dominated by the stationary points of S. For simplicity, let us assume for the moment that there is only one such stationary point, which we denote by $\bar{x}$,

$$\frac{\delta S}{\delta \bar{x}} = -\frac{d^2 \bar{x}}{dt^2} + V'(\bar{x}) = 0,$$ (2.9)

where the prime denotes differentiation with respect to x. Further, let us choose the x_ns to be eigenfunctions of the second variational derivative of S at $\bar{x}$,

$$-\frac{d^2 x_n}{dt^2} + V''(\bar{x}) x_n = \lambda_n x_n.$$ (2.10)

Then, in the small-$\hbar$ limit, the integral becomes a product of Gaussians, and we find

$$\langle x_f | e^{-HT/\hbar} | x_i \rangle = N e^{-S(\bar{x})/\hbar} \prod_n \lambda_n^{-\frac{1}{2}}[1 + O(\hbar)]$$

$$= N e^{-S(\bar{x})/\hbar}[\det(-\partial_t^2 + V''(\bar{x}))]^{-\frac{1}{2}}[1 + O(\hbar)].$$ (2.11)

(Of course, we are tacitly assuming here that all the eigenvalues are positive. We shall shortly see what to do when this is not the case.) If there are several stationary points, in general one has to sum over all of them.

Equation (2.9) is the equation of motion for a particle of unit mass moving in a potential *minus* V. Thus,

$$E = \frac{1}{2}\left(\frac{d\bar{x}}{dt}\right)^2 - V(\bar{x})$$ (2.12)

is a constant of the motion. This can be used to determine the qualitative features of the solutions of Eq. (2.9) by inspection.

As a simple example, consider the potential shown in Fig. 2(a). Let us choose $x_i = x_f = 0$. Figure 2(b) shows the inverted potential, $-V$. It is

Fig. 2

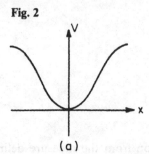

(a)

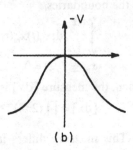

(b)

obvious from the figure that the only solution of Eq. (2.9) which obeys the boundary conditions is

$$\bar{x} = 0. \tag{2.13}$$

For this solution, $S = 0$. Thus, from Eq. (2.11),

$$\langle 0|e^{-HT/\hbar}|0\rangle = N[\det(-\partial_t^2 + \omega^2)]^{-\frac{1}{2}}[1 + O(\hbar)], \tag{2.14}$$

where

$$\omega^2 = V''(0). \tag{2.15}$$

In Appendix 1, I show that, for large T,

$$N[\det(-\partial_t^2 + \omega^2)]^{-\frac{1}{2}} = \left(\frac{\omega}{\pi\hbar}\right)^{\frac{1}{2}} e^{-\omega T/2}. \tag{2.16}$$

Thus, the ground-state energy is given by

$$E_0 = \tfrac{1}{2}\omega\hbar[1 + O(\hbar)]. \tag{2.17}$$

Also, the probability of the particle being at the origin when it is in its ground state is

$$|\langle x = 0|n = 0\rangle|^2 = (\omega/\pi\hbar)^{\frac{1}{2}}[1 + O(\hbar)]. \tag{2.18}$$

These are, of course, the correct semiclassical results. In the small-$\hbar$ limit, the particle is in a harmonic-oscillator ground-state concentrated at the origin and its energy is the ground-state energy of a harmonic oscillator.

2.2   *The double well and instantons*

We now turn to a less trivial problem,[5] the double well of Fig. 3(a). I will assume the potential is even, $V(x) = V(-x)$, and will denote its minima by $\pm a$. As before, I will add a constant to V, if necessary to make V vanish at its minima, and I will denote $V''(\pm a)$ by ω^2.

We will attempt to compute both

$$\langle -a|e^{-HT/\hbar}|-a\rangle = \langle a|e^{-HT/\hbar}|a\rangle, \tag{2.19a}$$

Fig. 3

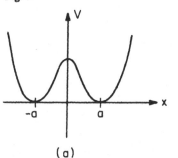

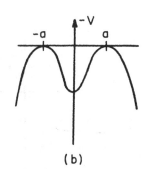

(a)   (b)

and

$$\langle a|e^{-HT/\hbar}|-a\rangle = \langle -a|e^{-HT/\hbar}|a\rangle, \qquad (2.19b)$$

by approximating the functional integral by its semiclassical limit, Eq. (2.11). Just as before, the first step is to find solutions of the classical Euclidean equation of motion, (2.9), consistent with our boundary conditions.

Of course, two such solutions are those in which the particle stays fixed on top of one or the other of the two hills in Fig. 3(b). However, there is another potentially interesting solution, one where the particle begins at the top of one hill (say the left one) at time $-T/2$, and moves to the top of the right hill at time $T/2$. Since we plan eventually to take T to infinity, we will focus on the form of the solution in this limit, where the particle attains the tops of the hills at times plus and minus infinity. In this case, we are dealing with a solution of the equation of motion with vanishing E; whence

$$dx/dt = (2V)^{\frac{1}{2}}. \qquad (2.20)$$

Equivalently,

$$t = t_1 + \int_0^x dx'(2V)^{-\frac{1}{2}}, \qquad (2.21)$$

where t_1 is an integration constant, the time at which x vanishes.

This solution is sketched in Fig. 4; it is called 'an instanton with center at t_1'. The name 'instanton' was invented by 't Hooft. The idea is that these objects are very similar in their mathematical structure to what are called solitons or lumps,[6] particle-like solutions of classical field theories: thus the '-on'. However, unlike lumps, they are structures in time (albeit Euclidean time): thus the 'instant-'. For the same reason, Polyakov suggested the name 'pseudoparticle', also used in the literature.

Of course, we can also construct solutions that go from a to $-a$,

Fig. 4

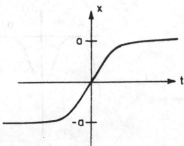

simply by replacing t by $-t$ in Eq. (2.21); these are called 'anti-instantons'.
Two properties of these solutions will be important to us:

(1) From Eq. (2.20), it is easy to derive a simple expression for S_0, the action of an instanton (or anti-instanton)

$$S_0 = \int dt[\tfrac{1}{2}(dx/dt)^2 + V] = \int dt(dx/dt)^2 = \int_{-a}^{a} dx(2V)^{\frac{1}{2}}. \qquad (2.22)$$

Note that this is the same as the integral that appears in the barrier-penetration formula, Eq. (1.7). We shall see shortly that this is no coincidence.

(2) For large t, x approaches a, and Eq. (2.20) can be approximated by

$$dx/dt = \omega(a - x). \qquad (2.23)$$

Thus, for large t,

$$(a - x) \propto e^{-\omega t}. \qquad (2.24)$$

Thus, instantons are, roughly speaking, well-localized objects, having a size on the order of $1/\omega$.

This is of critical importance, because it means that, for large T, the instanton and the anti-instanton are not the only approximate solutions of the equation of motion; there are also approximate solutions consisting of strings of widely separated instantons and anti-instantons. (You may be troubled by the sudden appearance in the argument of approximate solutions, approximate stationary points of S. If so, bear with me; I shall give a fuller explanation of this point later.)

I shall evaluate the functional integral by summing over all such configurations, with n objects (instantons or anti-instantons) centered at $t_1 \dots t_n$, where

$$T/2 > t_1 > t_2 \dots > t_n > -T/2. \qquad (2.25)$$

Fig. 5 shows one such configuration. T is assumed to be huge on the

Fig. 5

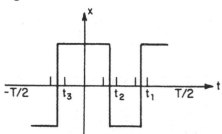

scale of the size of an instanton; thus the smooth curves of Fig. 4 appear as sharp jumps on the scale of Fig. 5. (The vertical marks on the time axis will be explained shortly.)

Now for the evaluation:

(1) For n widely separated objects, S is nS_0. This takes care of the exponential of the action.

(2) The evaluation of the determinant is a bit trickier. Let us consider the time evolution operator, e^{-HT}, as a product of operators associated with evolution between the points indicated by the vertical marks on the time axis in Fig. 5. If it were not for the small intervals containing the instantons and anti-instantons, V'' would equal ω^2 over the entire time axis, and thus we would obtain the same result we obtained for a single-well potential in Sect. 2.1,

$$\left(\frac{\omega}{\pi\hbar}\right)^{\frac{1}{2}} e^{-\omega T/2}. \tag{2.26}$$

The small intervals containing the instantons and anti-instantons correct this formula. Thus we obtain

$$\left(\frac{\omega}{\pi\hbar}\right)^{\frac{1}{2}} e^{-\omega T/2} K^n, \tag{2.27}$$

where K is defined by demanding that this formula give the right answer for one instanton. Later we shall obtain a more explicit expression for K.

(3) We must integrate over the locations of the centers:

$$\int_{-T/2}^{T/2} dt_1 \int_{-T/2}^{t_1} dt_2 \dots \int_{-T/2}^{t_{n-1}} dt_n = T^n/n!. \tag{2.28}$$

(4) We are not free to distribute instantons and anti-instantons arbitrarily. For example, if we start out at $-a$, the first object we encounter must be an instanton, the next one must be an anti-instanton, etc. Furthermore, if we are to end up back at $-a$, n must be even. Likewise, if we wish to end up at a, n must be odd.

274   *The uses of instantons*

Thus,

$$\langle -a|e^{-HT/\hbar}|-a\rangle = \left(\frac{\omega}{\pi\hbar}\right)^{\frac{1}{2}} e^{-\omega T/2} \sum_{\substack{even \\ n}} \frac{(Ke^{-S_0/\hbar}T)^n}{n!} [1+O(\hbar)],$$

(2.29)

while $\langle a|e^{-HT/\hbar}|-a\rangle$ is given by the same expression, summed over odd
*n*s. These sums are trivial:

$$\langle \pm a|e^{-HT/\hbar}|-a\rangle = \left(\frac{\omega}{\pi\hbar}\right)^{\frac{1}{2}} e^{-\omega T/2}\tfrac{1}{2}[\exp(Ke^{-S_0/\hbar}T)$$

$$\mp \exp(-Ke^{-S_0/\hbar}T)].$$ (2.30)

(From now on, to keep the page from getting cluttered, I will drop the
factors of $[1+O(\hbar)]$; remember that they are omnipresent though un-
written.)

Comparing this to Eq. (2.4), we see that we have two low-lying energy
eigenstates, with energies

$$E_\pm = \tfrac{1}{2}\hbar\omega \pm \hbar Ke^{-S_0/\hbar}.$$ (2.31)

If we call these eigenstates $|+\rangle$ and $|-\rangle$, we also see that

$$|\langle +|\pm a\rangle|^2 = |\langle -|\pm a\rangle|^2 = \langle a|-\rangle\langle -|-a\rangle$$

$$= -\langle a|+\rangle\langle +|-a\rangle = \tfrac{1}{2}\left(\frac{\omega}{\pi\hbar}\right)^{\frac{1}{2}}.$$ (2.32)

Of course, these are the expected results: the energy eigenstates are the
spatially even and odd combinations of harmonic oscillator states centered
at the bottoms of the two wells; the degeneracy of the two energy eigenvalues is broken only by barrier penetration (and thus the difference of
the energies is proportional to the barrier-penetration factor, $e^{-S_0/\hbar}$), and
the state of lower energy, which we have denoted by $|-\rangle$, is the spatially
even combination.

Our next task is to evaluate *K*. Before we do this, though, some comments should be made about what we have done so far:

(1) Really we have no right to retain the second term in Eq. (2.31). It
is not only exponentially small compared to the first term, it is exponentially small compared to the uncomputed $O(\hbar^2)$ corrections to the first
term. However, it is the leading contribution to the difference of the
energies, $E_+ - E_-$; a purist would retain it only in the expression for this
difference and not in the expressions for the individual energies.

(2) Our approximation has been based on the assumption that the
instantons and anti-instantons are all widely separated. As a consistency
check, we should verify that the major portion of our final result comes
from configurations where this is indeed the case.

This check is easy to carry out. For any fixed x, the terms in the exponential series, $\sum x^n/n!$, grow with n until n is on the order of x; after this point, they begin to decrease rapidly. Applying this to the sum in Eq. (2.29), we see the important terms are those for which

$$n \lesssim KTe^{-S_0/\hbar}. \tag{2.33}$$

That is to say, for small $\hbar$, the important terms in the sum are those for which n/T, the density of instantons and anti-instantons, is exponentially small, and thus the average separation is enormous. Note that this average separation is independent of T; our approximation is indeed a small-$\hbar$ approximation; the conditions for its validity are independent of T, as long as T is sufficiently large.

This approximation of summing over widely separated instantons is called the dilute-gas approximation, because of its similarity to the approximation of that name in statistical mechanics.

(3) Finally, I want to deliver the promised fuller explanation of the idea of an approximate stationary point of S. Let us begin by studying an integral over a single variable,

$$I = \int_0^T dt\, e^{-S(t)/\hbar}, \tag{2.34}$$

where S is a function of t monotonically decreasing to some asymptotic value, $S(\infty)$. Thus the integrand has no stationary points in the region of integration. Nevertheless, it is easy to find the approximate form of the integral for small $\hbar$ and large T:

$$I \approx Te^{-S(\infty)/\hbar}. \tag{2.35}$$

Speaking loosely, the integral is dominated by the stationary point at infinity. It is straightforward to generalize this phenomenon to multi-dimensional integrals: we assume an integrand whose graph has a sort of trough in it; the line along the bottom of the trough flattens out as we go to infinity. Speaking less pictorially, there is a line in the multi-dimensional space such that the integrand is a minimum with respect to variations perpendicular to the line and approaches some limiting value as one goes to infinity along the line. Of course, the line could itself be generalized to a hyperplane, a generalized 'bottom of the trough'. This is in fact the situation for our 'approximate stationary points'; the locations of the instantons and anti-instantons are the variables along the bottom of the trough; S becomes stationary (and equal to nS_0) only when they all go to infinity.

This concludes the comments; we now turn to the evaluation of K.

We must study the eigenvalue equation, Eq. (2.10), with $\bar{x}$ a single

instanton. Because of time translation invariance, this equation necessarily possesses an eigenfunction of eigenvalue zero,

$$x_1 = S_0^{-\frac{1}{2}} \, d\bar{x}/dt. \tag{2.36}$$

(The normalization factor comes from Eq. (2.22).) Were we to integrate
over the corresponding expansion coefficient, c_1, in Eq. (2.6), we would
obtain a disastrous infinity. Fortunately, we have already done this
integration, in the guise of integrating over the location of the center of
the instanton in Eq. (2.28). The change of $x(t)$ induced by a small change
in the location of the center, t_1, is

$$dx = (d\bar{x}/dt) \, dt_1. \tag{2.37}$$

The change induced by a small change in the expansion coefficient, c_1, is

$$dx = x_1 \, dc_1. \tag{2.38}$$

Hence,

$$(2\pi\hbar)^{-\frac{1}{2}} \, dc_1 = (S_0/2\pi\hbar)^{\frac{1}{2}} \, dt_1. \tag{2.39}$$

Thus, in evaluating the determinant, we should not include the zero
eigenvalue, but we should include in K a factor[7] of $(S_0/2\pi\hbar)^{\frac{1}{2}}$. Hence, the
one-instanton contribution to the transition matrix element is given by

$$\langle a|e^{-HT}|-a \rangle_{\text{one inst.}}$$

$$= NT(S_0/2\pi\hbar)^{\frac{1}{2}} e^{-S_0/\hbar} (\det'[-\partial_t^2 + V''(\bar{x})])^{-\frac{1}{2}}, \tag{2.40}$$

where det' indicates that the zero eigenvalue is to be omitted when computing the determinant. Comparing this to the one-instanton term in
Eq. (2.29), we find

$$K = (S_0/2\pi\hbar)^{\frac{1}{2}} \left| \frac{\det(-\partial_t^2 + \omega^2)}{\det'(-\partial_t^2 + V''(\bar{x}))} \right|^{\frac{1}{2}}. \tag{2.41}$$

This completes the computation.

Some remarks:

(1) To really sew things up, I should show that the formula we have
obtained for the energy splitting is the same as that obtained by the
traditional methods of wave mechanics. I do this in Appendix 2.

(2) I have been tacitly assuming that all the eigenvalues in Eq. (2.10) are
positive, other than the zero eigenvalue associated with x_1. It is easy to
prove that this is indeed the case. It is well-known that the eigenfunction
of a one-dimensional Schrödinger equation (like Eq. (2.10)) of lowest
eigenvalue has no nodes, the next-lowest eigenfunction has one node, etc.
Because the instanton is a monotone increasing function of t, x_1, proportional to the time derivative of the instanton, has no nodes. Thus
zero is the lowest eigenvalue and all the other eigenvalues are positive.

(3) K is proportional to $\hbar^{-\frac{1}{2}}$. This factor came from the zero eigenvalue

associated with time-translation invariance. Later in these lectures we will be analyzing theories that have larger invariance groups and for which the instantons have more than one zero eigenvalue associated with them. Clearly, for every zero eigenvalue there will be a factor of $\hbar^{-\frac{1}{2}}$. This rule for counting powers of $\hbar$ will be very important to us, for, as I explained in Sect. 1, counting powers of $\hbar$ is equivalent to counting powers of coupling constants.

2.3   *Periodic potentials*

Let us consider a periodic potential, like the one sketched in Fig. 6(a). (For simplicity, I have chosen the minima of V to be the integers.) If we ignore barrier penetration, the energy eigenstates are an infinitely degenerate set of states, each concentrated at the bottom of one of the wells. Barrier penetration changes this single eigenvalue into a continuous band of eigenvalues; the true energy eigenstates are the eigenstates of unit translations, the Bloch waves. Let us see how this old result can be obtained by instanton methods.

As we see from Fig. 6(b), the instantons are much the same as in the preceding problem. The only novelty is that the instantons can begin at any initial position, $x=j$, and go to the next one, $x=j+1$. Likewise, the anti-instantons can go from $x=j$ to $x=j-1$. Otherwise, everything is as before.

Thus, when doing the dilute-gas sum, we can sprinkle instantons and anti-instantons freely about the real axis; there is no constraint that instantons and anti-instantons must alternate. Of course, as we go along

Fig. 6

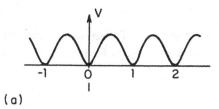

(a)

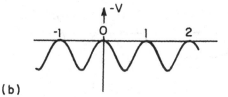

(b)

the line, each instanton or anti-instanton must begin where its predecessor ended. Furthermore, the total number of instantons minus the total number of anti-instantons must equal the change in x between the initial and final position eigenstates.

Thus we obtain

$$\langle j_+ | e^{-HT/\hbar} | j_- \rangle = \left(\frac{\omega}{\pi\hbar} \right)^{\frac{1}{2}} e^{-\omega T/2} \sum_{n=0}^{\infty} \sum_{\bar{n}=0}^{\infty} \frac{1}{n!\bar{n}!}$$
$$\times (Ke^{-S_0/\hbar}T)^{n+\bar{n}} \delta_{n-\bar{n}-j_+ +j_-} \qquad (2.42)$$

where n is the number of instantons and $\bar{n}$ the number of anti-instantons. If we use the identity

$$\delta_{ab} = \int_0^{2\pi} d\theta \, e^{i\theta(a-b)}/2\pi, \qquad (2.43)$$

the sum becomes two independent exponential series, and we find

$$\langle j_+ | e^{-HT/\hbar} | j_- \rangle = \left(\frac{\omega}{\pi\hbar} \right)^{\frac{1}{2}} e^{-\omega T/2} \int_0^{2\pi} e^{i(j_- - j_+)\theta} \frac{d\theta}{2\pi}$$
$$\times \exp[2KT \cos\theta \, e^{-S_0/\hbar}] \qquad (2.44)$$

Thus we find a continuum of energy eigenstates labeled by the angle θ. The energy eigenvalues are given by

$$E(\theta) = \tfrac{1}{2}\hbar\omega + 2\hbar K \cos\theta \, e^{-S_0/\hbar}. \qquad (2.45)$$

Also,

$$\langle \theta | j \rangle = \left(\frac{\omega}{\pi\hbar} \right)^{\frac{1}{4}} (2\pi)^{-\frac{1}{2}} e^{ij\theta}. \qquad (2.46)$$

Hearteningly, this is just the right answer.

2.4   Unstable states and bounces[8]
Galilean pastiche:

SAGREDO: Let me test my understanding of these instanton methods by studying the potential of Fig. 7(a). If I neglect barrier penetration, in the

Fig. 7

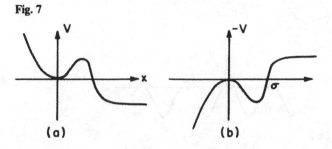

(a)   (b)

semiclassical limit, this potential has an energy eigenstate sitting in the bottom of the well. I wish to compute the corrections to the energy of this state due to barrier penetration. If I turn the potential upside down (Fig. 7(b)), I observe that the classical equation of motion has a solution in which the particle begins at the top of the hill at $x = 0$, bounces off the classical turning point σ, and returns to the top of the hill (Fig. 8). I will call this motion 'the bounce'. I will compute the transition matrix element between $x = 0$ and $x = 0$ by summing over configurations consisting of widely separated bounces, just as one sums over instantons and anti-instantons in the study of the double well. Indeed, the sum is the same as that for the double well (with the obvious redefinitions of S_0, ω^2, etc.), save that there is no restriction to an even or odd number of bounces. Thus I obtain the complete exponential series, rather than just the odd or even terms, and I find that

$$\langle 0|e^{-HT/\hbar}|0\rangle = \left(\frac{\omega}{\pi\hbar}\right)^{\frac{1}{2}} e^{-\omega T/2} \exp[KTe^{-S_0/\hbar}], \qquad (2.47)$$

and the energy eigenvalue is given by

$$E_0 = \tfrac{1}{2}\omega\hbar + \hbar Ke^{-S_0/\hbar}. \qquad (2.48)$$

SALVIATI: Alas, Sagredo, I fear you have erred in three ways. Firstly, the term you have computed is small compared to terms of order $\hbar^2$ which you have neglected, and thus you have no right to retain it. Secondly, I see by your sketch that the bounce has a maximum; therefore the eigenfunction x_1, which is proportional to the time derivative of the bounce, has a node. Thus it is not the eigenfunction of lowest eigenvalue, and there must be a nodeless eigenfunction, x_0, of a lower eigenvalue, that is to say, there must be a negative eigenvalue. Thus K, which is inversely proportional to the product of the square roots of the eigenvalues, is imaginary. Thirdly, the eigenvalue you attempt to compute is nowhere to be found in the spectrum of the Hamiltonian, because the state you are studying is rendered unstable by barrier penetration.

SAGREDO: Everything you say is correct, but I believe your criticisms show how to save the computation. An unstable state is one whose energy has an

Fig. 8

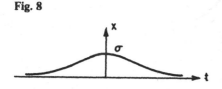

imaginary part; thus it is only to be expected that K should be imaginary. Furthermore, the term I have computed, though indeed small compared to neglected contributions to the real part of E_0, is the leading contribution to the imaginary part of E_0. Thus the correct version of Eq. (2.48) is

$$\text{Im } E_0 = \Gamma/2 = \hbar|K|e^{-S_0/\hbar}, \tag{2.49}$$

where Γ is, as usual, the width of the unstable state.

As you can see, the Tuscan twosome are as quick-witted as ever, although (also as ever) their arguments are sometimes a bit sloppy. Sagredo has missed a factor of $\frac{1}{2}$; the correct answer is

$$\Gamma = \hbar|K|e^{-S_0/\hbar}. \tag{2.50}$$

To show that this is the case requires a more careful argument than Sagredo's. The essential point is Salviati's observation that the energy of an unstable state is not an eigenvalue of H; in fact, it is an object that can only be defined by a process of analytic continuation. I will now perform such a continuation.

To keep things as simple as possible, let us consider not an integral over all function space, but an integral over some path in function space parametrized by a real variable, z,

$$J = \int dz (2\pi\hbar)^{-\frac{1}{2}} e^{-S(z)/\hbar}, \tag{2.51}$$

where $S(z)$ is the action along the path. In particular, let us choose the path sketched in Fig. 9. This path includes two important functions that occur in the real problem: $x(t)=0$, at $z=0$, and the bounce, at $z=1$. Furthermore, the path is such that the tangent vector to the path at $z=1$ is x_0. Thus the path goes through the bounce in the 'most dangerous direction', that direction with which the negative eigenvalue is associated, and $z=1$ is a maximum of S, as shown in Fig. 10. S goes to minus infinity as z goes to infinity because the functions spend more and more time in

Fig. 9

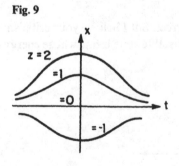

Fig. 10

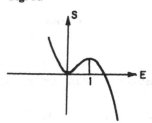

Fig. 11

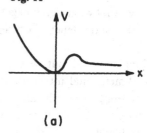 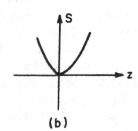

(a)   (b)

the region beyond the turning point, where V is negative; note that this implies that Eq. (2.51) is hopelessly divergent.

If $x=0$ were the absolute minimum of V, that is to say, if V were as shown in Fig. 11(a), we would have, for the same path, the situation shown in Fig. 11(b), and there would be no divergence in Eq. (2.48). Now let us suppose we analytically change V in some way such that we go from this situation back to the one of interest. To keep the integral convergent, we must distort the right-hand portion of the contour of integration into the complex plane. How we distort it depends on the details of the analytic passage from one potential to the other. In Fig. 12, I have assumed that it is distorted into the upper half plane. Following the standard procedure of the method of steepest descents, I have led the contour along the real axis to $z=1$, the saddle point, and then out along a line of constant imaginary part of S. The integral thus acquires an imaginary part; in the steepest-

Fig. 12

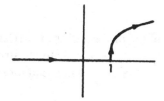

descent approximation,

$$\mathrm{im}\, J = \mathrm{Im} \int_1^{1+i\infty} dz (2\pi\hbar)^{-\frac{1}{2}} e^{-S(1)/\hbar} e^{-\frac{1}{2}S''(1)(z-1)^2/\hbar}$$

$$= \tfrac{1}{2} e^{-S(1)/\hbar} |S''(1)|^{-\frac{1}{2}}. \tag{2.52}$$

Note the factor of $\frac{1}{2}$; this arises because the integration is over only half of the Gaussian peak.

(If we had passed from one potential to the other in the conjugate manner, the contour would have been distorted into the lower half plane, and we would have obtained the opposite sign for the imaginary part. This is just a reflection of the well-known fact that what sign you get for the imaginary part of the energy of an unstable state depends on how you do your analytic continuation.)

Now, we have studied a one-dimensional integral, but we can always reduce our functional integral to a one-dimensional integral simply by integrating (in the Gaussian approximation) over all the variables orthogonal to our path. These directions involve only positive or zero eigenvalues near the stationary point and give us no trouble. In this manner we obtain Sagredo's answer, Eq. (2.49), except that the negative eigenvalue carries a factor of $\frac{1}{2}$ with it; that is to say, we obtain Eq. (2.50).

3   The vacuum structure of gauge field theories[9]

3.1   *Old stuff*

This subsection is a telegraphic compendium of formulae from gauge field theories. Its purpose is to establish notational conventions and possibly to jog your memory. If you do not already know the fundamentals of gauge field theory, you will not learn them here.[10]

Lie algebras. A representation of Lie algebra is a set of N anti-Hermitian matrices, T^a, $a = 1 \ldots N$, obeying the equations

$$[T^a, T^b] = c^{abc} T^c, \tag{3.1}$$

where the cs are the structure constants of some compact Lie group, G. It is always possible to choose the Ts such that $\mathrm{Tr}(T^a T^b)$ is proportional to δ^{ab}, although the constant of proportionality may depend on the representation. The Cartan inner product is defined by

$$(T^a, T^b) = \delta^{ab}. \tag{3.2}$$

Thus this is proportional to the trace of the product of the matrices.

So far I have not stated a convention that gives a scale to the structure constants and thus to the Ts. For SU(2), the case I will spend most time discussing, I will choose c^{abc} to be equal to ε^{abc}. Thus, for the isospinor representation,

$$T^a = -i\sigma^a/2, \tag{3.3}$$

where the σs are the Pauli spin matrices. In this case,

$$(T^a, T^b) = -2\,\mathrm{Tr}(T^aT^b). \tag{3.4}$$

Occasionally I will discuss SU(n), in particular SU(3). In this case I will choose the structure constants to agree with the preceding convention for the SU(2) subgroup composed of unitary unimodular transformations on two variables only. Thus, for SU(3), T^a is $-i\lambda^a/2$, where the λs are Gell-Mann's matrices.

Gauge fields. The gauge potentials are a set of vector fields, $A_\mu^a(x)$. It is convenient to define a matrix-valued vector field, $A_\mu(x)$, by

$$A_\mu = gA_\mu^a T^a, \tag{3.5}$$

where g is a constant called the gauge coupling constant. The field-strength tensor, $F_{\mu\nu}(x)$, is defined by

$$F_{\mu\nu} = \partial_\mu A_\nu - \partial_\nu A_\mu + [A_\mu, A_\nu]. \tag{3.6}$$

Pure gauge field theory is defined by the Euclidean action,

$$S = \frac{1}{4g^2}\int d^4x(F_{\mu\nu}, F_{\mu\nu}). \tag{3.7}$$

Sometimes I will write this in a shorthand form,

$$S = \frac{1}{4g^2}\int (F^2). \tag{3.8}$$

Gauge transformations. A gauge transformation is a function, $g(x)$, from Euclidean space into the gauge group, G. In equations,

$$g(x) = \exp \lambda^a(x)T^a, \tag{3.9}$$

where the λs are arbitrary functions. (Please do not confuse $g(x)$ with the coupling constant, g.) Under such a transformation,

$$A_\mu \to gA_\mu g^{-1} + g\partial_\mu g^{-1}, \tag{3.10}$$

and

$$F_{\mu\nu} \to gF_{\mu\nu}g^{-1}. \tag{3.11}$$

Thus, S is gauge-invariant. If $F_{\mu\nu}$ vanishes, then A_μ is a gauge-transform of zero; that is to say,

$$A_\mu = g\partial_\mu g^{-1}, \tag{3.12}$$

for some $g(x)$.

Covariant derivatives. The covariant derivative of the field strength tensor is defined by

$$D_\lambda F_{\mu\nu} = \partial_\lambda F_{\mu\nu} + [A_\lambda, F_{\mu\nu}].$$ (3.13)

Equation (3.7) leads to the Euclidean equations of motion

$$D_\mu F_{\mu\nu} = 0.$$ (3.14)

Given a field ψ that gauge-transforms according to

$$\psi \to g(x)\psi,$$ (3.15)

then the covariant derivative of ψ,

$$D_\mu\psi = \partial_\mu\psi + A_\mu\psi,$$ (3.16)

transforms in the same way.

3.2   The winding number

I propose to study Euclidean gauge field configurations of finite action (not necessarily solutions of the equations of motion).

Why?

The naive answer, sometimes given in the literature,[11] is that configurations of infinite action are unimportant in the functional integral, since, for such configurations, $e^{-S/\hbar}$ is zero. *This is wrong.* In fact, it is configurations of finite action that are unimportant; to be precise, they form a set of measure zero in function space. This has nothing to do with the divergences of quantum field theory; it is true even for the ordinary harmonic oscillator. (For a proof, see Appendix 3.) The only reason we are interested in configurations of finite action is that we are interested in doing semiclassical approximations, and a configuration of infinite action does indeed give zero if it is used as the center point of a Gaussian integral.

The convergence of the action integral is controlled by the behavior of A_μ for large r, where r is the radial variable in Euclidean four-space. To keep my arguments as simple as possible, I will assume that, for large r, A_μ can be expanded in an asymptotic series in inverse powers of r. (This assumption can be relaxed considerably without altering the conclusions.)[12] Thus, for the action to be finite, $F_{\mu\nu}$ must fall off faster than $1/r^2$ as r goes to infinity; that is to say, $F_{\mu\nu}$ must be $O(1/r^3)$. One's first thought is that this implies that A_μ is $O(1/r^2)$, but this is wrong: vanishing $F_{\mu\nu}$ does not imply vanishing A_μ, but merely that A_μ is a gauge transform of zero. Thus A_μ can be of the form

$$A_\mu = g\partial_\mu g^{-1} + O(1/r^2),$$ (3.17)

where g is a function from four-space to G of order one, that is to say, a function of angular variables only.

Thus, with every finite-action field configuration there is associated a group-element-valued function of angular variables, that is to say, a mapping of a three-dimensional hypersphere, S^3, into the gauge group, G. Of course, this assignment is not gauge-invariant. Under a gauge transformation, $h(x)$

$$A_\mu \to h A_\mu h^{-1} + h \partial_\mu h^{-1}. \tag{3.18}$$

Thus,

$$g \to hg + O(1/r^2). \tag{3.19}$$

If one could choose h to equal g^{-1} at infinity, one could transform g to one and eliminate it from Eq. (3.17). In general, though, this is not possible. The reason is that h must be a continuous function not just on the hypersphere at infinity, but throughout all four-space, that is to say, on a nested family of hyperspheres going all the way from r equals zero to r equals infinity. In particular, at the origin, h must be a constant, independent of angles. Thus, h at infinity can not be a general function on S^3, but must be one that can be obtained by continuous deformation from a constant function. Since any constant gauge transformation can trivially be obtained by continuous deformation from the identity transformation (all gauge groups are connected), we might as well say that h at infinity must be obtainable from $h = 1$ by a continuous deformation.

Given two mappings of one topological space into another, such that one mapping is continuously deformable into another, mathematicians say the two functions are 'homotopic' or 'in the same homotopy class'. What we have shown is that by a gauge transformation we can transform $g(x)$ into any mapping homotopic to $g(x)$, but we can not transform it into a function in another homotopy class. Thus, the gauge-invariant quantity associated with a finite-action field configuration is not a mapping of S^3 to G but a homotopy class of such mappings. Our task is to find these homotopy classes for physically interesting Gs.

To warm up for this task, let me consider a baby version of the problem for which the geometry is somewhat easier to visualize. I will work with the simplest of all gauge groups, U(1), the group of complex numbers of unit modulus. Thus the gauge field theory is ordinary electromagnetism. (However, I will still keep to the notational conventions established in Sect. 3.1; in particular, A_μ will be an imaginary quantity, i times the usual vector potential.) Also, I will work not in Euclidean four-space but in Euclidean two-space. I will still study fields obeying Eq. (3.17), although, of course, in two-space this condition is not a consequence of finiteness of the action. Because we are working in two-space, we have, instead of a hypersphere, S^3, an ordinary circle, S^1.

Now to work:

(1) G is the unit circle in the complex plane; thus, topologically, G is also S^1, and we have to study homotopy classes of mappings of S^1 into S^1. We will label the circle in space, the domain of our functions, in the standard way, by an angle θ ranging from 0 to 2π.

(2) It will be useful to define some standard mappings from S^1 to S^1. One is the trivial mapping,

$$g^{(0)}(\theta) = 1. \tag{3.20a}$$

Another is the identity mapping,

$$g^{(1)}(\theta) = e^{i\theta}. \tag{3.20b}$$

These are both part of a family of mappings,

$$g^{(\nu)}(\theta) = [g^{(1)}(\theta)]^\nu = e^{i\nu\theta}, \tag{3.20c}$$

where ν is an integer (positive, negative, or zero). ν is called the 'winding number', because it is the number of times we wind around G when we go once around the circle at infinity in two-space. (By convention, winding around minus once means winding around once in the negative direction.)

(3) Every mapping from S^1 to S^1 is homotopic to one of the mappings (3.20c). We do not have the mathematical machinery to prove this rigorously, but I hope I can make it plausible. Imagine taking a rubber band and marking on it in ink a sequence of values of θ running from 0 to 2π. We then wrap the band about a circle representing G, such that each value of θ lies above the point into which it is mapped. (Fig. 13 shows such a construction.) We can continuously deform the band, first to eliminate any folds, like the one on the top of the figure, and second to stretch the band so it lies uniformly on the circle. In this way we obtain some $g^{(\nu)}(\theta)$. (In the case shown, we obtain $g^{(1)}$.) Thus we can associate a winding number with every mapping. (Note that I have not yet shown that this number is uniquely defined.)

Fig. 13

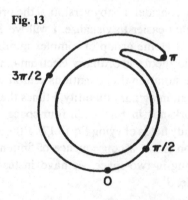

(4) I will now show that the winding number defined above is given by the integral formula

$$\nu = \frac{i}{2\pi} \int_0^{2\pi} d\theta g \, dg^{-1}/d\theta. \tag{3.21}$$

Firstly, by direct calculation, this gives the right answer for the standard mappings, Eq. (3.20c). Secondly, this quantity is invariant under continuous deformations. To prove this assertion it suffices to demonstrate invariance under infinitesimal deformations. A general infinitesimal deformation is of the form

$$\delta g = i(\delta\lambda)g, \tag{3.22}$$

where $\delta\lambda$ is some infinitesimal real function on the circle. Thus

$$\delta(g dg^{-1}/d\theta) = -id(\delta\lambda)/d\theta, \tag{3.23}$$

and the change in ν vanishes upon integration. (We now know that all of our standard mappings are in different homotopy classes and that the winding number is uniquely defined.)

(5) If

$$g(\theta) = g_1(\theta)g_2(\theta), \tag{3.24a}$$

then

$$\nu = \nu_1 + \nu_2. \tag{3.24b}$$

The proof is simple. The winding number is unchanged by continuous deformations. We can deform g_1 such that it is equal to one on the upper half of the circle $(0 \leqslant \theta \leqslant \pi)$ and g_2 such that it is equal to one on the lower half of the circle $(\pi \leqslant \theta \leqslant 2\pi)$. The integrand in Eq. (3.21) is then the sum of a part due to g_1 (vanishing on the upper semicircle) and a part due to g_2 (vanishing on the lower semicircle).

(6) Let us define

$$G_\mu = \frac{i}{2\pi} \varepsilon_{\mu\nu} A_\nu.^{13} \tag{3.25}$$

By Eqs. (3.17) and (3.21),

$$\nu = \lim_{r\to\infty} \int_0^{2\pi} r d\theta \hat{r}_\mu G_\mu, \tag{3.26}$$

where $\hat{r}_\mu$ is the radial unit vector. Thus, by Gauss's theorem,

$$\nu = \int d^2x \partial_\mu G_\mu. \tag{3.27}$$

Hence,

$$\nu = \frac{i}{4\pi} \int d^2x \varepsilon_{\mu\nu} F_{\mu\nu}. \tag{3.28}$$

I will now return to four-space, and take G to be SU(2). As we shall see, every argument will be a (mild) generalization of the arguments I have given for the baby problem.

(1) SU(2) is the group of unitary unimodular two-by-two matrices. It is well known that any such matrix can be uniquely written in the form

$$g = a + i\mathbf{b} \cdot \boldsymbol{\sigma}, \tag{3.29}$$

where $a^2 + |b|^2 = 1$. Thus, topologically, SU(2) is S^3, and we have to study homotopy classes of mappings from S^3 to S^3.

(2) It will be useful to define some standard mappings from S^3 to S^3. One is the trivial mapping,

$$g^{(0)}(x) = 1. \tag{3.30a}$$

Another is the identity mapping,

$$g^{(1)}(x) = (x_4 + i\mathbf{x} \cdot \boldsymbol{\sigma})/r. \tag{3.30b}$$

These are both part of a family of mappings,

$$g^{(\nu)}(x) = [g^{(1)}(x)]^\nu. \tag{3.30c}$$

where ν is an integer, called the winding number. (It is also sometimes called the Pontryagin index.) It measures the number of times the hypersphere at infinity is wrapped around G. (By convention, we say the hypersphere is wrapped around G in a negative sense if a right-handed triad of tangent vectors is mapped into a left-handed triad.)

(3) Every mapping from S^3 to S^3 is homotopic to one of our standard mappings (3.30c). We do not have the mathematical machinery to prove this assertion rigorously, but a plausibility argument can be constructed just as in the baby problem, with hyperspheres replacing circles. (If you have problems envisioning hyperspheres wrapped around hyperspheres, just accept the assertion on faith.) In this way we can associate a winding number with every mapping. (Note that I have not yet shown that this number is uniquely defined.)

(4) Let us define

$$\nu = \frac{1}{48\pi^2} \int d\theta_1 \, d\theta_2 \, d\theta_3 \, \varepsilon^{ijk} (g\partial_i g^{-1}, g\partial_j g^{-1} \, g\partial_k g^{-1}). \tag{3.31}$$

where θ_1, θ_2 and θ_3 are three angles that parametrize S^3. How these angles are chosen is irrelevant to Eq. (3.31); the Jacobian determinant that comes from changing the angles is canceled by the Jacobian determinant from the ε-symbol. Equation (3.31) is written using the Cartan inner product, that is to say, in a representation-independent way. Of course, for any particular representation of SU(2), we can rewrite Eq. (3.31) in terms of traces; for example, for the two-dimensional representation, by

Eq. (3.4),

$$v = -\frac{1}{24\pi^2} \int d\theta_1\, d\theta_2\, d\theta_3\, \mathrm{Tr}\varepsilon^{ijk} g\partial_i g^{-1}\, g\partial_j g^{-1}\, g\partial_k g^{-1}. \qquad (3.32)$$

I will show that this quantity is, firstly, a homotopy invariant, and secondly, agrees with the winding number as defined for our standard mappings. As before, a corollary of this proof will be that all of our standard mappings are in different homotopy classes and that the winding number is uniquely defined.

To show invariance under continuous deformations it suffices to show invariance under infinitesimal deformations. For any Lie group, a general infinitesimal transformation can be written as an infinitesimal right multiplication:

$$\delta g = g\delta\lambda^a(x)T^a \equiv g\delta T. \qquad (3.33)$$

Under this transformation,

$$\delta(g\partial_k g^{-1}) = -g(\partial_k \delta T)g^{-1}. \qquad (3.34)$$

The three derivatives in Eq. (3.32) make equal contributions to δv; thus,

$$\delta v \propto \int d\theta_1\, d\theta_2\, d\theta_3\, \varepsilon^{ijk}\, \mathrm{Tr} g\partial_i g^{-1}\, g\partial_j g^{-1}\, g(\partial_k \delta T)g^{-1}. \qquad (3.35)$$

If we use the identity,

$$0 = \partial_i(gg^{-1}) = g\partial_i g^{-1} + (\partial_i g)g^{-1}, \qquad (3.36)$$

this becomes

$$\delta v \propto \int d\theta_1\, d\theta_2\, d\theta_3\, \varepsilon^{ijk}\, \mathrm{Tr}\partial_i g^{-1}\, \partial_j g\partial_k \delta T, \qquad (3.37)$$

which vanishes upon integration by parts, because of the antisymmetry of the ε-symbol. This completes the proof of invariance under continuous deformations.

(5) Now to evaluate Eq. (3.32) for our standard mappings. The task is easiest for $g^{(1)}$, for the integrand is here obviously a constant, and we need evaluate it only at the north pole of the unit hypersphere, $x_4 = 1$, $x_i = 0$. At this point we might as well choose θ_i to equal x_i. Thus, from Eq. (3.30b),

$$g\partial_i g^{-1} = -\mathrm{i}\sigma_i, \qquad (3.38)$$

and

$$\mathrm{Tr}\varepsilon^{ijk} g\partial_i g^{-1}\, g\partial_j g^{-1}\, g\partial_k g^{-1} = -12. \qquad (3.39)$$

Since the area of a unit hypersphere is $2\pi^2$, we obtain the desired result, $v = 1$.

For the other standard mappings, the simplest way to proceed is to observe that if

$$g = g_1 g_2, \tag{3.40a}$$

then

$$v = v_1 + v_2. \tag{3.40b}$$

The argument is the same as for the baby problem, with semihyperspheres replacing semicircles.

(6) Let us define

$$G_\mu = 2\varepsilon_{\mu\nu\lambda\sigma}(A_\nu, \partial_\lambda A_\sigma + \tfrac{2}{3} A_\lambda A_\sigma). \tag{3.41}$$

A straightforward computation shows that

$$\partial_\mu G_\mu = \tfrac{1}{2}\varepsilon_{\mu\nu\lambda\sigma}(F_{\mu\nu}, F_{\lambda\sigma}). \tag{3.42}$$

The dual of an antisymmetric tensor (denoted by a tilde) is conventionally defined by

$$\tilde{F}_{\mu\nu} \equiv \tfrac{1}{2}\varepsilon_{\mu\nu\lambda\sigma}F_{\lambda\sigma}. \tag{3.43}$$

(The factor of $\tfrac{1}{2}$ is inserted in the definition so that $\tilde{\tilde{F}} = F$.) Equation (3.42) can thus be rewritten as

$$\partial_\mu G_\mu = (F_{\mu\nu}, \tilde{F}_{\mu\nu}) \equiv (F, \tilde{F}). \tag{3.44}$$

From the definition of $F_{\mu\nu}$,

$$G_\mu = \varepsilon_{\mu\nu\lambda\sigma}(A_\nu, F_{\lambda\sigma} - \tfrac{2}{3} A_\lambda A_\sigma). \tag{3.45}$$

This expression is useful in evaluating

$$\int d^4 x(F, \tilde{F}) = \int d^3 S \hat{r}_\mu G_\mu, \tag{3.46}$$

where $d^3 S$ is the element of area on a large hypersphere. The first term in Eq. (3.45) is $O(1/r^4)$ and makes no contribution to the integral; the second term simply gives (up to a multiplicative constant) the integral formula for the winding number, Eq. (3.31). Thus we obtain

$$\int d^4 x(F, \tilde{F}) = 32\pi^2 v. \tag{3.47}$$

Summary and generalizations. This has been a long analysis, and you may have lost track of what we were doing, so let me summarize the main results of this subsection. For a gauge field theory based on the group SU(2), every field configuration of finite action in four-dimensional Euclidean space has an integer associated with it, the Pontryagin index or winding number, v. It is not possible to continuously deform a configuration of one winding number into one of a different winding number while maintaining the finiteness of the action. We have two integral formulae for the winding number, one in terms of a surface integral over a large sphere, Eq. (3.31), and one in terms of a volume integral over all four-space, Eq. (3.47).

How much of this depends on the gauge group being SU(2)? Firstly, if the gauge group is U(1), it is easy to see that every mapping of S^3 into U(1) is continuously deformable into the trivial mapping (all of S^3 mapped into a single point). *Thus, for an Abelian gauge field theory, there is no analog of the winding number.* Secondly, for a general simple Lie group, G, there is a remarkable theorem due to Raoul Bott[14] that states that any continuous mapping of S^3 into G can be continuously deformed into a mapping into an SU(2) subgroup of G. *Thus, everything we have discovered for SU(2) is true for an arbitrary simple Lie group; in particular, it is true for SU(n).* I stress that 'everything' means *everything*. In particular, not a single numerical factor in the integral formulas for the winding number needs alteration, so long as we choose the normalization of the Cartan inner product appropriately (as we have). Finally, since a general compact Lie group is locally the direct product of an Abelian group and a string of simple groups, *for a general gauge field theory, there is an independent winding number for every simple factor group.*

3.3 Many vacua

We have learned a lot about classical gauge field theories; now it is time to confront the quantum theory. In principle, the Euclidean functional integral tells how to go from the classical theory to the quantum theory. As I explained in Sect. 2, we can use the functional integral to study the energy eigenstates of the theory; also, by adding appropriate source terms to the Hamiltonian (equivalently, to the Euclidean action) and then differentiating with respect to the sources at the end of the computation, we can study the expectation values of strings of operators, Euclidean Green's functions. However, for gauge field theories, there is a famous complication: to make the functional integral well-defined, we must impose a gauge-fixing condition.[15]

I will choose to work in axial gauge, $A_3 = 0$. I have several reasons for this choice. (1) It is possible to show[16] that every non-singular gauge field configuration can be put in axial gauge by a non-singular gauge transformation. It is by no means clear whether this is true for covariant gauges, for example. (2) In axial gauge the functional integral is directly equivalent to a canonical formulation of the theory;[17] there is no need of the ghost terms that occur in covariant gauges, or of the subsidiary conditions on the space of states that are needed in such gauges as $A_0 = 0$. (3) Most of the treatment in the literature of the phenomena we are about to discuss is in the gauge $A_0 = 0$. It is nice to show explicitly that the answers do not depend on this gauge choice. (4) Although axial gauge is terribly awkward for specific computations, once we have obtained functional-integral expressions for quantities of interest, we can use the standard Faddeev–Popov methods to transform these into some more convenient gauge.

In field theory, we normally plunge directly into infinite space. However, I will here study gauge field theory in a finite box of three-volume V, with definite boundary conditions, which I shall specify shortly. Just as in Sect. 2, I will also restrict the theory to a finite range of Euclidean time, T, with appropriate boundary conditions at initial and final times. Thus we are integrating over a box in Euclidean four-space, with boundary conditions on the (three-dimensional) walls of the box. Of course, I will eventually send both V and T to infinity. I again have reasons for this choice. (1) Certainly nothing is lost by beginning in a finite box; if the transition to infinite space goes smoothly, at worst we will have wasted a little time. (2) In some theories, we can gain information about the structure of the theory by seeing how things depend on the boundary conditions imposed on the walls of the box. For example, in a scalar field theory with spontaneous symmetry breakdown, the expectation value of the scalar field in the center of the box depends on the boundary conditions on the walls, no matter how large the box; this is one of the easiest ways to see that the theory has many vacua. (3) In the canonical quantization of the theory, it is necessary to eliminate A_0 from the action. To do this, it is necessary to find A_0 from $\partial_3^2 A_0$. In infinite space, this problem has many solutions; this ambiguity is usually resolved by applying *ad hoc* conditions on the behavior of A_0 at infinity. In a box with appropriate boundary conditions, this problem always has a unique solution.

There are many possible types of boundary conditions we could impose: we could fix some components of A_μ, some components of $F_{\mu\nu}$, some combinations of these, etc. A clue to a wise choice of boundary conditions is given by the surface term in the expression for the variation of the action. For example, for a free scalar field theory,

$$\delta S = \int d^3 S n^\mu \partial_\mu \phi \delta\phi + \cdots. \tag{3.48}$$

Here, $d^3 S$ is the element of surface area, n^μ is the normal vector to the surface, and the triple dots denote the usual volume integral of the Euler–Lagrange equations. From this expression we see that one way to make the surface terms vanish is to fix the value of ϕ on the walls of the box. Likewise, for a gauge field theory,

$$\delta S = \frac{1}{g^2} \int d^3 S n^\mu F_{\mu\nu} \delta A^\nu + \cdots. \tag{3.49}$$

From this expression we see that one way to make the surface term vanish is to fix the tangential components of A_μ on the surface. Note that there is no need to fix the normal component of A_μ; because $F_{\mu\nu}$ is antisymmetric, this makes no contribution to the surface integral.

We are not totally free to choose the tangential components of A_μ arbitrarily. Firstly, they must be chosen consistent with our gauge condition, $A_3 = 0$. Secondly, because we want to do semiclassical computations, we must choose our boundary conditions to be consistent with finiteness of the action, as the box goes to infinity. Equivalently, the boundary conditions must be consistent with the box being filled with a field configuration of a definite winding number. Furthermore, for fixed boundary conditions, this winding number is fixed, for only the tangential components of A_μ are needed to compute the normal component of G_μ. (See Eq. (3.41).)

Thus at least one relic of our boundary conditions remains no matter how large the box: we can not put an arbitrary finite-action field configuration in the box, but only one of a definite winding number. It turns out that the winding number is the *only* relic of the boundary conditions that survives as the box goes to infinity. The hand-waving argument for this is that the winding number is the *only* gauge-invariant quantity associated with the large-distance behavior of the fields. If you do not find this argument convincing, you will find a more careful one in Appendix 4.

Thus, for large boxes, we can forget about the boundary conditions in the functional integral and simply integrate over all configurations where the winding number, ν, has some definite value, n. I will denote the result of such an integration by $F(V, T, n)$. In equations,

$$F(V, T, n) = N \int [dA] e^{-S} \delta_{\nu n}. \tag{3.50}$$

where $[dA]$ denotes $[dA_1][dA_2][dA_4]$. Also, I have set $\hbar$ to one; we can always keep track of the powers of $\hbar$ by keeping track of the powers of g, as explained in Sect. 1.

$F(V, T, n)$ is a transition matrix element from some initial state to some final state (determined by our boundary conditions). What these states are will not be important to us. What is important is that for large times, T_1 and T_2,

$$F(V, T_1 + T_2, n) = \sum_{n_1 + n_2 = n} F(V, T_1, n_1) F(V, T_2, n_2). \tag{3.51}$$

This follows from Eq. (3.47), the expression for the winding number as the integral of a local density; this tells us that the way to put total winding number n in a large box is to put winding number n_1 in one part of the box and winding number n_2 in the remainder of the box, with $n = n_1 + n_2$. (Of course, such counting misses field configurations with significant action density on the boundary between the two sub-boxes, for there is no reason for the winding-number integral for each sub-box to be an integer for such configurations. However, we expect this to be a negligible surface effect for sufficiently large boxes.)

Pretty as it is, Eq. (3.51) is not what we would expect from a transition-matrix element that has a contribution from only a single energy eigenstate. Such an object would be a simple exponential, and would obey a multiplicative composition law for large times, not the convolutive composition law of Eq. (3.51). However, it is easy enough to turn convolutions into multiplications. The technique is called Fourier transformation:

$$F(V, T, \theta) \equiv \sum_n e^{in\theta} F(V, T, n)$$

$$= N \int [dA] e^{-S} e^{iv\theta}. \tag{3.52}$$

From Eq. (3.51),

$$F(V, T_1 + T_2, \theta) = F(V, T_1, \theta) F(V, T_2, \theta). \tag{3.53}$$

This is the correct composition law for a simple exponential. Thus we identify $F(V, T, \theta)$ as being (up to a normalization constant) the expectation value of e^{-HT} in an energy eigenstate, which we denote by $|\theta\rangle$ and call the θ vacuum.

$$F(V, T, \theta) \propto \langle \theta | e^{-HT} | \theta \rangle$$

$$= N' \int [dA] e^{-S} e^{iv\theta}. \tag{3.54}$$

where N' is a new normalization constant.

Our analysis has been simple and straightforward (I hope), but we have been led to a very unintuitive conclusion. Our original gauge field theory seems to have split up into a family of disconnected sectors, labeled by the angle θ, each with its own vacuum. Furthermore, in each of these sectors, the computational rules are the same as those we would have naively written down if we had not gone through any of this analysis, except that an extra term, proportional to $(F, \tilde{F})$, has been added to the Lagrangian density. Probably half the people who have played with gauge field theories have thought, at one time or another, of adding such a term, and they have discarded the possibility, because the added term is a total divergence (see Eq. (3.44)) and thus has no effect on the equations of motion and therefore 'obviously' has no effect on the physics of the theory. Of course, at this stage in our investigation, it is still possible that we have been fooling ourselves, that the extra term indeed has no effect on the physics, and that all the θ vacua we think we have discovered are simply duplicates of the same state. We shall eliminate this possibility immediately.

(I should remark that what we have done here closely parallels the treatment of a periodic potential in Sect. 2.3, except the arguments are somewhat more abstract and in a different order. The winding number is something like the total change in x (the difference between the number of instantons and the number of anti-instantons) in Sect. 2.3, and the θ vacua are something like the $|\theta\rangle$ eigenstates. The two big differences are that we found the analogs of the $|\theta\rangle$ states without pausing to talk about the analogs of the $|j\rangle$ states, and that we did the Fourier transform that untangled the energy spectrum before we saturated the functional integral with instantons. The first difference is unimportant; if I had wanted to, I could have added two extra paragraphs when I was talking about $F(V, T, n)$ and discussed the analogs of the $|j\rangle$ states. (They are called n vacua.) As for the instantons, they are the subject of the next subsection.)

3.4 Instantons: generalities

In the next subsection I shall explicitly construct instantons, finite-action solutions of the Euclidean gauge-field equations with $\nu = 1$. Most of the qualitative consequences of these solutions are independent of their detailed structure and follow merely from the fact of their existence. Therefore, in this subsection, I will simply assume that instantons exist and draw some conclusions from this assumption.

I will denote the action of an instanton by S_0. Because S_0 is finite, the instanton can not be invariant under spatial translations. Thus there exists at least a four-parameter family of instanton solutions; I will call these parameters 'the location of the center of the instanton'. The winding number is parity-odd. Thus there must also exist at least a four-parameter family of solutions with $\nu = -1$, the parity transforms of the instanton solutions, which I will call anti-instantons. Just as in Sect. 2, we can build approximate solutions consisting of n instantons and $\bar{n}$ anti-instantons, with their centers at arbitrary widely separated locations. These approximate solutions have $\nu = n - \bar{n}$.

Again as in Sect. 2, we approximate Eq. (3.54) by summing over all these configurations. Thus we obtain

$$\langle\theta|e^{-HT}|\theta\rangle \propto \sum_{n,\bar{n}} (Ke^{-S_0})^{n+\bar{n}}(VT)^{n+\bar{n}}e^{i(n-\bar{n})\theta}/(n!\,\bar{n}!)$$

$$= \exp(2KVTe^{-S_0}\cos\theta), \tag{3.55}$$

where K is a determinantal factor, defined as in Sect. 2. Thus, the energy of a θ vacuum is given by

$$E(\theta)/V = -2K \cos \theta \, e^{-S_0}. \tag{3.56}$$

Note that, as should be the case in a field theory, the different vacua are distinguished not by different energies, but by different energy densities. (Also note the similarity with the energy spectrum of a periodic potential, Eq. (2.45).)

We can go on and compute the expectation values of various operators. A particularly easy (and particularly instructive) computation is that of the expectation value of $(F, \tilde{F})$. By translational invariance,

$$\langle\theta|(F(x), \tilde{F}(x))|\theta\rangle = \frac{1}{VT} \int d^4x \, \langle\theta|(F, \tilde{F})|\theta\rangle. \tag{3.57}$$

Thus, by Eq. (3.47),

$$\langle\theta|(F, \tilde{F})|\theta\rangle = \frac{32\pi^2 \displaystyle\int [dA] \nu e^{-S} e^{i\nu\theta}}{VT \displaystyle\int [dA] e^{-S} e^{i\nu\theta}}$$

$$= -\frac{32\pi^2 i}{VT} \frac{d}{d\theta} \ln\left(\int [dA] e^{-S} e^{i\nu\theta}\right). \tag{3.58}$$

Hence there is no need to do a fresh summation over a dilute instanton–anti-instanton gas, since we have just evaluated the quantity in parentheses in Eq. (3.55). Thus in our approximation,

$$\langle\theta|(F, \tilde{F})|\theta\rangle = -64\pi^2 i K e^{-S_0} \sin \theta. \tag{3.59}$$

Some comments:

(1) The expectation value is independent of V and T, as it should be.

(2) The expectation value is an imaginary number, again as it should be. The reason is that

$$(F, \tilde{F}) = (F_{12}, F_{34}) + \text{permutations}. \tag{3.60}$$

When we continue from Euclidean space to Minkowski space, F_{12} remains F_{12}, but, just as x_4 becomes ix_0, so does F_{34} become iF_{30}. Thus, if we had obtained a real answer, we would have found that in Minkowski space (the real world) a Hermitian operator would have had an imaginary vacuum expectation value, a disaster.

(3) Both the vacuum energy density and the vacuum expectation value depend non-trivially on θ. Thus the θ-vacua are indeed all different from each other.

3.5   *Instantons: particulars*

$$\int d^4x(F, F) = \left[\int d^4x(F, F) \int d^4x(\tilde{F}, \tilde{F}) \right]^{\frac{1}{2}}$$

$$\geq \left| \int d^4x(F, \tilde{F}) \right|, \tag{3.61}$$

by the Schwartz inequality. Thus, for any winding number, we have an absolute lower bound on the action,

$$S \geq \frac{8\pi^2}{g^2} |v|. \tag{3.62}$$

Furthermore, equality is attained if and only if

$$F = \pm \tilde{F}, \tag{3.63}$$

where the positive (negative) sign holds for positive (negative) v.

This inequality was first derived by Belavin, Polyakov, Schwartz, and Tyupkin,[9] who used it to search for instantons. Their idea was to look for solutions of Eq. (3.63). If such solutions exist, they are minima of the action for fixed winding number, and thus stationary points of the action under local variations, that is to say, solutions of the field equations. Furthermore, since they have lower action than any other solutions of the same winding number (if other solutions exist), they dominate the functional integral, and, for our purposes, are the only solutions we need worry about. Finally, as a bonus, Eq. (3.63) is a first-order differential equation and considerably more tractable than the second-order field equations.

Let us begin the search with $v = 1$. We know that any field configuration with $v = 1$ can be gauge-transformed such that

$$A_\mu = g^{(1)} \partial_\mu [g^{(1)}]^{-1} + O(1/r^2), \tag{3.64}$$

where

$$g^{(1)} = \frac{x_4 + i\mathbf{x} \cdot \boldsymbol{\sigma}}{r}. \tag{3.65}$$

Equation (3.64) is rotationally invariant, in the sense that the effect of any four-dimensional rotation can be undone by an appropriate gauge transformation. This is a consequence of the statement that a rotation is a continuous deformation and thus does not change the winding number. There is also a short direct proof: Under a general rotation

$$g^{(1)} \to g g^{(1)} h^{-1}, \tag{3.66}$$

where g and h are elements of SU(2) determined by the rotation. (This is a standard formula; it is the usual way of demonstrating the isomorphism between SO(4) and SU(2)⊗SU(2).) Thus,

$$A_\mu \to g A_\mu g^{-1} + O(1/r^2).$$ (3.67)

This, as promised, can be undone by a gauge transformation, indeed, by a gauge transformation of the first kind, a constant gauge transformation.

This suggests that we search for a solution of Eq. (3.63) that is rotationally invariant in the same sense. That is to say, we make the Ansatz,

$$A_\mu = f(r^2) g^{(1)} \partial_\mu [g^{(1)}]^{-1},$$ (3.68)

where, to avoid a singularity, f must vanish at the origin. From here on it is straightforward plug-in-and-crank, which I will spare you. It turns out that we do indeed obtain a solution in this way, if

$$f = \frac{r^2}{r^2 + \rho^2},$$ (3.69)

where ρ is an arbitrary constant, called 'the size of the instanton'. The existence of solutions of arbitrary sizes is a necessary consequence of the scale invariance of the classical field theory. (This fact will occasion some embarrassment shortly.)

Once we have a solution to any field theory, we can obtain new solutions by applying the invariances of the theory. In the case at hand, these are generated by (1) scale transformations, (2) rotations, (3) the four-parameter group of spatial translations, (4) the four-parameter group of special conformal transformations, and (5) gauge transformations. Scale transformations simply change the size of the instanton; thus they just shift around the members of our one-parameter family of solutions but generate no new solutions. Rotations, as I have shown, can always be undone by gauge transformations. Spatial translations generate genuinely new solutions, and give us four more parameters, the 'location of the center of the instanton'. Although I do not have time to demonstrate it here, it turns out[18] that special conformal transformations can be undone by gauge transformations and translations.

Gauge transformations, as usual, require special consideration. It is easy to see that any non-trivial gauge transformation changes (3.68). Because $g^{(1)}$ is a function of angles only, the radial component of A_μ, A_r, vanishes. Thus, under a general non-singular gauge transformation, $g(x)$,

$$A_r \to g A_r g^{-1} + g \partial_r g^{-1} = g \partial_r g^{-1}.$$ (3.70)

Hence, if the gauge transformation is not to change A_μ, g must be independent of r. That is to say, its value everywhere must be its value at the origin; g must be a constant gauge transformation. But the only constant gauge transformation that leaves A_μ unchanged is the identity. (Remember, the effect of a constant gauge transformation is the same as that of a rotation.)

You might think that this discussion of gauge transformations is irrelevant. After all, when we do the quantum theory, we must work in a fixed gauge, such as axial gauge, and it is commonly said that once we have fixed the gauge we have no freedom to make gauge transformations. However, although commonly said, this is not strictly true; all standard gauges still allow constant gauge transformations.[19] This is as it should be. Constant gauge transformations act like ordinary symmetries; they put particles into multiplets (if there is no spontaneous symmetry breakdown), impose selection rules on scattering processes, etc. Thus, in a sensible formulation of the theory, they should remain as manifest symmetries of the Hamiltonian. Whether you accept this philosophy or not, the fact remains that constant gauge transformation applied to an instanton solution (transformed to obey the gauge conditions) will generate a different solution still obeying the gauge conditions. Thus we have found an eight-parameter family of solutions, one parameter from scale transformations, four from translations, and three from constant gauge transformations.

Are there other solutions with unit winding number? Atiyah and Ward[20] state that there are none. I can not give their proof here because I do not understand it. Nevertheless, mathematicians I trust say that their argument is not only legitimate but brilliant, so let us assume they are right and continue.

Solutions of higher winding number (if they exist) are of no interest to us. We have used approximate solutions consisting of n widely separated objects (instantons or anti-instantons) to evaluate the functional integral. These approximate solutions depend on $8n$ parameters, 8 for each object. Now suppose there are exact solutions that can be interpreted as n objects; that is to say, they depend on $8n$ (or fewer) parameters and become our approximate solutions when some of the parameters (the separations between the objects) become large. In this case, all we learn by knowing these exact solutions exist is that the dilute-gas approximation is better than we think it is – but we already know that it is good enough for our purposes. There might also be exact solutions that can not be interpreted in this way. To have a definite example, let me suppose there were a 'binstanton', a brand-new solution of winding number two. Then in evaluating the functional integral, we would have to sum over a dilute gas of instantons, anti-instantons, binstantons, and anti-binstantons. Thus,

Eq. (3.56) would be replaced by

$$E(\theta)/V = -2K \cos \theta e^{-S_0} - 2K' \cos 2\theta e^{-S'_0}, \qquad (3.71)$$

where the primed quantities are the action and determinantal factor for a binstanton. But S'_0 is twice S_0, so the new term is exponentially small compared to the old one and should be neglected.[21]

3.6   *The evaluation of the determinant and an infrared embarrassment*

We now know enough to go a long way towards explicitly evaluating the right-hand side of Eq. (3.56).

(1) S_0 is $8\pi^2/g^2$.

(2) We have an eight-parameter family of solutions and thus eight eigenmodes of eigenvalue zero in the small-vibration problem. Thus K contains a factor of $(1/\hbar^{\frac{1}{2}})^8$, or, equivalently $1/g^8$. Everything else in K is independent of $\hbar$, and thus independent of g.

(3) We have already done the integral over instanton location. The integral over constant gauge transformations is an integral over a compact group and thus gives only a constant numerical factor, the volume of SU(2). The integral over instanton sizes is potentially troublesome, since ρ can be anywhere between zero and infinity, so we will, for the moment, keep it as an explicit integral.

(4) Thus we obtain

$$E(\theta)/V = -\cos \theta \, e^{-8\pi^2/g^2} g^{-8} \int_0^\infty \frac{d\rho}{\rho^5} f(\rho M), \qquad (3.72)$$

where f is an unknown function and M is the arbitrary mass (more properly, arbitrary inverse wavelength) that is needed to define the renormalization prescription in a massless field theory. (I have avoided mentioning renormalization until now, but renormalization is essential in any computation that involves an infinite number of eigenmodes, as does this one. In Sect. 5 I will give a more detailed discussion of the ultraviolet divergences in determinantal factors and their removal by the usual one-loop renormalization counterterms.) The form of the integral is determined by dimensional analysis; an energy density has dimensions of $1/(\text{length})^4$.

(5) However, M and g are not independent parameters. Renormalization-group analysis[22] tells us that they must enter expressions for observable quantities only in the combination

$$\frac{1}{g^2} - \beta_1 \ln M + O(g^2), \qquad (3.73)$$

where β_1 is a coefficient which can be computed from one-loop perturbation theory. In the case at hand, β_1 is $11/12\pi^2$.

(6) This fixes the form of f. Thus,

$$E(\theta)/V = - A \cos \theta \; e^{-8\pi^2/g^2} g^{-8} \int_0^\infty \frac{d\rho}{\rho^5} (\rho M)^{8\pi^2\beta_1}[1 + O(g^2)],$$

(3.74)

where A is a constant independent of g, ρ, and M.

(7) To determine A requires a lot of hard work,[23] so I shall stop the calculation here. Even though we have not been able to carry things out to the end, it is remarkable how far we have been able to go with so little effort.

No doubt you have noticed that the integral we have derived is infrared-divergent. The origin of the divergence is clear from the derivation of the integral: the effective coupling constant (in the sense of the renormalization group) becomes large for large instantons, and this makes the integrand blow up. Thus the divergence is an embarrassment but not a catastrophe. It would be a catastrophe if we obtained a divergent answer in a regime in which we trusted our approximations. This is not the situation here; the divergence arises in the regime of large effective coupling constant, where all small-coupling approximations are certainly wrong. Phrased another way, the fact that the integrand has the wrong behavior for large ρ is overshadowed by the fact that it is the wrong integrand. Thus we are free to hope that strong-coupling effects (which we can not at the moment compute) introduce some sort of effective infrared cutoff in the integrand. This hope might be wrong, but it is not ruled out by anything we have done so far.

I admit that this argument is blatant hand-waving. However, it is not some new hand-waving special to instanton calculations, but the same old hand-waving that accompanies any discussion of the large-scale behavior of non-Abelian gauge field theories. For example, there is evidence that the observed hadrons are made of weakly coupled quarks. But if the quarks are weakly coupled, why can we not knock them out of the hadron? Well, in a gauge field theory the effective coupling constant grows at large distances, etc., much hand-waving, infrared slavery and quark confinement.

Everything that we have done for SU(2) can be extended straightforwardly to SU(3). To begin with, an SU(2) instanton solution can trivially be made into an SU(3) instanton solution; all that needs to be done is to say that three of the gauge fields, those associated with an

SU(2) subgroup, are of the form given, while the other five vanish. It is believed that these exhaust the set of solutions of Eq. (3.63) with unit winding number, although, unlike the SU(2) case, there is, to my knowledge, no rigorous proof of this statement. If this is indeed the case, there are only two minor differences between the SU(3) computation and the SU(2) one. (1) Instead of three parameters associated with constant gauge transformations, we have seven. (One of the eight SU(3) generators commutes with the SU(2) subgroup and does not change the solution.) Thus the factor of g^{-8} in Eq. (3.74) is replaced by one of g^{-12}. (2) β_1 has the proper value for an SU(3) gauge theory, $11/8\pi^2$.

4   The Abelian Higgs model in $1+1$ dimensions[24]

In this section I will discuss a field theory in which instanton effects drastically change the particle spectrum, the Abelian Higgs model in two-dimensional space-time.

In any number of dimensions, this is the theory of a complex scalar field with quartic self-interactions, minimally coupled to an Abelian gauge field with gauge coupling constant e, called the electric charge. In our notation, the theory is defined by the Euclidean Lagrangian density,

$$\mathscr{L} = \frac{1}{4e^2}(F, F) + D_\mu \psi^* D_\mu \psi + \frac{\lambda}{4}(\psi^* \psi)^2 + \frac{\mu^2}{2}\psi^* \psi, \qquad (4.1)$$

where λ is a positive number and μ^2 may be either positive or negative. To this must be added renormalization counterterms; however, renormalization will play no part in our computations, and, to keep things as simple as possible, I will not distinguish between bare and renormalized parameters.

Perturbation theory tells us that for weak coupling the qualitative properties of the theory depend critically on the sign of μ^2:

(1) If μ^2 is positive, the theory is simply the electrodynamics of a charged scalar meson. The mass spectrum consists of the charged meson, its antiparticle, and a massless vector meson, the photon. The force between widely separated external charges is the ordinary Coulomb force. These statements require some modification in two dimensions. Firstly, because there are no transverse directions, there is no photon. Secondly, because the Coulomb force is independent of distance, it is impossible to separate a meson and an antimeson; in contemporary argot, the charged particles are confined. The spectrum of the theory consists of a sequence of meson–antimeson bound states, rather like the spectrum of positronium, except

that these states are all stable, since they can not decay through the emission of (nonexistent) photons.

(2) If μ^2 is negative, the Higgs phenomenon takes place. In the ground state of the theory,

$$|\langle\psi\rangle|^2 = -\mu^2/\lambda \equiv a^2. \qquad (4.2)$$

The particle spectrum consists of a massive neutral scalar meson and a massive neutral vector meson. The force between widely separated external charges falls off exponentially rapidly. These statements require no modification in two dimensions.

In the remainder of this section, I will argue that the preceding sentence is a lie; contrary to the predictions of perturbation theory, the qualitative properties of the model for negative μ^2 are the same as those for positive μ^2; the two-dimensional Abelian Higgs model does not display the Higgs phenomenon. To be precise, I will show that, for negative μ^2, the theory admits instantons, and, when the effects of these instantons are taken into account, the long-range force between external charges is independent of their separations. Also, I will be able to argue, from the behavior of the long-range force, that the theory contains (confined) charged particles. There is a quantitative difference between positive and negative μ^2, though: for positive μ^2, the strength of the long-range force is independent of $\hbar$; for negative μ^2, the strength of the long-range force is exponentially small in $\hbar$, the mark of an instanton effect.

Just as in Sect. 3, we must begin the analysis by classifying classical field configurations of finite action. Of course, before doing this, we must add a constant to the Lagrangian density so the minimum of the action is zero. Thus we write

$$\mathscr{L} = \frac{1}{4e^2}(F, F) + |D_\mu\psi|^2 + \frac{\lambda}{4}(|\psi|^2 - a^2)^2. \qquad (4.3)$$

This is the sum of three positive terms. In order that the third term not make a divergent contribution to the action, it is necessary that $|\psi|$ approach a as r goes to infinity. However, there is no restriction on the phase of ψ. In equations,

$$\lim_{r\to\infty} \psi(r, \theta) = g(\theta)a, \qquad (4.4)$$

where g is a complex number of unit modulus, an element of U(1). In order that the second term not make a divergent contribution to the action, it is necessary that

$$A_\mu = g\partial_\mu g^{-1} + O(1/r^2). \qquad (4.5)$$

(Remember, in our conventions, A_μ is an imaginary field.) The first term now automatically makes a finite contribution to the action.

The lovely thing about Eq. (4.5) is that it is identical to Eq. (3.17); that is to say, the problem of classifying finite-action configurations is the baby problem of Sect. 3.2. Thus the finite-action configurations are characterized by an integer, v, the winding number, just as they are for four-dimensional gauge field theories. By Eq. (3.28), the integral expression for the winding number is

$$v = \frac{i}{4\pi} \int d^2x \, \varepsilon_{\mu\nu} F_{\mu\nu}. \tag{4.6}$$

Equivalently,

$$v = \frac{i}{2\pi} \oint A_\mu \, dx_\mu, \tag{4.7}$$

where the integral is over the circle at infinity.

Although I will not bother to explicitly display them here, it turns out that the Euclidean field equations have solutions with unit winding number, instantons, again just like four-dimensional gauge theories.[25] The only relevant difference, for our purposes, is that the Higgs model is not scale invariant; thus the instantons have a fixed size and the problems associated with integrating over scale transformations do not arise. Otherwise, though, everything is much the same as it was before, and we can copy step-by-step our earlier analysis and uncover the vacuum structure of the theory.

Thus, just as before, we have a family of θ-vacua, with energy densities given by

$$E(\theta)/L = -2Ke^{-S_0} \cos \theta. \tag{4.8}$$

Here L is the volume of (one-dimensional) space, S_0 is the action of an instanton, and K is a determinantal factor. Also, by copying the derivation of Eq. (3.59), we find that

$$\langle \theta | \varepsilon_{\mu\nu} F_{\mu\nu} | \theta \rangle = 8\pi K e^{-S_0} \sin \theta. \tag{4.9}$$

As before, this has the right reality properties; when we continue to Minkowski space, we pick up a factor of i that cancels the factor of i in our definition of A_μ. We see from this equation that the θ-vacua are characterized by a constant expectation value of the electric field F_{01}. In two dimensions, unlike four, such a constant 'background field' is not in conflict with Lorentz invariance.[26]

Now that we understand the vacuum structure, let us compute the force between widely separated external charges. To be more precise, let us introduce into the system two static charges of equal magnitude, q, and opposite sign, separated by a distance L', and let us compute (for large L') Δ, the change in the energy of a θ-vacuum caused by these charges. The standard method of computing Δ uses Wilson's loop integral,[27]

$$W = \exp\left(-\frac{q}{e} \oint A_\mu \, dx_\mu \right), \tag{4.10}$$

where the integration is over the rectangular path shown in Fig. 14. According to Wilson, the vacuum energy shift is given by

$$\Delta = -\lim_{T' \to \infty} \frac{1}{T'} \ln \langle \theta | W | \theta \rangle. \tag{4.11}$$

In our case,

$$\langle \theta | W | \theta \rangle = \frac{\int [dA][d\psi^*][d\psi] W e^{-S} e^{i\nu\theta}}{\int [dA][d\psi^*][d\psi] e^{-S} e^{i\nu\theta}}, \tag{4.12}$$

and our task is to compute these two functional integrals in our standard dilute-gas approximation, for large L' and T' (and, of course, for even larger L and T, the spatial and temporal extent of the universe). In Eq. (4.8) we have already calculated the denominator. To calculate the numerator, let us divide the sum over instantons and anti-instantons into two independent sums: one over objects lying inside the loop and one over objects lying outside the loop. By this division we neglect contributions coming from configurations in which instantons and anti-instantons overlap the loop, but, for large L, T, L', and T', this is a very small portion of the available configurations and can reasonably be neglected. (Of course, if our calculation gives zero for its answer, then these configura-

Fig. 14

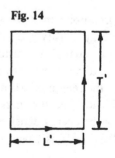

tions will be the most important ones and we will have to go back and compute them.) The functional integrand splits neatly into the product of an 'outside' term and an 'inside' term: $S = S^{\text{outside}} + S^{\text{inside}}$, $v = v^{\text{outside}} + v^{\text{inside}}$, while

$$W = \exp(2\pi i q v^{\text{inside}} / e). \tag{4.13}$$

Thus, for the outside objects, we have the same sum as for the denominator, except that the available volume of Euclidean two-space is not LT but $LT - L'T'$. For the inside objects, we also have the same sum, except that the available volume is $L'T'$, and θ is replaced by $\theta + 2\pi q/e$.

Thus,

$$\ln\langle\theta|W|\theta\rangle = 2Ke^{-S_0}[(LT - L'T')\cos\theta + L'T'\cos(\theta + 2\pi q/e) - LT\cos\theta], \tag{4.14}$$

where the first term comes from the outside sum, the second from the inside sum, and the third from the denominator. Hence,

$$\Delta = 2L'Ke^{-S_0}[\cos\theta - \cos(\theta + 2\pi q/e)]. \tag{4.15}$$

This is proportional to L', the separation between the external charges; thus there is a constant force between external charges at large separation. As announced, there is no quantitative difference between positive and negative μ^2. However, there is a qualitative difference. For positive μ^2, the strength of the force is proportional to q^2 for small $\hbar$; for negative μ^2, it is exponentially small in $\hbar$. (Remember, if we had not chosen our units so $\hbar$ was one, S_0 would have been $S_0/\hbar$.)

There is a simple physical interpretation of this result. For small θ and small q/e,

$$\Delta = L'Ke^{-S_0}[(\theta + q/e)^2 - \theta^2], \tag{4.16}$$

$$E(\theta) = LKe^{-S_0}\theta^2 + \text{constant}, \tag{4.17}$$

and,

$$\langle\theta|F_{12}|\theta\rangle = 4\pi Ke^{-S_0}\theta. \tag{4.18}$$

These expressions have an obvious interpretation: In a θ-vacuum, there is a background electric field, and an energy density proportional to the square of this field. Because we are in one spatial dimension, the external charges act like condenser plates in three dimensions; they induce a constant field proportional to their charge in the region between them, which is added to the pre-existing background field. Thus the energy shift is the separation multiplied by the difference of the energy density of the new field and that of the old. Equation (4.14) is just this trivial picture complicated by nonlinear terms in the expression for the energy density as a function of the field.

One aspect of these nonlinear complications is of physical import: Eq. (4.14) is periodic in q with period e. This is explicable if the theory contains charged particles of charge e. If this is the case, there is a process that can change the charge on our condenser plates by $\pm e$: a particle–antiparticle pair can materialize in the region between the plates, and the particle can fly to one plate and the antiparticle to the other. This process will occur whenever it is energetically favorable. For sufficiently large L', this is equivalent to saying that it will occur whenever it lowers the energy density, because the energetic cost of making a pair is independent of L', and the energetic gain of lowering the energy density is proportional to L'. Thus qs that are equal modulo e lead to identical physics; no matter which one you start out with, pairs are made until the charge on the plates reaches its optimum value, the one that gives minimum energy density.

What if we were to do a parallel computation in a four-dimensional gauge field theory, with non-Abelian external charges? Would we also obtain a force independent of separation? Alas, we would not. There is an L' in Eq. (4.15) because there is an $L'T'$ in Eq. (4.14), that is to say, because even an instanton deep within the loop has a non-negligible effect on the loop integral. This is precisely what does not happen in four dimensions. At large distances from an instanton, A_μ is $g\partial_\mu g^{-1}$, plus terms that fall off far too rapidly to affect the loop integral. However, the loop integral is gauge-invariant, and we can always gauge-transform g such that it is constant everywhere except within a small cone emerging from the instanton perpendicular to the plane of the loop. Whatever confines quarks, it is not instantons.

5 't Hooft's solution of the U(1) problem

5.1 *The mystery of the missing meson*

The U(1) problem is an apparent contradiction between two pieces of accepted wisdom. One is wisdom of the 1970s, that hadronic physics is quantum chromodynamics. The other is wisdom of the 1960s, that hadronic physics is approximately invariant under chiral SU(2)⊗SU(2). Let me remind you of the meaning of these two propositions.

Quantum chromodynamics is a field theory whose dynamical variables are an octet of SU(3) gauge fields and a family of SU(3) triplet Dirac bispinor fields, called quarks. In Minkowski space, the Lagrangian density is

$$\mathcal{L} = -\frac{1}{4g^2}(F_{\mu\nu}, F^{\mu\nu}) + \sum_f \bar{\psi}_f(iD_\mu\gamma^\mu - m_f)\psi_f, \qquad (5.1)$$

where f, called the flavour index, labels the various triplets. The usual exact and approximate symmetries of hadron physics (charge, isospin, Gell-Mann's SU(3), etc.) act only on the flavour indices; all physical hadrons are supposed to be singlets under the gauge group. (This last statement is sometimes called quark confinement; it is still far from proved, although there are some suggestive arguments.) ψ_1 and ψ_2 form an isodoublet, the non-strange quarks; ψ_3 is the strange quark; ψ_4 is the charmed quark; there may or may not be additional flavors.

Chiral SU(2)$\otimes$SU(2) is the group generated by the strangeness-conserving weak-interaction currents and their parity transforms. Its diagonal subgroup is conventional isospin. This group is very close to being an exact symmetry of the strong interactions; it is a much better symmetry than SU(3) and roughly as good a symmetry as isospin. However, were this symmetry to be exact, only the isospin subgroup would be a manifest symmetry; the remainder of the group would be a Nambu–Goldstone symmetry, with three massless Goldstone bosons, the pions. The smallness of the pion mass (on a hadronic mass scale) is a measure of the goodness of the symmetry. This is the picture that stands in back of all the stunningly successful soft-pion computations of the mid 1960s.

Now for the apparent contradiction. In quantum chromodynamics, the limit of perfect SU(2)$\otimes$SU(2) symmetry is the limit in which the non-strange quarks are massless. In this limit, the Lagrangian (4.1) obviously has a further chiral U(1) symmetry; it is invariant under

$$\psi_f \rightarrow e^{-i\alpha\gamma_5}\psi_f, \quad (f=1,2) \qquad (5.2)$$

where α is a real number. The associated conserved current is

$$j_\mu^5 = \sum_{f=1}^{2} \bar{\psi}_f\gamma_\mu\gamma_5\psi_f. \qquad (5.3)$$

I emphasize that the appearance of this additional chiral symmetry is very special to quantum chromodynamics; for example, the σ model has no such additional symmetry in the chiral limit.

Now, either this additional symmetry is manifest or it is spontaneously broken. If it were manifest, all non-massless hadrons would occur in parity doublets. This is not the case; thus it must be spontaneously broken. But if it is spontaneously broken, Goldstone's theorem tells us there must be an associated isoscalar pseudoscalar Goldstone boson. This is the U(1) problem: *what happened to the fourth Goldstone boson?*

One's first thought is that the missing meson is the eta, but this is wrong. The chiral U(1) symmetry is broken by the same mass term that breaks chiral SU(2)⊗SU(2), and thus the fourth Goldstone boson should have roughly the same mass as the pions. The eta is far too heavy. This can be made more precise: using conventional soft-pion methods, Weinberg[28] has shown that a U(1) Goldstone boson must have a mass less than $\sqrt{3}m_\pi$. The eta grossly disobeys this inequality. Also, if we consider the approximation in which the strange quark mass also vanishes, and in which we have perfect chiral SU(3)⊗SU(3) symmetry, the eta takes its place with the pions in an octet of Goldstone bosons. But in this limit we still have an additional U(1) symmetry and we still have a missing meson.

(This should be all that I need to say about the eta. However, there is some confusion abroad on this point, and thus I emphasize that *there is no connection between the eta and the* U(1) *problem*. The eta is a red herring; it is just another hadron; it is no more a relic of a U(1) Goldstone boson than is the N**.)

It may seem that I have posed an insoluble problem; this is because I have lied to you. In fact, j_μ^5 is not a conserved current; it is afflicted with the famous Adler–Bell–Jackiw anomaly.[29] In the limit of N massless quarks,

$$\partial^\mu j_\mu^5 = \frac{N}{32\pi^2}\, \varepsilon^{\mu\nu\lambda\sigma}(F_{\mu\nu}, F_{\lambda\sigma}). \tag{5.4}$$

(Note the similarity between the right-hand side of this equation and the Pontryagin density. This will be important to us later.)

You might think that this is the end of the story; if the current is not conserved, there is no U(1) symmetry to worry about. Alas, life is not so simple. In Sect. 3, we showed that the Euclidean counterpart of the right-hand side of Eq. (5.4) could be written as the divergence of a (gauge-variant) function of A_μ and $F_{\mu\nu}$. It is easy to see that the same construction works in Minkowski space. Thus, if we define

$$J_\mu^5 = j_\mu^5 - \frac{N}{16\pi^2}\, \varepsilon_{\mu\nu\lambda\sigma}(A^\nu, F^{\lambda\sigma} - \tfrac{2}{3}A^\lambda A^\sigma), \tag{5.5}$$

this current is gauge-variant but conserved.

If we work in a covariant gauge (and why should we not?), the added term commutes with the quark fields at equal times. Thus we can derive, for Green's functions made of one J_μ^5 and a string of gauge-invariant quark multilinears, chiral U(1) Ward identities of the usual form. And since these are of the usual form, they lead to the usual conclusion: chiral U(1) is a symmetry; either Green's functions made of quark multilinears alone are U(1) symmetric, or there are Goldstone poles in Green's functions for one J_μ^5 and a string of quark multilinears.

Is there no way out? Well, there is one. The Hilbert space of a gauge field theory quantized in a covariant gauge is notoriously full of negative-norm timelike photons and similar gauge phantoms, states that never couple to gauge-invariant operators. Could it be that the Goldstone boson is such a phantom? No, this is not possible; the formulation of the question is wrong. If the Goldstone boson does not couple at all to gauge-invariant operators, it cannot produce a pole in a Green's function for one J_μ^5 and a string of gauge-invariant operators.

The proper formulation of the question was found by Kogut and Susskind,[30] who had the bright idea of looking at the Schwinger model, massless spinor electrodynamics, in $1 + 1$ dimensions in a covariant gauge. The Schwinger model is an exactly soluble theory that has properties very close to those we have been discussing. In particular, there is a gauge-invariant axial current with an anomalous divergence and a gauge-variant conserved axial current, and, most important, there is chiral symmetry breakdown without Goldstone poles in gauge-invariant Green's functions. What Kogut and Susskind found in the covariant-gauge Schwinger model were two free massless fields, ϕ_+ and ϕ_-. ϕ_+ creates quanta of positive norm and has the usual propagator; ϕ_- creates quanta of negative norm and has minus the usual propagator. (Remember, a covariant gauge is full of negative-norm states from the very beginning.) All gauge-invariant quantities couple to the sum of these fields, $\phi_+ + \phi_-$; this has zero propagator and produces no singularities. Thus gauge-invariant Green's functions are free of Goldstone poles. However, the gauge-variant conserved current couples to the gradient of the difference, $\partial_\mu(\phi_+ - \phi_-)$. Thus, when one considers a Green's function for one gauge-variant current and a string of gauge-invariant fields, the relative minus sign in the coupling cancels the relative minus sign in the propagators, and Goldstone poles appear where they should. This set-up is called a Goldstone dipole. (The terminology is a bit misleading, because there are only single poles in Green's functions, but I shall stick with it anyway.)

Thus according to Kogut and Susskind, the proper formulation of our question is, is the U(1) symmetry of quantum chromodynamics spontaneously broken via a Goldstone dipole? You might think that this is a question that could be asked seriously only by a field theorist driven mad by spending too many years in too few dimensions. Nevertheless, as 't Hooft[9] brilliantly showed, the answer is yes. The remainder of this section is an explanation of his computation.

5.2   *Preliminaries: Euclidean Fermi fields*

Before we can treat quantum chromodynamics by functional integration, we must know how to integrate over Euclidean Fermi fields. This section is a description of the theory of such integration, with all mathematical fine points ruthlessly suppressed.[31] I will develop the theory by defining Fermi integration as a 'natural' generalization of Bose integration. At the end, I will justify my definitions by showing that they lead to formulae equivalent to those obtained by conventional canonical quantization.

Let us begin by defining our integration variables. For Bose theories, we integrate over *c*-number Euclidean fields. These are objects that commute with each other at arbitrary separations; they can be thought of as the classical (vanishing *ħ*) limit of quantum Bose fields. This suggests that the proper variables for a Fermi theory should be classical Fermi fields, objects which *anticommute* with each other at arbitrary separations. Thus, for example, for the theory of a single Dirac field, we would expect our integration variables to be two Euclidean bispinors, $\bar{\psi}$ and ψ, obeying

$$\{\psi(x), \psi(y)\} = \{\bar{\psi}(x), \bar{\psi}(y)\} = \{\psi(x), \bar{\psi}(y)\} = 0, \tag{5.6}$$

for all Euclidean points *x* and *y*.

The last of these relations is crucial, for it implies that $\bar{\psi}$ can not be in any sense the adjoint of ψ times some matrix. For if this were so, the last relation (multiplied by the inverse matrix) would state that the sum of two positive semi-definite objects, $\psi\psi^\dagger$ and $\psi^\dagger\psi$, was zero. This would only be possible if ψ vanished, not a happy situation for a prospective integration variable. Thus if we are to have any hope of founding a sensible integration theory, we must treat ψ and $\bar{\psi}$ as *totally independent variables*.

This independence is the main novelty of Euclidean Fermi fields; the rest of the construction is straightforward. We define the Euclidean *γ*-matrices to be four Hermitian matrices obeying

$$\{\gamma_\mu, \gamma_\nu\} = 2\delta_{\mu\nu}. \tag{5.7}$$

We use these to define the O(4) transformation law for ψ in the usual way, and define $\bar{\psi}$ to transform like the adjoint of ψ. We define γ_5, a Hermitian matrix, by

$$\gamma_5 = \gamma_1\gamma_2\gamma_3\gamma_4. \tag{5.8}$$

Thus, $\bar{\psi}\psi$ is a scalar, $\bar{\psi}\gamma_5\psi$ a pseudoscalar, $\bar{\psi}\gamma_\mu\psi$ a vector, etc.

The Euclidean action for a free Dirac field is

$$S = -\int d^4x\, \bar{\psi}(i\partial_\mu\gamma_\mu - im)\psi. \tag{5.9}$$

The minus sign is pure convention; we could always absorb it into ψ if we wanted to. (Remember, we are free to transform ψ without touching $\bar{\psi}$.) The i in front of the mass term is not conventional. It is there to insure that the Euclidean propagator is proportional to $(\not{p}+im)/(p^2+m^2)$; if it were not for the i, we would have tachyon poles. If m vanishes, Eq. (5.9) is invariant under chiral transformations,

$$\psi \rightarrow e^{-i\alpha\gamma_5}\psi, \quad \bar{\psi} \rightarrow \bar{\psi}e^{-i\alpha\gamma_5}. \tag{5.10}$$

The quark part of the Euclidean action for quantum chromodynamics is obtained from Eq. (5.9) by replacing ordinary derivatives by covariant derivatives.

So much for the integrand; now for the integration. For Bose fields, we defined functional integration as iterated integration over ordinary numbers. Therefore, let us begin by defining integration for a function of a single anticommuting quantity, a. (Of course, for a single quantity, the anticommutation algebra degenerates to a single equation, $a^2 = 0$.)

We want to define

$$\int da f(a), \tag{5.11}$$

for an arbitrary function, f. We want this to have the usual linearity property: the integral of a linear combination of two functions should be the linear combination of the integrals. In addition, we would like the integral to be translation-invariant

$$\int da f(a+b) = \int da f(a), \tag{5.12}$$

where b is an arbitrary anticommuting quantity. I will now show that these conditions determine the integral, up to a normalization factor.

The reason is that there are only two linearly independent functions of a, 1 and a; all higher powers vanish. We will choose our normalization such that

$$\int da\, a = 1. \tag{5.13}$$

From this, and Eq. (5.12),

$$\int da\, 1 = 0. \tag{5.14}$$

For functions of many anticommuting variables, we define multiple integrals as iterated single integrals. Thus, for example, a complete integration table for the four linearly independent functions of two anticommuting variables, a and $\bar{a}$, is

$$\int da\, d\bar{a} \left\{ \begin{matrix} \bar{a}a \\ \bar{a} \\ a \\ 1 \end{matrix} \right\} = \left\{ \begin{matrix} 1 \\ 0 \\ 0 \\ 0 \end{matrix} \right\}. \tag{5.15}$$

As an application of this table, I will evaluate

$$\int da\, d\bar{a}\, e^{\lambda \bar{a}a} = \int da\, d\bar{a}(1 + \lambda \bar{a}a)$$

$$= \lambda. \tag{5.16}$$

We can now define integration over Fermi fields exactly as we defined integration over Bose fields in Sect. 2. We introduce two arbitrary complete orthonormal sets of c-number functions, ψ_r and $\bar{\psi}_r$,

$$\int d^4x\, \psi_r^\dagger \psi_s = \int d^4x \bar{\psi}_r \bar{\psi}_s^\dagger = \delta_{rs}. \tag{5.17}$$

We expand the Fermi fields in terms of these functions,

$$\psi = \sum_r a_r \psi_r, \quad \bar{\psi} = \sum_r \bar{a}_r \bar{\psi}_r, \tag{5.18}$$

and define

$$[d\psi][d\bar{\psi}] = \prod_r da_r\, d\bar{a}_r. \tag{5.19}$$

As an application let me evaluate

$$\int [d\psi][d\bar{\psi}] e^{-S}, \tag{5.20a}$$

where

$$S = -\int d^4x\, \bar{\psi} A \psi, \tag{5.20b}$$

and A is some linear operator, possibly depending on external c-number fields. For simplicity, let me assume that A commutes with $A^\dagger$. (This is the case for a quark in an external gauge field.) Then we can choose the ψ_rs to be the eigenfunctions of A,

$$A\psi_r = \lambda_r \psi_r, \tag{5.21}$$

and we can choose $\bar{\psi}_r$ to be $\psi_r^\dagger$. Thus

$$S = -\sum_r \lambda_r \bar{a}_r a_r, \tag{5.22}$$

and

$$\int [d\psi][d\bar{\psi}] e^{-S} = \prod_r \lambda_r$$

$$= \det A. \tag{5.23}$$

Note that this is the inverse of the answer we would have obtained had we done the identical integral with ψ and $\bar{\psi}$ complex Bose fields.

I will now show that Eq. (5.22) is the correct answer, that it is identical to the normal field-theoretic expression for the vacuum-to-vacuum transition amplitude in a theory of a quantized Dirac field interacting with external c-number fields. In this theory, this amplitude is the sum of all Feynman graphs with no external Fermi lines. This in turn is the exponential of the sum of all connected (that is to say, one-loop) graphs. Now, if ψ were a Bose field, we know that the amplitude would be the inverse determinant, because we trust functional integration for Bose fields. But the only effect of replacing bosons by fermions is to multiply the one-loop graphs by minus one. This inverts the exponential of the one-loop graphs, that is to say, it turns the inverse determinant into the determinant.

In any theory in which the Fermi fields enter the action at most bilinearly, we can always integrate over the Fermi fields, using Eq. (5.23), before we integrate over the Bose fields. In diagrammatic language, we can always sum the Fermi loops before we integrate over virtual bosons. Thus, because our definition of Fermi integration gives the right answer for a Dirac field in an external c-number field, it also gives the right answer for a Dirac field interacting with a quantum Bose field. In particular, it gives the right answer for quantum chromodynamics.

5.3   Preliminaries: chiral Ward identities

In this section is a discussion of the chiral Ward identities for a theory of a set of quantum Dirac fields interacting with c-number gauge fields. In the sequel, we shall use these identities in several different cases; thus it is useful to have them written down in their most general form, at hand when we need them.

Let ψ be a set of Euclidean Dirac fields, assembled into a big vector, which transforms according to some representation of SU(n), not necessarily irreducible, generated by a set of matrices, T^a. Let us define the constant C by

$$\mathrm{Tr}\, T^a T^b = - C\delta^{ab}. \tag{5.24}$$

Thus, for example, for a set of N fields each transforming according to the n-dimensional representation of SU(n),

$$C = N/2. \tag{5.25}$$

We wish to study the theory of these fields interacting with given c-number gauge fields,

$$S = -\mathrm{i} \int \mathrm{d}^4x \ \bar{\psi}(\gamma_\mu D_\mu - M)\psi, \tag{5.26}$$

where D_μ is the covariant derivative defined by Eq. (3.16), and M is the mass matrix for the Dirac fields, assumed to be SU(n)-invariant. Let $\phi^{(r)}$, $r = 1 \ldots m$, be a set of local multilinear functions of the Dirac fields. The Euclidean Green's functions for these objects are defined by

$$\langle \phi^{(1)}(x_1) \ldots \phi^{(m)}(x_m) \rangle^A = \frac{\int [d\psi][d\bar{\psi}] e^{-S} \phi^{(1)}(x_1) \ldots \phi^{(m)}(x_m)}{\int [d\psi][d\bar{\psi}] e^{-S}}$$

(5.27)

where I have inserted the superscript A to remind you that we are working in an external gauge field.

Now let us perform an infinitesimal change of variables in the numerator of Eq. (5.27),

$$\delta\psi = -i\gamma_5 \psi \delta\alpha, \quad \delta\bar{\psi} = -i\bar{\psi}\gamma_5 \delta\alpha,$$ (5.28a)

where $\delta\alpha$ is an infinitesimal function of Euclidean space. Since the ϕs are functions of the Dirac fields, they will change under the change of variables; we define $\partial\phi^{(r)}/\partial\alpha$ by

$$\delta\phi^{(r)} = (\partial\phi^{(r)}/\partial\alpha)\delta\alpha.$$ (5.28b)

Thus, for example, $\partial\bar{\psi}\psi/\partial\alpha$ is $-2i\bar{\psi}\gamma_5\psi$. A change of variables does not change the integral; thus, taking the variational derivative with respect to $\delta\alpha$, we find

$$\partial^\mu \langle j_\mu^5(y)\phi^{(1)}(x_1) \ldots \phi^{(m)}(x_m) \rangle^A$$
$$+ \langle \bar{\psi}M\gamma_5\psi(y)\phi^{(1)}(x_1) \ldots \phi^{(m)}(x_m) \rangle^A$$
$$+ \delta^{(4)}(y-x_1)\langle \partial\phi^{(1)}(x_1)/\partial\alpha \ldots \phi^{(m)}(x_m) \rangle^A$$
$$+ \cdots$$
$$+ \delta^{(4)}(y-x_m)\langle \phi^{(1)}(x_1) \ldots \partial\phi^{(m)}(x_m)/\partial\alpha \rangle^A = 0,$$ (5.29)

where j_μ^5 is $\bar{\psi}\gamma_\mu\gamma_5\psi$.

These are, of course, just the Euclidean version of the Ward identities we would have obtained in Minkowski space by studying the divergence of j_μ^5, and, of course, they are wrong, for they take no account of the Adler–Bell–Jackiw anomaly. I do not have the time here to recapitulate the theory of the anomaly, and I will simply state the correct version of Eq. (5.29): the zero on the right-hand side is replaced by

$$-\frac{iC}{8\pi^2}(F(y), \tilde{F}(y))\langle \phi^{(1)}(x_1) \ldots \phi^{(m)}(x_m) \rangle^A.$$ (5.30)

We can obtain a very useful equation by integrating the corrected Ward identity over y. The first term on the left vanishes by integration by parts; the theory contains no massless particles that could give a non-vanishing surface term. Also, on the right we can use

$$\int d^4y(F, \tilde{F}) = 32\pi^2 v.$$ (3.47)

Thus we obtain

$$2 \left\langle \int d^4y \, \bar{\psi}M\gamma_5\psi(y)\phi^{(1)}(x_1) \dots \phi^{(m)}(x_m) \right\rangle^A$$

$$+ \frac{\partial}{\partial \alpha} \langle \phi^{(1)}(x_1) \dots \phi^{(m)}(x_m)\rangle^A$$

$$= -4iCv\langle\phi^{(1)}(x_1) \dots \phi^{(m)}(x_m)\rangle^A.$$ (5.31)

Now all our artillery is at the ready; we can begin our assault on quantum chromodynamics.

5.4   QCD (*baby version*)

I will begin by analyzing a baby version of quantum chromo-dynamics, in which the gauge group is SU(2), and in which there is only a single isodoublet quark, of mass zero. In equations,

$$S = \int d^4x \left[\frac{1}{4g^2}(F, F) - i\bar{\psi}D_\mu\gamma_\mu\psi\right].$$ (5.32)

After we have worked out the baby theory, we will go on to the real thing.

Most of the analysis of Sect. 3 is essentially unaltered by the presence of a quark. In particular, all of our old instanton solutions are still solutions of the Euclidean equations of motion (with the quark fields set equal to zero). Thus we still have all the θ-vacua, and formulae like

$$E(\theta)/V = -2K \cos \theta \, e^{-S_0},$$ (3.56)

and

$$\langle\theta|(F, \tilde{F})|\theta\rangle = -64\pi^2iKe^{-S_0} \sin \theta,$$ (3.59)

remain unaltered. The only effect of the quarks is to insert into the definition of K a term proportional to

$$\det\left[\frac{i\slashed{D}}{i\slashed{\partial}}\right] = \det\left[\frac{i(\partial_\mu + A_\mu)\gamma_\mu}{i\partial_\mu\gamma_\mu}\right],$$ (5.33)

where A_μ is the field of an instanton.

This is a trifling alteration, but it is a tremendous trifle, for, as we shall see, $i\slashed{D}$ has a vanishing eigenvalue. Thus the determinant vanishes, as does $E(\theta)/V$ and $\langle\theta|(F, \tilde{F})|\theta\rangle$!

The vanishing eigenvalue can be demonstrated either by a short explicit computation or by a long indirect argument. I will choose the second method. Despite what you might think, this is not a perverse choice. (Well, not totally perverse.) The indirect argument will have some byproducts that will be very useful to us later.

For simplicity, I will assume (falsely) that $i\slashed{D}$ has a purely discrete spectrum,[32]

$$i\slashed{D}\psi_r = \lambda_r\psi_r. \tag{5.34}$$

Because $i\slashed{D}$ is Hermitian, all the λs are real. Because γ_5 anticommutes with γ_μ,

$$i\slashed{D}\gamma_5\psi_r = -\lambda_r\gamma_5\psi_r. \tag{5.35}$$

Thus non-vanishing eigenvalues always occur in pairs of opposite sign. Eigenfunctions of vanishing eigenvalue, on the other hand, can always be chosen to be eigenfunctions of γ_5,

$$\gamma_5\psi_r = \chi_r\psi_r, \quad (\lambda_r = 0) \tag{5.36}$$

Because $\gamma_5^2 = 1$, $\chi_r = \pm 1$. I will denote the number of eigenfunctions of these two types by $n_\pm$.

I will now prove the remarkable sum rule,[33]

$$n_- - n_+ = \nu. \tag{5.37}$$

Thus, not only is there a zero eigenvalue in the field of an instanton, there is a zero eigenvalue in *any* gauge field of non-zero winding number, whether or not it is a solution of the Euclidean equations of motion.

The proof rests on the chiral Ward identities for the quantum theory of a *massive* quark interacting with an external gauge field.

$$S = -i \int d^4x\, \bar{\psi}(\slashed{D} - m)\psi. \tag{5.38}$$

If we take the case of *no* ϕs, Eq. (5.31) becomes

$$-2i\nu = 2\left\langle \int d^4y\, \bar{\psi}m\gamma_5\psi \right\rangle^A$$

$$= \frac{2 \int [d\psi][d\bar{\psi}]e^{-S} \int d^4y\, \bar{\psi}m\gamma_5\psi}{\int [d\psi][d\bar{\psi}]e^{-S}}. \tag{5.39}$$

(Remember, in the case at hand, $C = \frac{1}{2}$.) To evaluate the functional integrals, we need the eigenfunctions and eigenvalues of $i(\slashed{D} - m)$. The eigenfunctions are those of $\slashed{D}$, and the eigenvalues are simply shifted by $-im$,

$$i(\slashed{D} - m)\psi_r = (\lambda_r - im)\psi_r. \tag{5.40}$$

If we expand the fields in the ψ_rs, the functional integrals become trivial, and we obtain

$$-2i\nu = \frac{2m \sum_r \int d^4y \, \psi_r^\dagger \gamma_5 \psi_r \prod_{s \neq r} (\lambda_s - im)}{\prod_r (\lambda_r - im)}$$

$$= 2m \sum_r \int d^4y \, \psi_r^\dagger \gamma_5 \psi_r (\lambda_r - im)^{-1}. \qquad (5.41)$$

Because eigenfunctions of a Hermitian operator with different eigenvalues are orthogonal,

$$\int d^4y \, \psi_r^\dagger \gamma_5 \psi_r = 0 \quad \text{if } \lambda_r \neq 0, \qquad (5.42)$$

while

$$\int d^4y \, \psi_r^\dagger \gamma_5 \psi_r = \chi_r \quad \text{if } \lambda_r = 0. \qquad (5.43)$$

Thus,

$$-2i\nu = 2i(n_+ - n_-). \qquad (5.44)$$

This is the desired result.

It turns out that the instanton obeys the sum rule by having one eigenfunction of vanishing eigenvalue with $\chi = -1$ and none with $\chi = +1$. (This also can be seen indirectly, without dirtying one's hands with explicit computations; see Appendix 5.) We shall never need the explicit form of the eigenfunction, but, just for completeness, I shall write it down here. For an instanton with center at X and size ρ,

$$\psi_0(x - X, \rho) = \rho[\rho^2 + (x - X)^2]^{-\frac{3}{2}}u, \qquad (5.45)$$

where u is a constant spinor. Likewise, for an anti-instanton, there is one eigenfunction of vanishing eigenvalue with $\chi = +1$, the parity transform of Eq. (5.45). For n widely separated instantons and anti-instantons, there are n such eigenfunctions, one centered about each object. (More properly, I should say that there are n approximate eigenfunctions with approximately vanishing eigenvalues, but, for the dilute-gas approximation, the qualifications are irrelevant.)

What is important for our purposes is that the sum rule implies that any field configuration with non-vanishing winding number has at least one eigenfunction of vanishing eigenvalue and thus a vanishing Fermi determinant. Thus, not just in the dilute gas approximation, but to *all* orders in the semiclassical expansion, all the θ vacua have the same energy and they all have a vanishing expectation value for $(F, \tilde{F})$.

A phenomenon this general must have a deep cause. We can discover this cause if we consider the chiral Ward identities for vanishing quark mass. There is a technical obstacle to this; for vanishing quark mass, the denominator in Eq. (5.27) vanishes, at least for fields with $v \neq 0$. This is easily surmounted; we define denominator-free Green's functions,

$$\langle\langle \phi^{(1)}(x_1) \ldots \rangle\rangle^A \equiv \int [d\psi][d\bar{\psi}] e^{-S} \phi^{(1)}(x_1) \ldots . \qquad (5.46)$$

By the same reasoning as before, these obey the Ward identities,

$$\left[\frac{\partial}{\partial \alpha} + 2iv \right] \langle\langle \phi_1(x_1) \ldots \rangle\rangle^A = 0, \qquad (5.47)$$

i.e. Eq. (5.31) without the mass term. The Green's functions of our baby version of chromodynamics are given by

$$\langle \theta | \phi^{(1)}(x_1) \ldots | \theta \rangle = \frac{\int [dA] e^{-S_g} e^{iv\theta} \langle\langle \phi^{(1)}(x_1) \ldots \rangle\rangle^A}{\int [dA] e^{-S_g} e^{iv\theta} \langle\langle 1 \rangle\rangle^A}, \qquad (5.48)$$

where S_g is the gauge-field part of the action. By Eq. (5.47),

$$\left[\frac{\partial}{\partial \alpha} + 2 \frac{\partial}{\partial \theta} \right] \langle \theta | \phi^{(1)}(x_1) \ldots | \theta \rangle = 0. \qquad (5.49)$$

Thus, the effect of a chiral U(1) transformation can be undone by a change of θ. That is to say, chiral U(1) transformations turn one θ-vacuum into another; chiral U(1) symmetry is spontaneously broken, and the θ-vacua are the many vacua that appear when a symmetry suffers spontaneous breakdown. This is startling; after all, when we first met the θ-vacua in Sect. 3, they had no connection with chiral symmetry – there was no chiral symmetry for them to be connected with! Nevertheless, it is an inevitable result of our analysis, and it explains why all the θ-vacua have the same energy density and the same expectation value of $(F, \tilde{F})$; it is because these quantities are chiral U(1) invariants.

(Parenthetical remark: the factor of 2 in Eq. (5.49) is worth comment. It tells us that when we make a chiral rotation by π we return to the same θ-vacuum. This is as it should be.

$$e^{-i\pi\gamma_5} = -1. \qquad (5.50)$$

Thus a chiral rotation by π has the same effect on the fields as a spatial rotation by 2π; we would be very unhappy if this symmetry suffered spontaneous breakdown.)

There is one possible loophole in the argument I have given. It remains a logical possibility that, for every Green's function, the derivative with respect to α and the derivative with respect to θ both vanish. If this happened, we would have, not spontaneous symmetry breakdown, but manifest symmetry, and the θ-vacua would be mathematical artifacts, superfluous duplicates of a single vacuum.

I will now eliminate this possibility by computing, in the dilute-gas approximation

$$\langle\theta|\sigma_{\pm}(x)|\theta\rangle = \frac{\int [\mathrm{d}A][\mathrm{d}\psi][\mathrm{d}\bar{\psi}]\mathrm{e}^{-S}\mathrm{e}^{\mathrm{i}\nu\theta}\sigma_{\pm}(x)}{\int [\mathrm{d}A][\mathrm{d}\psi][\mathrm{d}\bar{\psi}]\mathrm{e}^{-S}\mathrm{e}^{\mathrm{i}\nu\theta}}, \qquad (5.51)$$

where

$$\sigma_{\pm} = \tfrac{1}{2}\bar{\psi}(1\pm\gamma_5)\psi. \qquad (5.52)$$

These are chiral eigenfields,

$$\partial\sigma_{\pm}/\partial\alpha = \mp 2\mathrm{i}\sigma_{\pm}. \qquad (5.53)$$

Thus, if we obtain a non-zero answer, we will know that spontaneous symmetry breakdown has occurred.

The computation will parallel closely that of the vacuum energy of a pure gauge field theory in Sect. 3. Indeed, as the calculation proceeds, we will accumulate all the terms that led to our earlier expression for the determinantal factor, K, as an integral over instanton size, ρ,

$$K = 2g^{-8} \int_0^\infty \frac{\mathrm{d}\rho}{\rho^5} f(\rho M), \qquad (5.54)$$

where M is the renormalization mass. As these old terms come up, I shall call them to your attention, but I will not bother to write them down; I will keep explicit track only of new terms that modify the integrand in Eq. (5.54).

There is one important novelty in the dilute-gas approximation. For n widely separated instantons and anti-instantons, $\mathrm{i}\slashed{D}$ has n vanishing eigenvalues. Thus the integral over Fermi fields will vanish unless the integrand contains

$$\prod_{\lambda_r = 0} \bar{a}_r a_r. \qquad (5.55)$$

Such a term can appear only if we are computing a Green's function involving at least $2n$ Dirac fields. Hence, for any fixed Green's function, the potentially infinite sum over instantons and anti-instantons terminates.

I will first do the σ_- computation:

In the denominator of Eq. (5.51), the only configuration that does not have a surplus of vanishing eigenvalues is one of no instantons and no anti-instantons, that is to say, the classical vacuum, $A_\mu = 0$. Thus the denominator is simply the product of a Bose determinant and a Fermi determinant. The same Bose determinant appeared in the denominator in our earlier computation. The Fermi determinant, $\det(i\not{D})$, is a new factor. In the numerator, we need a configuration with $\nu = 1$, by Eq. (5.47). The only one that does not have a surplus of vanishing eigenvalues is one instanton and no anti-instantons. Let us do the Fermi integral first; this gives

$$\tfrac{1}{2}\psi_0^\dagger(x - X, \rho)(1 - \gamma_5)\psi_0(x - X, \rho) \prod_{\lambda_r \neq 0} \lambda_r$$

$$= \psi_0^\dagger \psi_0(x - X, \rho) \det'(i\not{D}),$$ (5.56)

where det', as always, denotes a determinant with vanishing eigenvalues removed. The Bose integral gives a determinant and a bunch of collective-coordinate factors identical to those that go into K. Because $\det'(i\not{D})$ does not depend on X, the integration over the instanton location is trivial,

$$\int d^4X \, \psi_0^\dagger \psi_0(x - X) = 1.$$ (5.57)

Finally, we have a factor of $e^{-8\pi^2/g^2}$ from the instanton action, and a factor of $e^{i\theta}$ from the $e^{i\nu\theta}$.

The σ_+ computation is almost identical to the σ_- one; the only difference is that the relevant configuration is one anti-instanton, and thus, instead of a factor of $e^{i\theta}$, we have one of $e^{-i\theta}$.

Putting all this together, we find

$$\langle\theta|\sigma_\pm(x)|\theta\rangle = e^{-8\pi^2/g^2}e^{\mp i\theta}g^{-8} \, 2 \int_0^\infty \frac{d\rho}{\rho^5} \, f(\rho M) \frac{\det'(i\not{D})}{\det(i\partial)}.$$ (5.58)

(In case you have lost track of the meaning of my symbols, I remind you that $i\not{D}$ is the Dirac operator in the field of an instanton of size ρ.)

Just as before, we can use dimensional analysis to study the integrand in this formula. The eigenvalues of $i\not{D}$ have the dimensions of 1/length. One eigenvalue has been removed from the primed determinant; thus the ratio det'/det has dimensions of length, and must be of the form

$$\frac{\det'(i\not{D})}{\det(i\partial)} = \rho h(\rho M),$$ (5.59)

where h is an unknown function. Note that this gives the right dimensions for the expectation values of $\sigma_\pm$, 1/(length)^3.

From here on the argument is a rerun of that of Sect. 3: we can use the renormalization group to determine the form of the integrand up to an arbitrary multiplicative constant, be embarrassed in the infrared, wave our hands about new physics giving an effective infrared cutoff, etc.

We now know spontaneous symmetry breakdown occurs. Are there

Goldstone bosons? Let us look for them in

$$\langle\theta|\sigma_+(x)\sigma_-(0)|\theta\rangle. \tag{5.60}$$

By reasoning which should now be familiar to you, only two field configurations are relevant: $A_\mu = 0$, and one instanton plus one anti-instanton. The first of these just gives the usual one-loop perturbation theory expression; this has a two-quark cut, but no Goldstone pole. The second just gives the product $\langle\theta|\sigma_+|\theta\rangle\langle\theta|\sigma_-|\theta\rangle$. This also has no Goldstone pole. By similar methods one can investigate other gauge-invariant Green's functions, such as $\langle\theta|j^5_\mu\sigma_\pm|\theta\rangle$ or $\langle\theta|j^5_\mu j^5_\nu|\theta\rangle$, and again find no Goldstone poles, but really there is no need to do these computations. If Goldstone bosons appear anywhere, they should appear in (5.60), and they do not.

In the last sentence, I should have said not 'appear anywhere', but 'appear among the physical states', that is to say, as singularities in gauge-invariant Green's functions. The situation is very different if we study a gauge-variant Green's function such as

$$\langle\theta|J^5_\mu(x)\sigma_-(0)|\theta\rangle = \langle\theta|j^5_\mu(x)\sigma_-(0)|\theta\rangle$$

$$+ \frac{i}{16\pi^2}\langle\theta|G_\mu(x)\sigma_-(0)|\theta\rangle, \tag{5.61}$$

where G_μ is defined in Eq. (3.41). As I have said, the first of the terms on the right has no Goldstone pole, but, as I will show, the second does. The argument is simple. In a covariant gauge, there is a Goldstone pole if and only if

$$\int d^4x\, \partial_\mu\langle\theta|G_\mu(x)\sigma_-(0)|\theta\rangle \neq 0. \tag{5.62}$$

If we use the identity,

$$\int d^4x\, \partial_\mu G_\mu = 32\pi^2\nu, \tag{5.63}$$

and the fact that the only configurations that contribute to (5.62) have $\nu = 1$, we find

$$\int d^4x\, \partial_\mu\langle\theta|G_\mu(x)\sigma_-(0)|\theta\rangle = 32\pi^2\langle\theta|\sigma_-|\theta\rangle \neq 0. \tag{5.64}$$

On the other hand, for $\langle\theta|J_\mu J_\nu|\theta\rangle$, the contributing configurations have vanishing ν, and thus there is no Goldstone pole.

To summarize, we have found in the dilute-gas approximation: spontaneous breakdown of chiral U(1) symmetry, no Goldstone poles in gauge-invariant Green's functions, no Goldstone poles in the propagator of a gauge-variant conserved current, and a Goldstone pole in the Green's function for one gauge-variant current and one gauge-invariant operator. This is the Goldstone dipole of Kogut and Susskind.

5.5   QCD (*the real thing*)

Real quantum chromodynamics in the chiral SU(2)$\otimes$SU(2) limit differs from our baby version in two respects. Firstly, we have triplet quarks with gauge group SU(3) rather than doublet quarks with gauge group SU(2). Secondly, we have two massless quarks, rather than one. (I will ignore the massive quarks; they are irrelevant to the U(1) problem.)

Replacing an SU(2) doublet by an SU(3) triplet makes hardly any change. If this were the only difference, we would still have instantons, and the constant C of Eq. (5.25) would still be $\frac{1}{2}$; the only thing we would need to change in Sect. 5.4 would be the integral over instanton size, where g^{-8} would become g^{-12}.

In contrast, replacing one massless triplet by two makes a profound change. C is doubled, and thus the sum rule (5.37) is changed to

$$n_- - n_+ = 2\nu. \tag{5.65}$$

Hence, i$\not{D}$ in an instanton field has two vanishing eigenvalues rather than one. (We do not really need a fancy sum rule to see this; we have two independent quark fields, so every eigenvalue occurs twice, once for ψ_1 and once for ψ_2.) Thus, two fields no longer suffice to take care of all the vanishing eigenvalues, and all quark bilinears have zero expectation values.

This is no obstacle to demonstrating the spontaneous breakdown of chiral U(1) symmetry; we just have to study quadrilinears rather than bilinears. For example, the same computation that before gave a non-vanishing expectation value for $\bar{\psi}_1(1-\gamma_5)\psi_1$ will now give a non-vanishing expectation value for $\bar{\psi}_1(1-\gamma_5)\psi_1\bar{\psi}_2(1-\gamma_5)\psi_2$.

There is a reason for this. We have found spontaneous breakdown of chiral U(1), but not of chiral SU(2)$\otimes$SU(2); the θ-vacua are all invariant under chiral SU(2)$\otimes$SU(2). (There are two ways to see this. (1) There are too few θ-vacua for them to be anything but invariant; for spontaneous breakdown of chiral SU(2)$\otimes$SU(2) we need at least a three-parameter family of vacua. (2) Chiral U(1) transformations are connected to θ by the anomalous divergence of the isosinglet axial current; the isotriplet axial current is anomaly-free.) All Lorentz-invariant quark bilinears transform according to the representation $(\frac{1}{2}, \frac{1}{2})$ of SU(2)$\otimes$SU(2), and must have vanishing expectation values. However, there are quadrilinear SU(2)$\otimes$SU(2) singlets, such as

$$\tfrac{1}{2}\varepsilon_{ij}\varepsilon_{kl}\bar{\psi}_i(1-\gamma_5)\psi_k\bar{\psi}_j(1-\gamma_5)\psi_l$$
$$=\bar{\psi}_1(1-\gamma_5)\psi_1\bar{\psi}_2(1-\gamma_5)\psi_2-\bar{\psi}_1(1-\gamma_5)\psi_2\bar{\psi}_2(1-\gamma_5)\psi_1. \tag{5.66}$$

These operators can have non-vanishing expectation values.

The doubling of C also changes Eq. (5.49) to

$$\left[\frac{\partial}{\partial\alpha}+4\,\frac{\partial}{\partial\theta}\right]\langle\theta|\phi_1(x_1)\ldots|\theta\rangle=0. \tag{5.67}$$

Thus a chiral rotation by $\pi/2$, rather than π,

$$\psi_{1,2}\to-i\gamma_5\psi_{1,2}, \tag{5.68}$$

returns us to the same θ-vacuum. Again, this is an effect of unbroken $SU(2)\otimes SU(2)$. If we multiply this by the $SU(2)\otimes SU(2)$ transformation,

$$\psi_1\to-i\gamma_5\psi_1,\quad\psi_2\to i\gamma_5\psi_2, \tag{5.69}$$

we obtain

$$\psi_1\to\psi_1,\quad\psi_2\to-\psi_2, \tag{5.70}$$

which should not be spontaneously broken.

Of course, we do not want unbroken $SU(2)\otimes SU(2)$ in quantum chromodynamics; we want spontaneous breakdown; we want pions. However, there is no reason to be disturbed that pions have not emerged from our computations. Our methods are semiclassical, valid in the limit of vanishing $\hbar$, in principle capable only of revealing those phenomena that occur for arbitrarily weak coupling. We have learned that the breakdown of chiral $SU(2)\otimes SU(2)$ is not such a phenomenon. This is no surprise. What is a surprise (and a wonderful surprise) is that the breakdown of chiral $U(1)$ *is* such a phenomenon.

5.6   Miscellany[34]

There are some topics that I do not have the time to discuss in the detail they deserve but which I can not resist mentioning:

(1) For most theories with spontaneous symmetry breakdown, symmetry is restored at sufficiently high temperatures. Is this true here? This is an easy question to answer. Finite-temperature Green's functions are given by functional integrals over a Euclidean time inversely proportional to the temperature, with periodic time boundary conditions for Bose fields and antiperiodic ones for Fermi fields. Thus, as the temperature goes up, instantons of any given size eventually get squeezed out; there is no way to fit them into the available region of Euclidean space. However, no matter how high the temperature, there are always instantons so small that they barely notice the time boundary conditions. Thus, although asymmetries go to zero as a (calculable) power of the inverse temperature, symmetry is never fully restored. For extremely high temperatures, the only relevant instantons are so small that the effective coupling constant is extremely weak; thus we could make numerical computations of extreme accuracy, but only in a regime that is totally inaccessible to

experiment. I stress that this persistence of symmetry breakdown is a reflection of the scale invariance of classical chromodynamics, not of any property of instanton effects in general. For example, in the model of Sect. 4, there is a definite instanton size, and thus, at sufficiently high temperatures, all instanton effects disappear.

(2) Callan, Dashen, and Gross[1] have recently proposed a detailed picture of the dynamic structure of quantum chromodynamics. To explain their ideas, let me restrict myself to chromodynamics with two massless quarks, and let me imagine the universe cooling down from a very high temperature. Then, according to Callan, Dashen, and Gross:

(a) At very high temperatures, when the effective coupling constant is very small, chiral U(1) is spontaneously broken by instantons, but chiral $SU(2) \otimes SU(2)$ is still a good symmetry, and quarks are still unconfined. (Of course, this part is the standard picture which I have described in detail.)

(b) At somewhat lower temperatures, the effective coupling constant grows larger, and chiral $SU(2) \otimes SU(2)$ suffers spontaneous breakdown.[35] This is also an instanton effect, but an indirect one that can not be seen in the dilute gas approximation. Nevertheless, the effective coupling constant, although not tiny, is still small enough so that weak-coupling approximations are fairly reliable. (This part looks good to me.) Quarks are still unconfined.

(c) At still lower temperatures, and still larger effective couplings, new field configurations, called 'merons', become important in the functional integral. These produce a long-range force that confines the quarks.[36] (I can see nothing wrong with this idea in principle, but the details of the argument involve a stupendous amount of hand-waving. This part is just a suggestion (although a very clever suggestion) that may or may not someday become a theory of confinement.)

If you will excuse me for beating a dead horse one more time, this picture shows very sharply how misleading it is to say that 'instantons give the U(1) Goldstone boson a mass'. This implies that quarks get their masses through spontaneous symmetry breakdown, with the appearance of four Goldstone bosons, and then instantons come to the rescue. This is not what happens.

(To be fair, I should modify the last sentence and say, 'This is not what happens in the picture of Callan, Dashen, and Gross.' A skeptic might imagine replacing paragraph (c) above by, 'At still lower temperatures, and still larger effective couplings, new field configurations become important which restore chiral U(1) invariance. At a yet later stage, this suffers spontaneous breakdown and a Goldstone boson appears.' To my knowledge, there is no chromodynamic computation that offers the slightest evidence for this disgusting alternative, but it is not logically excluded.)[37]

(3) I have stressed several times that spontaneous breakdown of U(1) (without Goldstone bosons) is independent of spontaneous breakdown of SU(2)⊗SU(2) (with Goldstone bosons). In a recent paper, Crewther[38] has argued ingeniously that these phenomena are not just independent; they are inconsistent. This would be bad news if it were true, but I do not believe that it is; I think Crewther's arguments are invalid. However, since Crewther and I are at this moment entering our fourth month of correspondence on this matter, and since neither of us has yet convinced the other of the error of his ways, I will say no more about this.

(4) In all the θ-vacua, except for $\theta = 0$ or π, CP-noninvariant operators have non-vanishing expectation values. Thus it seems that in most of the θ-vacua we have observable strong CP violation. Of course, this is an illusion; the θ-vacua are transformed into each other by the U(1) group, and thus all experiments must yield the same results in any vacuum. Phrased more explicitly, for every θ-vacuum there is a discrete symmetry under which the vacuum is invariant, the product of CP and an appropriate U(1) transformation, and we are free to redefine CP to be this transformation.

All this is for massless quarks. The situation changes drastically when the quarks have masses, either because we have put them in by hand, or because they have Yukawa couplings to weak-interaction Higgs mesons. Now we no longer have U(1) symmetry; there is a potential clash between the definition of CP selected by θ and that selected by the quark mass operator, and there is the disastrous possibility of strong CP violation.

(Let me dispose of a red herring. You might think that all this might be said of a theory in which U(1) breaks down in the ordinary way, with Goldstone bosons, as in the U(1) σ model. In this case, there is no problem; as soon as we add a U(1) violating interaction, no matter how weak, the order parameter, the analog of θ, automatically aligns itself with the perturbation. *This is not what happens here.* The easiest (and unfortunately also the least convincing) way of seeing this is to remember that when all the dust of Sect. 3 settled, θ emerged as effectively a coupling constant, the coefficient of a term in the action. Thus we would no more expect θ to change discontinuously in response to an external perturbation than we would expect g to.)

Several mechanisms have been suggested for avoiding this disaster.[39] At the moment I favor an up quark with vanishing bare mass, that is to say, with vanishing coupling to the Higgs fields. In this case, we still have a

U(1) symmetry, chiral U(1) acting on the up quark only, and thus we have no *CP* problem. Unfortunately, this conflicts with current-algebra estimates of the up mass; these all agree that it is somewhere between $\frac{1}{2}$ and $\frac{2}{3}$ of the down mass. However, all these estimates are based on soft-kaon and soft-eta computations, and these are notoriously less accurate than soft-pion computations. For example, only soft-pion methods are needed to compute the slope of $\eta \to 3\pi$, in good agreement with experiment; soft-eta methods are needed to compute the rate, off by a factor of three.[40] So perhaps a massless up quark is not such a silly idea. Still, I would be happier if I had a more elegant solution, and one with more predictive power.

6 The fate of the false vacuum[41]

6.1 *Unstable vacua*

In Sect. 2.4 I explained how to use instanton methods to study a particle theory with a false (that is to say, unstable) ground state. In this section I will apply these methods to a field theory with a false ground state, that is to say, a false vacuum.

For simplicity, I will restrict myself to the theory of a single scalar field in four-dimensional space-time, with dynamics defined by the Euclidean action

$$S = \int d^4x [\tfrac{1}{2}(\partial_\mu \phi)^2 + U(\phi)], \tag{6.1}$$

where U is a function of the form shown in Fig. 15. Note that U possesses two relative minima, ϕ_+ and ϕ_-, but only ϕ_- is an absolute minimum. In analogy to Sect. 2.4, I have used my freedom to add a constant to U to insure that $U(\phi_+)=0$. The state of the classical field theory for which $\phi = \phi_-$ is the unique classical state of lowest energy, and, at least for weak coupling, corresponds to the unique vacuum state of the quantum theory. The state of the classical field theory for which $\phi = \phi_+$ is also a stable classical equilibrium state. However, in the quantum theory it is rendered unstable by barrier penetration; it is a false vacuum.

Fig. 15

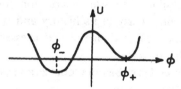

Even without any knowledge of instantons and bounces, it is easy to understand the qualitative features of the decay of the false vacuum. The decay closely parallels the nucleation processes of statistical physics, like the crystallization of a supersaturated solution or the boiling of a superheated fluid. Imagine Fig. 15 to be a plot of the free energy of a fluid as a function of density. The false vacuum corresponds to the superheated fluid phase and the true vacuum to the vapor phase. Thermodynamic fluctuations are continually causing bubbles of vapor to materialize in the fluid. If the bubble is too small, the gain in volume energy caused by the materialization of the bubble is more than compensated for by the loss in surface energy, and the bubble shrinks to nothing. However, once in a while a bubble is formed large enough so that it is energetically favorable for the bubble to grow. Once this occurs, there is no need to worry about fluctuations anymore; the bubble expands until it converts the available fluid to vapor (or coalesces with another bubble).

An identical picture describes the decay of the false vacuum, with quantum fluctuations replacing thermodynamic ones. Once in a while a bubble of true vacuum will form large enough so that it is energetically favorable for the bubble to grow. Once this happens, the bubble spreads throughout the universe, a cancer of space, converting false vacuum to true.

Thus the thing to compute is not a decay probability per unit time, Γ, but a decay probability per unit time per unit volume, Γ/V, for the probability per unit time that in a given volume a critical bubble will form is proportional to the volume (at least if the volume is much bigger than the bubble).

Of course, such a computation would be bootless were it not for cosmology. An infinitely old universe must be in a true vacuum, no matter how slowly the false vacuum decays. However, the universe is not infinitely old, and, at the time of the big bang, the universe might well have been in the false vacuum. For example, in the Weinberg–Salam model, if the mass of the Higgs meson exceeds Weinberg's lower bound, the asymmetric vacuum, in which we live, has a lower energy than the symmetric vacuum. However, if the Higgs mass is less than $\sqrt{2}$ times the lower bound, the symmetric vacuum is a local minimum of the potential, a possible false vacuum. Now we know that at high temperatures (i.e. in the early universe), symmetry breaking disappears in this model; the symmetric vacuum is the true ground state. Thus it is possible to envision a situation in which the universe gets into the false vacuum early in its history and is stuck there as it cools off; in such a situation, knowledge of Γ/V is essential if we wish to describe the future of the universe.

(I stress that I am just using the Weinberg–Salam model as an example.

I have chosen it because it is familiar and concrete, but in some ways it is a bad choice for our purposes. Firstly, the model involves, not one scalar field, but many scalar and vector fields. Secondly, the vacuum stability features I have described are not properties of the classical potential, $U(\phi)$, but require consideration of one-loop corrections. Thus the formalism I am going to develop is not applicable to this case. As long as we are talking about this model, though, you might be tempted to consider the possibility that the Higgs mass is less than Weinberg's lower bound, that we are living in the false vacuum. As Linde[41] has pointed out, this is silly; if this were the case, there would be no way for the universe to get into the false vacuum in the first place.)

The relevant parameter for cosmology is that cosmic time for which the product of Γ/V and the volume of the past light cone is of order unity. If this time is on the order of microseconds, the universe is still hot when the false vacuum decays, even on the scale of high-energy physics, and a zero-temperature computation of Γ/V is inapplicable. If this time is on the order of years, the decay of the false vacuum will lead to a sort of secondary big bang, with interesting cosmological consequences. If this time is on the order of billions of years, we have occasion for anxiety.

6.2   *The bounce*

We know from Sect. 2.4 how to compute Γ/V. We must find the bounce, $\bar{\phi}$, a solution of the Euclidean equations of motion,

$$\partial_\mu \partial_\mu \bar{\phi} = U'(\bar{\phi}), \tag{6.2}$$

that goes from the false ground state at time minus infinity to the false ground state at time plus infinity,

$$\lim_{x_4 \to \pm \infty} \bar{\phi}(\mathbf{x}, x_4) = \phi_+. \tag{6.3}$$

To these boundary conditions we can add another. It is easy to see that if the action of the bounce is to be finite,

$$\lim_{|\mathbf{x}| \to \infty} \bar{\phi}(\mathbf{x}, x_4) = \phi_+. \tag{6.4}$$

Once we have found the bounce, it is trivial to compute Γ/V. To leading order in $\hbar$,

$$\Gamma/V = K e^{-S_0}, \tag{6.5}$$

where S_0 is $S(\bar{\phi})$ and K is a determinantal factor, defined as in Sect. 2.4.

I will shortly construct the bounce. Before I do so, though, I want to make some comments:

(1) We already see the power of our method. The problem of barrier penetration in a system with an infinite number of degrees of freedom has

been reduced to a study of the properties of a single classical partial differential equation.

(2) The factor of V in the expression for Γ arises automatically in our method. No non-trivial solution of Eqs. (6.2)–(6.4) is translation invariant. Thus we must integrate over the location of the bounce. This gives us a factor of V, just as did the integration over instanton location in Sect. 3.

(3) It might be that there are many solutions to Eqs. (6.2)–(6.4). We are only interested in the solutions of minimum action, for these make the dominant contribution to the functional integral.

(4) We are not interested in the trivial solution, $\phi = \phi_+$. For this solution, $\delta^2 S/\delta\phi^2$ has no negative eigenvalues, and thus makes no contribution to the vacuum decay probability.

(5) If we imbed $\bar{\phi}$ in a one-parameter family of functions,

$$\phi_\lambda(x) = \bar{\phi}(x/\lambda), \tag{6.6}$$

then,

$$S(\phi_\lambda) = \tfrac{1}{2}\lambda^2 \int d^4x (\partial_\mu \bar{\phi})^2 + \lambda^4 \int d^4x\, U(\bar{\phi}). \tag{6.7}$$

Because $\bar{\phi}$ is a solution of the equations of motion, this must be stationary at $\lambda = 1$. Thus,

$$\int d^4x (\partial_\mu \bar{\phi})^2 = -4 \int d^4x\, U(\bar{\phi}), \tag{6.8}$$

and

$$S_0 = \frac{1}{4} \int d^4x (\partial_\mu \bar{\phi})^2 > 0. \tag{6.9}$$

This is reassuring. Since U is somewhere negative, one might worry about the possibility that S_0 was negative, which would lead to a very strange dependence of the decay probability on $\hbar$. This possibility has now been eliminated. Also,

$$d^2S/d\lambda^2 = -\frac{1}{2} \int d^4x (\partial_\mu \bar{\phi})^2 < 0. \tag{6.10}$$

Thus, at $\bar{\phi}$, $\delta^2 S/\delta\phi^2$ has at least one negative eigenvalue, and $\bar{\phi}$ does contribute to the decay probability. Of course, if there were more than one negative eigenvalue, we would have to rethink the analysis of Sect. 2.4. However, as I shall show eventually, this does not happen; there is only one negative eigenvalue.

Now for the construction of the bounce: Eqs. (6.2)–(6.4) are O(4) invariant. Thus it is not unreasonable to guess that the bounce might also be O(4) invariant, that is to say, that $\bar{\phi}$ might depend only on the distance from some point in Euclidean space. Recently, Glaser, Martin, and I were able to show that this guess is right, under mild conditions on U; there always exists an O(4)-invariant bounce and it always has strictly lower action than any O(4)-noninvariant bounce.[42] The rigor of our proof is matched only by its tedium; I would not lecture on it to my worst enemy. However, it is possible to give a sloppy argument for the first part (existence) although, unfortunately, not for the second (action minimization).

I will now give this argument.

If we choose the center of symmetry to be the origin of coordinates, then O(4) symmetry is the statement that $\bar{\phi}$ is a function only of the radial variable, r. Thus Eq. (6.2) becomes

$$\frac{d^2\bar{\phi}}{dr^2} + \frac{3}{r}\frac{d\bar{\phi}}{dr} = U'(\bar{\phi}), \tag{6.11}$$

while Eqs. (6.3) and (6.4) both become

$$\lim_{r\to\infty} \bar{\phi}(r) = \phi_+. \tag{6.12}$$

Also,

$$\left.\frac{d\bar{\phi}}{dr}\right|_{r=0} = 0. \tag{6.13}$$

Otherwise, $\bar{\phi}$ would be singular at the origin.

The key to the argument is the observation that if we interpret $\bar{\phi}$ as a particle position and r as time, Eq. (3.9) is the mechanical equation for a particle moving in a potential *minus* U and subject to a somewhat peculiar viscous damping force with Stokes's law coefficient inversely proportional to the time. The particle is released at rest at time zero, Eq. (6.13); we wish to show that if the initial position is properly chosen, the particle will come to rest at time infinity at ϕ_+, that is to say, on top of the right-hand hill in Fig. 16.

I shall demonstrate this by showing that if the particle is released to the right of ϕ_-, and is sufficiently close to ϕ_-, it will overshoot and pass ϕ_+ at some finite time. On the other hand, if it is released sufficiently far

Fig. 16

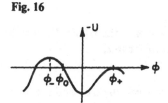

to the right of ϕ_-, it will undershoot and never reach ϕ_+. Thus (arguing in the worst tradition of nineteenth century British mathematics) by continuity there must be an intermediate initial position for which it just comes to rest at ϕ_+.

To demonstrate undershoot is trivial. If the particle is released to the right of ϕ_0, it does not have enough energy to climb the hill to ϕ_+. The damping force does not affect this argument, because viscous damping always diminishes the energy.

To demonstrate overshoot requires a little more work. For ϕ very close to ϕ_-, we may safely linearize Eq. (6.11),

$$\left(\frac{d^2}{dr^2}+\frac{3}{r}-\mu^2\right)(\bar{\phi}-\phi_-)=0, \tag{6.14}$$

where μ^2 is $U''(\phi_-)$. The solution to Eq. (6.14) is

$$\bar{\phi}-\phi_-=2[\bar{\phi}(0)-\phi_-]I_1(\mu r)/\mu r. \tag{6.15}$$

Thus, if we choose ϕ to be initially sufficiently close to ϕ_-, we can arrange for it to stay arbitrarily close to ϕ_- for arbitrarily large r. But for sufficiently large r, the viscous damping force can be neglected, since it is inversely proportional to r. But if we neglect viscous damping, the particle overshoots.   Q.E.D.

We have made great progress. We have reduced the partial differential equation for the bounce to an ordinary differential equation. But we can go even farther; in the limit of small energy-density difference between the true and false vacuum, we can obtain an explicit expression for the bounce and for S_0, as I shall now show.

6.3   The thin-wall approximation

Let $U_+(\phi)$ be an even function of ϕ,

$$U_+(\phi)=U_+(-\phi), \tag{6.16}$$

with minima at some points $\pm a$,

$$U'_+(\pm a)=0. \tag{6.17}$$

Also, let us define

$$\mu^2=U''_+(\pm a). \tag{6.18}$$

Now let us add to U_+ a small term that breaks the symmetry,

$$U=U_+ + \varepsilon(\phi-a)/2a, \tag{6.19}$$

where ε is a positive number. This defines a theory of the sort we have been discussing. To lowest non-trivial order in ε,

$$\phi_\pm=\pm a, \tag{6.20}$$

and ε is the energy-density difference between the true and the false vacua.

from the mechanical analogy of Sect. 6.2. In order not to lose too much energy, we must choose $\phi(0)$, the initial position of the particle, very close to ϕ_-. The particle then stays close to ϕ_- until some very large time, $r = R$. Near time R, the particle moves quickly through the valley in Fig. 16, and slowly comes to rest at ϕ_+ at time infinity. Translating from the mechanical analogy back into field theory, the bounce looks like a large four-dimensional spherical bubble of radius R, with a thin wall separating the false vacuum without from the true vacuum within.

To go on, we need more information about the wall of the bubble. For r near R, we can neglect the viscous damping term and we can also neglect the ε-dependent term in U. We thus obtain

$$d^2\bar\phi/dr^2 = U'_+(\bar\phi). \tag{6.21}$$

This is the classical equation of motion for a particle in a symmetric double-welled potential, the equation we studied in Sect. 2.2, the equation that had one-dimensional instantons for its solutions. Indeed, a one-dimensional instanton centered at R is the solution we need here, for such a function goes from $-a$ to a as r increases through R, just what we want. This is our approximate description of the bounce.

The only thing missing from this description is the value of R. This is easily obtained by a variational computation:

$$S = 2\pi^2 \int_0^\infty r^3 \, dr[\tfrac{1}{2}(d\bar\phi/dr)^2 + U]. \tag{6.22}$$

We can divide this integral into three regions: the outside of the bubble, the skin of the bubble, and the inside of the bubble. Within the accuracy of our approximation, in the outside region, $\phi = \phi_+$ and $U = 0$; thus we get no contribution from this part of the integral. In the inside region, $\phi = \phi_-$ and $U = -\varepsilon$; thus from this part of the integral we get

$$-\tfrac{1}{2}\pi^2 R^4 \varepsilon. \tag{6.23}$$

Over the skin, r is approximately R, and, over this small region, the ε-dependent terms in U are negligible; thus from this part of the integral we get

$$2\pi^2 R^3 \int dr[\tfrac{1}{2}(d\bar\phi/dr)^2 + U_+] = 2\pi^2 R^3 S_1, \tag{6.24}$$

where S_1 is the action of a one-dimensional instanton,

$$S_1 = \int_{-a}^{a} (2U_+)^{\frac{1}{2}} \, d\phi. \tag{6.25}$$

334   *The uses of instantons*

Varying with respect to R, we find

$$dS/dR = 0 = -2\pi^2 R^3 \varepsilon + 6\pi^2 R^2 S_1. \tag{6.27}$$

Hence,

$$R = 3S_1/\varepsilon. \tag{6.28}$$

This completes the approximate description of the bounce. We also know S_0:

$$S_0 = 27\pi^2 S_1^4/2\varepsilon^3. \tag{6.29}$$

I have described what we have done as an approximation that is valid in the limit of small ε. Now that we have gone through the computation, we can phrase the condition for the validity of the approximation more precisely: the approximation is good if the radius of the bubble is much larger than the thickness of the bubble wall; R must be much larger than $1/\mu$, or, equivalently,

$$3S_1\mu \gg \varepsilon. \tag{6.30}$$

6.4   *The fate of the false vacuum*

In a particle problem like that of Sect. 2.4, we can describe the decay process in the language of the old quantum theory. The particle sits at the bottom of the potential well until, at some random time, it makes a quantum jump to the other side of the barrier, materializing at the point labeled σ in Fig. 7. At this point, the potential energy of the particle is the same as it was at the bottom of the well; thus its kinetic energy must vanish; equivalently, it has zero velocity. These conditions give the initial-value data for the subsequent motion of the particle, which is totally governed by classical mechanics. Like all descriptions of quantum-mechanical processes in the language of the old quantum theory, this one must be taken with a large grain of salt; it will certainly lead us astray if we try to use it to describe measurements made just outside the potential barrier. Nevertheless, it is very useful as an asymptotic description, for discussing what happens far from the barrier and long after the time the system decays. For example, this is the description we all use when we discuss the macroscopic detection of an alpha particle emitted by an unstable nucleus.

This description can readily be extended to a system with many degrees of freedom. The point σ becomes the point in multi-dimensional configuration space where all velocities vanish; that is to say, it is the midpoint of the bounce. Thus, for the field theory we have been studying, the description of the vacuum decay process in the language of the old quantum theory is: the classical field makes a quantum jump (say at time zero) to the state defined by

$$\phi(x_0 = 0, \mathbf{x}) = \bar{\phi}(\mathbf{x}, x_4 = 0), \tag{6.31a}$$

and

$$\partial_0 \phi(x_0 = 0, \mathbf{x}) = 0. \tag{6.31b}$$

Afterwards, it evolves according to the classical Minkowskian field equation,

$$(\nabla^2 - \partial_0^2)\phi = U'(\phi). \tag{6.32}$$

The first of these equations implies that the same function, $\bar{\phi}(r)$, that gives the shape of the bounce in four-dimensional Euclidean space also gives the shape of the bubble at the moment of its materialization in ordinary three-space. Indeed, it does more; because the Minkowskian field equation is simply the analytic continuation of the Euclidean field equation back to real time, the desired solution of Eqs. (6.31) and (6.32) is simply the analytic continuation of the bounce:

$$\phi(x_0, \mathbf{x}) = \bar{\phi}(r = [|\mathbf{x}|^2 - x_0^2]^{\frac{1}{2}}). \tag{6.33}$$

(As a consequence of Eq. (6.13), $\bar{\phi}$ is an even function of r, so we need not worry about which branch of the square root to take.)

We can immediately draw some very interesting consequences of Eq. (6.33):

(1) O(4) invariance of the bounce becomes O(3, 1) invariance of the solution of the classical field equations. In other words, the growth of the bubble, after its materialization, looks the same to any Lorentz observer.

(2) In the case of small ε, discussed in Sect. 6.3, there is a thin wall, localized at $r = R$, separating true vacuum from false. As the bubble expands, this wall traces out the hyperboloid

$$|\mathbf{x}|^2 - x_0^2 = R^2. \tag{6.34}$$

Typically, we would expect R to be a microphysical number, on the order of a fermi, give or take ten orders of magnitude. This means that by macrophysical standards, once the bubble materializes it begins to expand almost instantly with almost the velocity of light.

(3) As a consequence of this rapid expansion, if a bubble were expanding toward us at this moment, we would have essentially no warning of its approach until its arrival. This is shown graphically in Fig. 17. The heavy curve is the bubble wall, Eq. (6.34). A stationary observer, O, cannot tell a bubble has formed until he intercepts the future light cone, W, projected from the wall at the time of its formation. A time R later, that is to say, on the order of 10^{-10}–10^{-30} sec later, he is inside the bubble and dead. (In the true vacuum, the constants of nature, the masses and couplings of

Fig. 17

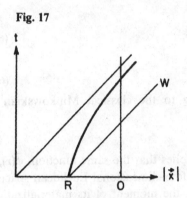

the elementary particles, are all different from what they were in the false vacuum, and thus the observer is no longer capable of functioning biologically, or even chemically.) Since even 10^{-10} sec is considerably less than the response time of a single neuron, there is literally nothing to worry about; if a bubble is coming toward us, we shall never know what hit us.

(4) The rapidly expanding bubble wall obviously carries a lot of energy. How much? A section of bubble wall at rest carries energy S_1 per unit area. Because any part of the bubble wall at any time is obtained from any other part by a Lorentz transformation, a section of wall expanding with velocity v carries energy $S_1/(1-v^2)^{\frac{1}{2}}$ per unit area. Thus, at a time when the radius of the bubble is $|\mathbf{x}|$, the energy of the wall is

$$E_{\text{wall}} = 4\pi|\mathbf{x}|^2 S_1/(1-v^2)^{\frac{1}{2}}. \tag{6.35}$$

By Eq. (6.34),

$$v = \mathrm{d}|\mathbf{x}|/\mathrm{d}t = (1 - R^2/|\mathbf{x}|^2)^{\frac{1}{2}}. \tag{6.36}$$

Thus,

$$E_{\text{wall}} = 4\pi|\mathbf{x}|^3 S_1/R = 4\pi\varepsilon|\mathbf{x}|^3/3. \tag{6.37}$$

Thus, in the thin-wall approximation, all the energy released by converting false vacuum to true goes to accelerate the bubble wall. This refutes the naive expectation that the decay of the false vacuum would leave behind it a roiling sea of mesons. In fact, the expansion of the bubble leaves behind only the true vacuum.

6.5   *Determinants and renormalization*

I said earlier that the determinantal factor K in Eq. (6.5) was defined as in the particle problem of Sect. 2.4. This is basically true, but there are three technical differences. (1) In particle mechanics, we had only one infinitesimal translation, and thus one zero eigenvalue, to worry about; here we have four. (2) It was critical in the analysis of Sect. 2.4 that the second variational derivative of the action at the bounce had one and only one negative eigenvalue. Is the same true here? (3) Whenever we study a relativistic field theory, we must deal with ultraviolet divergences and renormalization. Of course, this last remark also applied to the gauge field theories of Sect. 3, where I swept renormalization problems under the rug. However, we now have a problem with a much simpler renormalization structure (only a single scalar field to worry about, no problems with gauge invariance and gauge-fixing terms, etc.), so it is worth confronting renormalization head-on.

I will deal with these three problems in the order in which I have stated them.

(1) Vanishing eigenvalues. Because we have four infinitesimal translations, we have four eigenfunctions with eigenvalue zero, proportional to $\partial_\mu \bar{\phi}$. We must determine the constant of proportionality, that is to say, the normalization of the eigenfunctions. This is easy to do. By the spherical symmetry of the bounce,

$$\int d^4x \, \partial_\mu \bar{\phi} \, \partial_\nu \bar{\phi} = \tfrac{1}{4}\delta_{\mu\nu} \int \partial_\lambda \bar{\phi} \, \partial_\lambda \bar{\phi}$$

$$= \delta_{\mu\nu} S_0, \qquad\qquad (6.38)$$

by Eq. (6.9).

Thus, as far as zero eigenvalues go, the only difference between the problem at hand and the particle problem of Sect. 2.4 is that we have four factors of $(S_0/2\pi)^{\frac{1}{2}}$ rather than one. Hence,

$$K = \frac{S_0^2}{4\pi^2} \left| \frac{\det'[-\partial_\mu\partial_\mu + U''(\bar{\phi})]}{\det[-\partial_\mu\partial_\mu + U''(\phi_+)]} \right|^{-\frac{1}{2}}, \qquad\qquad (6.39)$$

assuming we have no problems with negative eigenvalues or renormalization.

(2) Negative eigenvalues. We already know that $\delta^2 S/\delta\phi^2$ evaluated at the bounce has at least one negative eigenvalue. Can there be more than one? To answer this question I will have to steal some information from the paper by Glaser, Martin, and me that I referred to earlier.[42] There we showed that the bounce could be characterized as the absolute minimum of S for fixed

$$V = \int d^4x \, U. \qquad\qquad (6.40)$$

This implies that there can not be two independent eigenvectors with negative eigenvalues; for, if there were, we could form a linear combination of the eigenvectors tangent to the surface of constant V, and the bounce would not even be a local minimum of S with fixed V, let alone an absolute minimum.

(3) Renormalization. Until now all of our dynamics has been expressed in terms of unrenormalized quantities. We must now recast our formulae in terms of renormalized quantities. We begin with S itself,

$$S = S_R + \sum_{n=1}^{\infty} S^{(n)}. \tag{6.41}$$

Here S_R is the renormalized action, a functional of exactly the same form as S, but with all unrenormalized quantities replaced by their renormalized counterparts, and $S^{(n)}$ is the action induced by standard renormalization counterterms computed from the sum of all n-loop graphs. To avoid excessive clutter in my equations, I will redefine ϕ to be the renormalized field, U to be the polynomial that occurs in S_R, $\bar{\phi}$ to be the bounce as computed from S_R, and S_0 to be $S_R(\bar{\phi})$.

The renormalization counterterms serve to remove all ultraviolet divergences from all one-particle irreducible Green's functions. Equivalently, they serve to remove all ultraviolet divergences from the effective action, $\gamma(\phi)$, the generating functional of these Green's functions. To one-loop order,[43]

$$\exp \gamma(\phi) = \exp[S_R(\phi) + S^{(1)}(\phi)] \det[-\partial_\mu \partial_\mu + U''(\phi)]^{\frac{1}{2}}. \tag{6.42}$$

It will be important to us shortly that (for renormalizable Us) the right-hand side of this equation is free of ultraviolet divergences for arbitrary ϕ.

Now let us imagine computing Γ/V iteratively, first treating S_R as if it were the total action, and then taking account of the renormalization counterterms perturbatively. If we had not set $\hbar$ equal to one, $S^{(n)}$ would have been proportional to $\hbar^n$. Thus, to the order in which we are working, the only counterterm we need consider is $S^{(1)}$.

The first thing we must realize is that the counterterms may destroy our convention that $S(\phi_+)$ vanishes. We can take care of this trivially by replacing S_0 in Eq. (6.5) by the difference $S_0 - S(\phi_+)$.

Secondly, adding new terms to S_R will change the stationary points of S. In particular, it will change the bounce. Let us write

$$\bar{\phi} \to \bar{\phi} + \Delta\bar{\phi}. \tag{6.43}$$

Then

$$S(\bar{\phi}) \to S_0 + \int d^4x \, \frac{\delta S_R}{\delta \bar{\phi}} \Delta\bar{\phi} + S^{(1)}(\bar{\phi}) + \ldots, \tag{6.44}$$

where the triple dots indicate terms that are negligible in the order in which we are working. The second term vanishes because the bounce is a stationary point of S_R. Thus, for our purposes,

$$S(\bar{\phi}) = S_0 + S^{(1)}(\bar{\phi}).$$ (6.45)

By the same reasoning,

$$S(\phi_+) = S^{(1)}(\phi_+).$$ (6.46)

Putting all this together, we find

$$\frac{\Gamma}{V} = \frac{S_0^2}{4\pi^2} \exp[-S_0 - S^{(1)}(\bar{\phi}) + S^{(1)}(\phi_+)] \left| \frac{\det'[-\partial_\mu\partial_\mu + U''(\bar{\phi})]}{\det[-\partial_\mu\partial_\mu + U''(\phi_+)]} \right|^{-\frac{1}{2}}$$ (6.47)

The point of this exercise is not the simplicity of this formula. Equation (6.47) is an ugly mess, and I know no way of evaluating it for even the simplest theories without using a computer. Rather, the point is that ordinary renormalization works for instanton computations. As a good renormalized expression should be, Eq. (6.47) is free of ultraviolet divergences; each determinant is paired with an exponential of $S^{(1)}$, just as in Eq. (6.42). (That one of the factors is a primed determinant is irrelevant; omitting any finite number of eigenvalues has no effect on the ultraviolet divergence.)

6.6 Unanswered questions

This concludes what I know about the fate of the false vacuum. There remain many interesting unanswered questions:

(1) I have discussed the expansion of a bubble of true vacuum into false vacuum. What if the initial state of the world is not the false vacuum, but some state of non-zero particle density built on the false vacuum? What happens when a bubble wall encounters a particle?

(2) I have discussed spontaneous decay of the false vacuum. However, there is also the possibility of induced decay. In particular, in a collision of two particles of very high energy, there might be a non-negligible cross-section for the production of a bubble. How can one estimate this cross-section?

(3) If we assume that the universe starts out in a false vacuum, at some time in its expansion bubbles begin to form. Because the formation of bubbles is totally Lorentz-invariant, the average distance between bubbles at their time of formation must be of the same order of magnitude as the time at which bubbles begin to appear. Because bubble walls expand with the speed of light, after a time interval of the same order of magnitude, bubble walls begin to collide. What happens then? Can such events be accommodated in the history of the early universe?

The preceding paragraphs are taken verbatim from a paper I wrote at the end of 1976. I still do not know the answers to any of these questions; maybe you will be able to do better than I.

Appendix 1: How to compute determinants[44]

We wish to study the equation

$$(-\partial_t^2 + W)\psi = \lambda\psi,$$ (A.1.1)

where W is some bounded function of t. Let us define $\psi_\lambda(t)$ as the solution of this equation obeying the boundary conditions

$$\psi_\lambda(-T/2) = 0, \quad \partial_t\psi_\lambda(-T/2) = 1.$$ (A.1.2)

The operator $-\partial_t^2 + W$ (acting on the space of functions vanishing at $\pm T/2$) has an eigenvalue, λ_n, if and only if

$$\psi_{\lambda_n}(T/2) = 0.$$ (A.1.3)

As in the text, we define

$$\det(-\partial_t^2 + W) = \prod_n \lambda_n.$$ (A.1.4)

Now, let $W^{(1)}$ and $W^{(2)}$ be two functions of t, and let $\psi_\lambda^{(1,2)}$ be the associated solutions of Eq. (A.1.1). I will prove that

$$\det\left[\frac{-\partial_t^2 + W^{(1)} - \lambda}{-\partial_t^2 + W^{(2)} - \lambda}\right] = \frac{\psi_\lambda^{(1)}(T/2)}{\psi_\lambda^{(2)}(T/2)}.$$ (A.1.5)

Proof. The left-hand side of this formula is a meromorphic function of λ, with a simple zero at each $\lambda_n^{(1)}$ and a simple pole at each $\lambda_n^{(2)}$. By elementary Fredholm theory, it goes to one as λ goes to infinity in any direction except along the positive real axis. The right-hand side is a meromorphic function with exactly the same zeros and poles. By elementary differential-equation theory, it also goes to one in the same limit. Thus the ratio of the two sides is an analytic function of λ that goes to one as λ goes to infinity in any direction except along the positive real axis. That is to say, it is one.   Q.E.D.

If we define a quantity N by

$$\frac{\det(-\partial_t^2 + W)}{\psi_0(T/2)} = \pi\hbar N^2,$$ (A.1.6)

then, by Eq. (A.1.5), N is independent of W. I will use this expression to define the normalization constant N in the functional integral. (Note that no explicit definition of this quantity was given in the text, so I am perfectly free to define it as I wish here.) Thus we have the desired formula for evaluating Gaussian functional integrals,

$$N[\det(-\partial_t^2 + W)]^{-\frac{1}{2}} = [\pi\hbar\psi_0(T/2)]^{\frac{1}{2}}.$$ (A.1.7)

As a specific example, for the harmonic oscillator, $W = \omega^2$,

$$\psi_0 = \omega^{-1} \sinh \omega(t + T/2), \tag{A.1.8}$$

from which Eq. (2.16) immediately follows.

Appendix 2: The double well done doubly well[45]

In this appendix I shall show that the formulae derived in the text for the splitting of the ground-state energies in a double-well potential, Eqs. (2.31) and (2.41), are equivalent to the results of ordinary wave mechanics. To do this, I will have to both evaluate the determinants that appear in Eq. (2.41) (using the method of Appendix 1) and do the wave-mechanical computation. To keep my equations as simple as possible, I will choose my units such that $\omega = 1$.

Evaluating determinants

We have to evaluate a primed determinant, one with the zero eigenvalue omitted. I will do this by evaluating the full determinant on a finite interval, $[-T/2, T/2]$, dividing this by its smallest eigenvalue, λ_0, and then letting T go to infinity.

Thus we must construct solutions of

$$[-\partial_t^2 + U''(\bar{x})]\psi_\lambda = \lambda\psi. \tag{A.2.1}$$

We already know one solution with $\lambda = 0$,

$$x_1 = S_0^{-\frac{1}{2}} \, d\bar{x}/dt \to Ae^{-|t|}, \quad t \to \pm\infty. \tag{A.2.2}$$

The constant A is determined by the integral expression for the instanton, Eq. (2.21),

$$t = \int_0^{\bar{x}} dx(2V)^{-\frac{1}{2}} = -\ln[S_0^{-\frac{1}{2}}A^{-1}(a - \bar{x})] + O(a - \bar{x}). \tag{A.2.3}$$

Equation (A.2.1) must have a second solution with $\lambda = 0$, which I denote by y_1. It will be convenient to normalize y_1 such that its Wronskian with x_1 is given by

$$x_1\partial_t y_1 - y_1\partial_t x_1 = 2A^2. \tag{A.2.4}$$

Thus,

$$y_1 \to \pm Ae^{|t|}, \quad t \to \pm\infty. \tag{A.2.5}$$

We can now construct ψ_0 of Appendix 1. For large T,

$$\psi_0(t) = (2A)^{-1}(e^{T/2}x_1 + e^{-T/2}y_1). \tag{A.2.6}$$

Hence,

$$\psi_0(T/2) = 1. \tag{A.2.7}$$

This takes care of the determinant. To find the lowest eigenvalue, we must find $\psi_\lambda(t)$ for small λ. This can be done by a standard method: we turn Eq. (A.2.1) into an integral equation and iterate once. This can readily be seen to yield

$$\psi_\lambda(t) = \psi_0(t) - \lambda(2A^2)^{-1} \int_{-T/2}^{t} dt' [y_1(t)x_1(t') - x_1(t)y_1(t')]\psi_0(t'), \quad (A.2.8)$$

plus terms of order λ^2, which we neglect. By Eq. (A.2.6),

$$\psi_\lambda(T/2) = 1 - \lambda(4A^2)^{-1} \int_{-T/2}^{T/2} dt[e^T x_1^2 - e^{-T} y_1^2]. \quad (A.2.9)$$

For large T, the second term in this expression is bounded, and thus negligible compared to the first term. Thus, for large T,

$$\psi_\lambda(T/2) = 1 - \lambda(4A^2)^{-1}e^T, \quad (A.2.10)$$

because x_1 is properly normalized.

Thus the lowest eigenvalue is given by

$$\lambda_0 = (4A^2)e^{-T}, \quad (A.2.11)$$

and, for large T,

$$\frac{\det'[-\partial_t^2 + U''(\bar{x})]}{\det[-\partial_t^2 + \omega^2]} = \frac{\psi_0(T/2)}{\lambda_0 e^T/2} = \frac{1}{2A^2}. \quad (A.2.12)$$

Reassuringly, this is non-zero and T-independent.

Plugging this in to Eqs. (2.31) and (2.41), we find that the lowest energy levels are given by

$$E_\pm = \hbar/2 \pm A(\hbar S_0/\pi)^{\frac{1}{2}} e^{-S_0/\hbar}. \quad (A.2.13)$$

Solving the Schrödinger equation
We wish to study the solutions of

$$-\tfrac{1}{2}\hbar^2 \partial_x^2 \psi + V\psi = E\psi. \quad (A.2.14)$$

As long as x is not near the bottoms of the wells, we can use standard WKB solutions. Near the bottom of each well, though, there are two turning points. These are *not* separated by many wavelengths, so we can not use the standard connection formulae for a linear turning point. Fortunately, near the bottom of a well, in a region that includes both turning points, we may safety approximate V by a harmonic-oscillator potential. Thus, for example, for x near a, we may write

$$-\tfrac{1}{2}\hbar^2 \partial_x^2 \psi + \tfrac{1}{2}(x-a)^2 \psi = E\psi. \quad (A.2.15)$$

Our strategy will be to match WKB solutions of Eq. (A.2.14) outside the wells to solutions of Eq. (A.2.15) in the bottoms of the wells. Furthermore, since we know the solutions are either even or odd, we can restrict ourselves to positive x, and only have to do this awkward matching for the right-hand well.

I will begin by constructing the even and odd WKB solutions for $0 \leqslant x < a$. If we define

$$k(x) = [2(V - E)]^{\frac{1}{2}}, \tag{A.2.16}$$

then these are

$$\psi_{\pm} = k^{-\frac{1}{2}} \left[\exp \hbar^{-1} \int_0^x k \, dx' \pm \exp - \hbar^{-1} \int_0^x k \, dx' \right]. \tag{A.2.17}$$

For the solutions we are interested in, E is itself of order $\hbar$. Thus we may ignore E in the factor of $k^{-\frac{1}{2}}$, and expand to first order in the exponential,

$$k = (2V)^{\frac{1}{2}} - E(2V)^{-\frac{1}{2}}. \tag{A.2.18}$$

To match on to the solutions of Eq. (A.2.15), we need the form of the WKB solutions as x enters the regime of validity of the quadratic approximation to V, $V = (a - x)^2/2$. In this regime $k(x)$ is just $(a - x)$, while we can compute the E-independent term in the integral by

$$\int_0^x dx (2V)^{\frac{1}{2}} = \int_0^a dx (2V)^{\frac{1}{2}} - \int_a^x dx (2V)^{\frac{1}{2}}$$
$$= \tfrac{1}{2} S_0 - \tfrac{1}{2}(a - x)^2. \tag{A.2.19}$$

For the E-dependent term in the integral, we can use Eq. (A.2.3). Thus we obtain

$$\psi_{\pm} = (a - x)^{-\frac{1}{2}} \{ \exp \hbar^{-1} [\tfrac{1}{2} S_0 - \tfrac{1}{2}(a - x)^2 + E \ln S_0^{-\frac{1}{2}} A^{-1}(a - x)]$$
$$\pm \exp - \hbar^{-1} [\tfrac{1}{2} S_0 - \tfrac{1}{2}(a - x)^2 + E \ln S_0^{-\frac{1}{2}} A^{-1}(a - x)] \} \tag{A.2.20}$$

If we write

$$E = \hbar(\tfrac{1}{2} + \varepsilon), \tag{A.2.21}$$

then Eq. (A.2.20) becomes

$$\psi_{\pm} = \{ e^{S_0/2\hbar} S_0^{-\frac{1}{2}} A^{-\frac{1}{2}} \exp[-(a - x)^2/2\hbar]$$
$$\pm (a - x)^{-1} e^{-S_0/2\hbar} S_0^{\frac{1}{2}} A^{\frac{1}{2}} \exp[(a - x)^2/2\hbar] \} [1 + O(\varepsilon)]. \tag{A.2.22}$$

We will hold this expression in reserve while we go on to study the solutions of Eq. (A.2.15).

We already know one solution of Eq. (A.2.15), for $\varepsilon = 0$,

$$\psi_1 = \exp[-(a - x)^2/2\hbar]. \tag{A.2.23}$$

Of course, there is another (odd, increasing) solution, ϕ_1. This does not have a simple form in terms of elementary functions, but its asymptotic form, for $|x-a| \gg \hbar$, is easily computed by the WKB approximation, or just read off from Eq. (A.2.22),

$$\phi_1 = (a-x)^{-1} \exp[(a-x)^2/2\hbar]. \tag{A.2.24}$$

It will turn out that this is all that we need. Note that I have normalized ϕ_1 such that the Wronskian of the two solutions is

$$\phi_1 \partial_x \psi_1 - \psi_1 \partial_x \phi_1 = 2/\hbar. \tag{A.2.25}$$

We wish to solve Eq. (A.2.15) for small ε. By the same argument as led to Eq. (A.2.8),

$$\psi = \psi_1 - \varepsilon \int_x^\infty dx' \, \psi_1(x')[\psi_1(x')\phi_1(x) - \phi_1(x')\psi_1(x)]. \tag{A.2.26}$$

I have chosen here the solution that vanishes as x goes to plus infinity. Thus, this is the appropriate solution for matching with the decreasing WKB solution in the region $(x-a) \gg \hbar$. Thus, the only matching left to do is in the region $(a-x) \gg \hbar$.

In this region, we can use

$$\int_{-\infty}^\infty dx \, \psi_1^2 = (\pi\hbar)^{\frac{1}{2}}. \tag{A.2.27}$$

to write

$$\psi = \exp[-(a-x)^2/2\hbar][1 + O(\varepsilon)] - \varepsilon(\pi\hbar)^{\frac{1}{2}}(a-x)^{-1} \exp[(a-x)^2/2\hbar]. \tag{A.2.28}$$

As it should be, this is proportional to Eq. (A.2.22), if we choose

$$\varepsilon = e^{-S_0/\hbar} A(S_0/\pi\hbar)^{\frac{1}{2}}. \tag{A.2.29}$$

This is the desired result, and it is identical to the result of the dilute-gas approximation, Eq. (A.2.13).

Almost identical methods to these can be used to check the dilute-gas formula for the width of an unstable state, Eq. (2.50). You might find it an instructive exercise to see that things work out in this case also.

Appendix 3: Finite action is zero measure[46]

In this appendix I will show that, even for a one-dimensional harmonic oscillator, motions of finite action form a set of measure zero in function space.

If we define eigenvalues λ_n and expansion coefficients c_n as in Sect. 2.1, then, for a harmonic oscillator, the quadratic approximation to the action is exact,

$$S = \frac{1}{2} \sum_n \lambda_n c_n^2. \tag{A.3.1}$$

If we introduce new variables, $b_n = c_n(\lambda_n/\hbar)^{\frac{1}{2}}$, then

$$S = \frac{\hbar}{2} \sum_n b_n^2. \tag{A.3.2}$$

Let us define a slightly unconventional normalization constant, N', by

$$N'[dx] = \prod_n (2\pi)^{-\frac{1}{2}} db_n. \tag{A.3.3}$$

This has been chosen such that

$$N' \int [dx] e^{-S/\hbar} = 1. \tag{A.3.4}$$

How much of this integral comes from motions of finite action? The integrand is positive, and every motion of finite action lies in a cube of side L

$$|b_n| \leqslant L \quad \text{for all } n, \tag{A.3.5}$$

for sufficiently large L. Thus, the finite-action contribution to the integral must be less than

$$\overline{\lim_{L \to \infty}} \prod_{n=1}^{\infty} (2\pi)^{-\frac{1}{2}} \int_{-L}^{L} db_n \, e^{-b_n^2} = 0. \tag{A.3.6}$$

Q.E.D.

Appendix 4: Only winding number survives

This appendix is the promised (in Sect. 3.3) demonstration that for a sufficiently large box, the only relic of the boundary conditions imposed on the walls of the box is the winding number.

Consider a rectangular box in Euclidean four-space, with sides $L_1 \ldots L_4$. I will label the eight hyperplanes that bound the box by their normal vectors; thus I will refer to the upper 1-wall, the lower 1-wall, the upper 2-wall, etc. (Upper and lower here refer to greater and lesser values of the appropriate coordinate.)

On the walls of the box the tangential components of A_μ are given in a way consistent with finiteness of the action, that is to say, consistent with

$$A_\mu = g \partial_\mu g^{-1}. \tag{A.4.1}$$

Thus, giving the tangential components of A_μ on the walls is equivalent to giving g on the walls (up to an irrelevant multiplicative constant). The gauge condition $A_3 = 0$ still allows arbitrary x_3-independent gauge transformations. I will use the freedom to make such a transformation to transform g to one on the lower 3-wall. Because the vanishing of A_3 implies the vanishing of $\partial_3 g$, g is automatically one on all walls except the upper 3-wall. On this wall, g is given as a function of three variables, $g(x_1, x_2, x_4)$, equal to one on the boundary of the wall. (I stress that the only function of this gauge transformation is to simplify my subsequent arguments. Since the functional integral is gauge-invariant, anything I can prove with this gauge convention I could prove without it; it is just that the arguments would be clumsier.)

Now let us imbed our original box, with boundary conditions given by $g_1(x_1, x_2, x_4)$ in a larger box, with the same lowermost corner (chosen to be the origin of coordinates), and with the same sides L_1, L_2, and L_4, but with third side $L_3 + \Delta$. Let the boundary conditions on the larger box be given by some function $g_2(x_1, x_2, x_4)$.

Theorem. If g_1 and g_2 are in the same homotopy class, then any field configuration defined inside the original box consistent with its boundary conditions can be extended to a field configuration defined inside the larger box, consistent with its boundary conditions and the gauge condition $A_3 = 0$, at the cost of an increase in action of order $1/\Delta$.

Before I prove this theorem I will make some comments:

(1) The theorem would certainly not be true if g_1 and g_2 were in different homotopy classes. In this case, to get from g_1 to g_2, we would have to put at least one instanton in the new volume; this would increase the action by at least $8\pi^2/g^2$, independent of the value of Δ.

(2) We are free to choose Δ to be proportional to, say, $L_3^{\frac{1}{2}}$. Thus, for a very large box, the fractional change in the volume of the box is negligible, as is the change in the action. In the language of statistical physics, changing the boundary conditions while keeping the winding number fixed is just a surface effect, not a volume effect.

(3) There is an apparent paradox that may have bothered you. For any fixed configuration of instantons and anti-instantons, $g(x_1, x_2, x_4)$ is fixed. How then can we get all configurations consistent with a fixed winding number with a single set of boundary conditions? The theorem supplies the answer. We do not get all these configurations; we get only a small portion of them. However, we do get 'close relatives' of all of them, configurations that differ only by a small distortion very close to the upper 3-wall. The difference caused by this small distortion is negligible for a sufficiently large box.

Now for the proof. By assumption, g and g' are in the same homotopy class. Thus there is a continuous function of four variables, $g(x_1, x_2, s, x_4)$, with $0 \leqslant s \leqslant 1$, such that

$$g(x_1, x_2, 0, x_4) = g, \quad g(x_1, x_2, 1, x_4) = g_2. \qquad (A.4.2)$$

Let $g(x)$ be a function defined in the added volume by

$$g(x) = g(x_1, x_2, (x_3 - L_3)/\Delta, x_4). \tag{A.4.3}$$

If we could choose

$$A_\mu = g\partial_\mu g^{-1}, \tag{A.4.4}$$

then we could effect the desired transition at *no* cost in added action. Unfortunately, this is impossible; Eq. (A.4.4) is inconsistent with the gauge condition $A_3 = 0$. However,

$$\begin{aligned} A_\mu &= g\partial_\mu g^{-1}, \quad \mu \neq 3 \\ &= 0, \qquad \mu = 3, \end{aligned} \tag{A.4.5}$$

is consistent with the gauge condition and will effect the transition.

We must compute the action associated with Eq. (A.4.5). If we make a gauge transformation by g^{-1}, Eq. (A.4.5) becomes

$$\begin{aligned} A_\mu &= 0, \qquad \mu \neq 3, \\ &= g^{-1}\partial_\mu g, \quad \mu = 3. \end{aligned} \tag{A.4.6}$$

(A gauge transformation does not change the action.) From Eq. (A.4.6), we see that A_3 is proportional to $1/\Delta$. The only non-vanishing components of $F_{\mu\nu}$ are $F_{\mu 3}$, also proportional to $1/\Delta$. Thus the Lagrangian density is proportional to $1/\Delta^2$. However, the volume of integration is only proportional to Δ.   Q.E.D.

Appendix 5: No wrong-chirality solutions[47]

In this appendix I will show that, if

$$F_{\mu\nu} = \tilde{F}_{\mu\nu}, \tag{A.5.1}$$

then the only normalizable solution of both

$$D_\mu \gamma_\mu \psi = 0, \tag{A.5.2}$$

and

$$\gamma_5 \psi = \psi, \tag{A.5.3}$$

is $\psi = 0$.

From Eq. (A.5.2),

$$D_\nu \gamma_\nu D_\mu \gamma_\mu \psi = D_\mu D_\mu \psi + \tfrac{1}{2} F_{\mu\nu} \gamma_\mu \gamma_\nu \psi = 0. \tag{A.5.4}$$

Also,

$$F_{\mu\nu} \gamma_\mu \gamma_\nu \gamma_5 = -\tilde{F}_{\mu\nu} \gamma_\mu \gamma_\nu. \tag{A.5.5}$$

Thus,

$$D_\mu D_\mu \psi = 0. \tag{A.5.6}$$

Multiplying by $\psi^\dagger$ and integrating, we find

$$\int d^4x \, D_\mu \psi^\dagger D_\mu \psi = 0. \tag{A.5.7}$$

Hence

$$D_\mu \psi = 0, \tag{A.5.8}$$

for all μ. If we go to axial gauge, this implies, in particular, that ψ is independent of x_3. The only such normalizable function is $\psi = 0$.   Q.E.D.

Notes and references

1. These topics are all drawn from the classic part of the theory. 'Classic', in this context, means work done more than six months ago. A good summary of the more recent research of one of the most active groups in this field is C. Callan, R. Dashen, and D. Gross, *Phys. Rev.* **D17**, 2717 (1978).
2. Polyakov's early work is summarized in A. M. Polyakov, *Nucl. Phys.* **B121**, 429 (1977).
3. See, for example, R. Feynman and A. Hibbs, *Quantum Mechanics and Path Integrals* (McGraw-Hill, New York, 1965).
4. See the note on notation at the end of Sec. 1.
5. It was Polyakov[2] who recognized the double well as the prototypical instanton problem.
6. For a review of lumps, see Chapter 6 in this volume.
7. This is, of course, nothing but the standard prescription for handling collective coordinates in soliton problems. See J. L. Gervais and B. Sakita, *Phys. Rev.* **D11**, 2943 (1975).
8. The treatment here follows that of C. Callan and S. Coleman, *Phys. Rev.* **D16**, 1762 (1977). The idea of handling unstable states this way goes back to Langer's analysis of the droplet model in statistical mechanics (J. S. Langer, *Ann. Phys.* (N.Y.) **41**, 108 (1967)). The factor of $\frac{1}{2}$, of which much is made below, occurs in Langer's analysis and was explained to me by Michael Peskin.
9. The order of my exposition will not be the historical order of discovery. Here is the way it happened. The topological structure of finite-action Euclidean gauge-field configurations was uncovered and the instanton solutions discovered by A. A. Belavin, A. M. Polyakov, A. S. Schwartz, and Yu. S. Tyupkin, *Phys. Lett.* **59B**, 85 (1975). The importance of the instantons was realized by G. 't Hooft (*Phys. Rev. Lett.* **37**, 8 (1976); *Phys. Rev.* **D14**, 3432 (1976)) who used them to solve the U(1) problem. (I won't get to this until Sect. 5.) 't Hooft's work was clarified and extended by R. Jackiw and C. Rebbi (*Phys. Rev. Lett.* **37**, 172 (1976)) and by C. Callan, R. Dashen, and D. Gross (*Phys. Lett.* **63B**, 334 (1976)), who discovered the properties of pure gauge field theories discussed in this section.
10. For a review of gauge field theories, see Chapter 5 in this volume.
11. And sometimes given by me. I thank Arthur Wightman for awakening me from my dogmatic slumbers.
12. It suffices to assume that the gauge field is without (gauge-invariant) singularities if we make a stereographic projection. of four-space onto a four-sphere. I would love to found the analysis on finiteness of the action, without even this assumption about the behavior of the fields at infinity, but I have not been able to do so.

13. Sign convention. In n-space, $\varepsilon_{1\ldots n} = 1$. Symbols with upper indices are defined by raising with the appropriate (Euclidean or Minkowskian) metric.
14. R. Bott, *Bull. Soc. Math. France* **84**, 251 (1956).
15. See Note 10.
16. See Note 6.
17. At least in a box; see the next paragraph.
18. R. Jackiw and C. Rebbi, *Phys. Rev.* **D14**, 517 (1976).
19. Even this is true for axial gauge (in infinite space) only if we add additional gauge conditions.[6]
20. M. Atiyah and R. Ward, *Comm. Math. Phys.* **55**, 117 (1977).
21. Although I have just argued that this knowledge is irrelevant to our immediate purposes, an enormous amount *has* been learned recently about solutions to the Euclidean gauge-field equations. In fact, 'binstantons' do not exist, but $8|\nu|$-parameter families of solutions with winding number ν do. For a review (with references to the original literature) see R. Jackiw, C. Nohl, and C. Rebbi, in *Particles and Fields*, Proceedings of the 1977 Banff Summer Institute, edited D. H. Boal and A. N. Kamal, Plenum (1978).
22. For a review of the renormalization group applied to gauge theories, see Note 10.
23. Done by 't Hooft, a hard worker (second paper cited in Ref. 9). 't Hooft's computation has been somewhat simplified. See A. Belavin and A. M. Polyakov, *Nucl. Phys.* **B123**, 429 (1977); F. Ore, *Phys. Rev.* **D16**, 2577 (1977); S. Chahda, A. D'Adda, P. di Vecchia, and F. Nicodemi, *Phys. Lett.* **72B**, 103 (1977).
24. The analysis reported here is based on C. Callan, R. Dashen, and D. Gross, *Phys. Lett.* **66B**, 375 (1977). The fact that the Abelian Higgs model in two dimensions does not display the Higgs phenomenon was discovered independently by two of my graduate students, Frank De Luccia and Paul Steinhardt. They did not write up their results because I did not believe them. I take this occasion to apologize to them for my stupidity.
25. The problem is identical to that of constructing flux tubes in superconductors. See Ref. 6, and references cited therein.
26. Indeed, θ-vacua, with precisely the same interpretation (but derived in a completely different way), occur in the massive Schwinger model, quantum electrodynamics of charged fermions in $1+1$ dimensions. (See S. Coleman, R. Jackiw, and L. Susskind, *Ann. Phys.* (N.Y.) **93**, 267 (1975), and S. Coleman, *Ann. Phys.* (N.Y.) **101**, 239 (1976).) The arguments that work for the Schwinger model also work for the Higgs model when μ^2 is positive, so we also obtain θ-vacua in this case despite the absence of instantons.
27. K. Wilson, *Phys. Rev.* **D10**, 2445 (1974). The standard expression for W has a factor of $-iq$ where I have one of $-q/e$; the source of the difference is the factor of ie hidden in my definition of A_μ.
28. S. Weinberg, *Phys. Rev.* **D11**, 3583 (1975). This paper, titled 'The U(1) problem', gives a characteristically lucid description of the situation just before 't Hooft's breakthrough. (As a major unanswered question, Weinberg lists 'How does the underlying gluon-gauge invariance enforce the equal coupling of the positive- and negative-metric Goldstone bosons to gauge-invariant operators?')
29. S. L. Adler, *Phys. Rev.* **177**, 2426 (1969). J. S. Bell and R. Jackiw, *Nuovo Cimento* **60**, 47 (1969). W. Bardeen, *Phys. Rev.* **184**, 1848 (1969).
30. J. Kogut and L. Susskind, *Phys. Rev.* **D11**, 3594 (1976).
31. For more details on Fermi integration, see F. A. Berezin, *The Method of Second Quantization* (Academic Press, New York and London, 1966).

350   *The uses of instantons*

32. The easiest way to fix this up is to stereographically project Euclidean four-space onto a four-sphere; $\not{D}$ is then projected into an operator with a pure discrete spectrum. This changes the determinant, but only by a factor that is independent of the gauge field. Since, as we shall see, our final results will only depend on ratios of determinants, this change is irrelevant.
33. To my knowledge, this sum rule was first derived by A. S. Schwarz, *Phys. Lett.* **67B**, 172 (1977). The derivation in the literature closest to the one given here is that of L. Brown, R. Carlitz, and C. Lee, *Phys. Rev.* **D16**, 417 (1977).
34. This section is mainly afterthoughts; I did not know most of these things at the time these lectures were given.
35. A related picture of how instantons break SU(2)⊗SU(2) is advanced by D. Caldi, *Phys. Rev. Lett.* **39**, 121 (1977).
36. An (apparently) very different picture of how merons effect confinement has been advanced by J. Glimm and A. Jaffe, *Phys. Rev.* **D18**, 463 (1978). G. 't Hooft has advocated completely different configurations (*Nucl. Phys.* **B138**, 1 (1978)).
37. The preceding paragraph is the product of conversations with Michael Peskin, who has observed that a group of two-dimensional models analyzed by C. Callan, R. Dashen, and D. Gross (*Phys. Rev.* **D16**, 2526 (1977)) display (in a certain sense) a restoration of chiral symmetry at large scales, the first half of the above scenario.
38. R. Crewther, *Phys. Lett.* **70B**, 349 (1977).
39. R. D. Peccei and H. R. Quinn, *Phys. Rev. Lett.* **38**, 1440 (1977); *Phys. Rev.* **D16**, 1791 (1977). F. Wilczek, *Phys. Rev. Lett.* **40**, 279 (1978). S. Weinberg, *Phys. Rev. Lett.* **40**, 223 (1978) and 'Instantons without axions' (unpublished).
40. These ideas are the product of discussions with S. Glashow and D. Nanopoulos.
41. The key paper on this subject is M. B. Voloshin, I. Yu. Kobzarev, and L. B. Okun, *Yad. Fiz.* **20**, 1229 (1974) (*Sov. J. Nucl. Phys.* **20**, 644 (1975)). The instanton approach to the problem was developed in S. Coleman, *Phys. Rev.* **D15**, 2929 (1977), and C. Callan and S. Coleman.[8] (Large portions of the text of this section are plagiarized from these two papers.) Similar ideas were developed independently by M. Stone, *Phys. Rev.* **D14**, 3568 (1976) and *Phys. Lett.* **67B**, 186 (1977). P. Frampton was the first to study these phenomena in the Weinberg–Salam model (*Phys. Rev. Lett.* **37**, 1378 (1976)); however, Frampton's conclusions have been criticized severely (and, I think, correctly) by A. Linde (*Phys. Lett.* **70B**, 306 (1977); **92B**, 119 (1980); *Rept. Prog. Phys.* **42**, 389 (1979).
42. S. Coleman, V. Glaser, and A. Martin, *Comm. Math. Phys.* **58**, 211 (1978).
43. See, for example, R. Jackiw, *Phys. Rev.* **D9**, 1686 (1974).
44. Formulae related to the one developed here can be found throughout the literature. Two references out of many: J. H. Van Vleck, *Proc. Nat. Acad. Sci.* **14**, 178 (1928). R. Dashen, B. Hasslacher, and A. Neveu, *Phys. Rev.* **D10**, 4114 (1974). The derivation given here was developed in conversations with Ian Affleck.
45. This appendix reports on computations done with C. Callan. A somewhat different attack on the problem (with the same conclusions) is E. Gildener and A. Patrascioiu, *Phys. Rev.* **D16**, 423 (1977).
46. I give no reference not because these results are novel but because they are a standard part of the theory of Weiner integrals.
47. This appendix is a transcription of an argument of Brown *et al.*[33]

8

1/N

(1979)

1 Introduction

More variables usually means greater complexity, but not always. There exist families of field theories with symmetry group SO(N) (or SU(N)) that become simpler as N becomes larger. More precisely, the solutions to these theories possess an expansion in powers of $1/N$. This expansion is the subject of these lectures.

There are two reasons to study the $1/N$ expansion.

(1) It can be used to analyze model field theories. This is important. Most of us have a good intuition for the phenomena of classical mechanics. We were not born with this intuition; we developed it toiling over problems involving rigid spheres that roll without slipping and similar extreme but instructive simplifications of reality. One reason we have such a poor intuition for the phenomena of quantum field theory is that there are so few simple examples; essentially all we have to play with is perturbation theory and a handful of soluble models. The $1/N$ expansion enables us to enlarge this set.

In Section 2 I develop the $1/N$ expansion for ϕ^4 theory and apply it to two-dimensional models with similar combinatoric structures, the Gross–Neveu model and the CP^{N-1} model. These models display (in the leading $1/N$ approximation) such interesting phenomena as asymptotic freedom, dynamical symmetry breaking, dimensional transmutation, and non-perturbative confinement; they are worth studying.

(2) It is possible that the $1/N$ expansion, with N the number of colors, might fruitfully be applied to quantum chromodynamics. In the real world, N is 3, so an expansion in powers of $1/N$ may not seem like such a good idea. This objection is without force, as is shown by the following wisecrack by Ed Witten:

$$\frac{e^2}{4\pi} = \frac{1}{137} \Leftrightarrow e = 0.30.$$

351

352   1/N

Of course, this does not show that the $1/N$ expansion in QCD will necessarily be as good an approximation as perturbation theory in QED, but
it does show that there is no reason to reject it *a priori*.

Unfortunately, it is not possible to make a decisive test of the approximation, because no one knows how to compute even the first term in the
expansion in closed form. However, it is possible to argue that this first
term, whatever its detailed form, has many properties that are also shared
by the real world, and which are otherwise underived from field theory.
These include the saturation of scattering amplitudes by an infinite number
of narrow resonances, the essential feature of dual-resonance models.
I discuss these matters in Section 3.

Although united here, these two classes of applications have very different standings. The work on model field theories is modest but solid, a
permanent part of our knowledge. In contrast, the work on chromodynamics is ambitious but conjectural. It is possible that it will lead to
great breakthroughs; it is possible that it will fizzle out, like so many
hopeful programs before it.

I should warn you that these lectures are introductory rather than
encyclopedic. Much more could be said about every topic I discuss.
I have not gone into more depth in part because of lack of time and in
part because of lack of competence. I am not an expert in these matters;
one reason I decided to lecture on them this summer was to force myself
to learn them.

Much of what I do know I have learned from conversations with Roman
Jackiw, Hugh Osborn, Howard Schnitzer, Gerard 't Hooft, Ken Wilson,
and Edward Witten. My debt to Witten is enormous; most of the second
half of Section 3 is plagiarism of his ideas.

(Note to the reader: If you are only interested in chromodynamic
applications, I suggest you read just the first five paragraphs of Section 2,
and then proceed directly to Section 3.)

2   **Vector representations, or, soluble models**
2.1 ϕ^4 theory (half-way)[1]
        I will begin the development of the $1/N$ expansion with a theory
that is (I hope) familiar to you, the $O(N)$ version of ϕ^4 theory. I warn you in
advance that I will stop the discussion half-way, after I have worked out all
the combinatorics but before I have evaluated any Feynman integrals.
This is not because the details of the model are not interesting and instructive (they are, especially in the Nambu–Goldstone mode), but because
I am using this theory only as a warm-up, and want to get on to the even
more interesting and instructive Gross–Neveu and $\mathbb{C}P^{N-1}$ models.

The dynamical variables of the theory are a set of N scalar fields, ϕ^a, $a = 1 \ldots N$, with dynamics defined by the Lagrange density,

$$\mathscr{L} = \tfrac{1}{2}\partial_\mu \phi^a \partial^\mu \phi^a - \tfrac{1}{2}\mu_0^2 \phi^a \phi^a - \tfrac{1}{8}\lambda_0 (\phi^a \phi^a)^2, \tag{2.1}$$

where the sum on repeated indices is implied. Since I am going to stop the investigation before evaluating any Feynman integrals, I might as well keep the dimension of space-time arbitrary (but less than or equal to four).

To get an idea of what is going on, I have written down in Fig. 1 the first few diagrams (in ordinary perturbation theory) for the scattering of two mesons of type a into two mesons of type b ($a \neq b$). The first diagram displayed, the Born term, is $O(\lambda_0)$. The second diagram is $O(\lambda_0^2 N)$, because there are N possible choices for the internal index, c. The third diagram, in contrast, is only $O(\lambda_0^2)$; the internal indices are fixed and there is no sum to do.

The explicit factor of N in the second diagram makes the large-N limit seem nonsensical, but this is easily rectified. All we need do is define

$$g_0 \equiv \lambda_0 N, \tag{2.2}$$

and declare that we wish to study the limit of large-N with fixed g_0 (not fixed λ_0). The first diagram is now $O(g_0/N)$; as we shall see, this is the leading non-trivial order in $1/N$. The second diagram is $O(g_0^2/N)$, the same order in $1/N$. The third diagram is $O(g_0^2/N^2)$, next order in $1/N$ and negligible compared to the two preceding diagrams in the large-N limit.

This is the first step in constructing the $1/N$ expansion. We must decide what parameters to hold fixed as N becomes large. If we make the wrong choice, we can obtain either a trivial theory (only the Born term survives) or one without a $1/N$ expansion (there are graphs proportional to positive powers of N). Of course, we have not yet shown that the second possibility does not occur in the theory at hand. However, there are clearly an infinite

Fig. 1

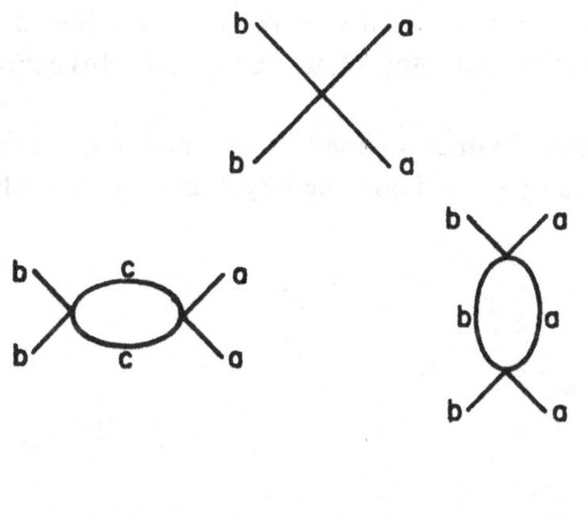

Fig. 2

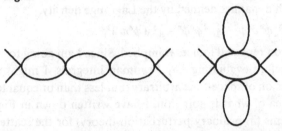

number of graphs proportional to $1/N$, times various powers of g_0; two of them are shown in Fig. 2. (To keep the graphs from being hopeless jumbles, I have left out the index labels; I hope you can figure out where they go.)

To keep all these diagrams straight, and to show that there are no diagrams proportional to positive powers of N, is a combinatoric challenge. We can simplify life considerably by introducing an auxiliary field, σ, and altering the Lagrange density:

$$\mathcal{L} \to \mathcal{L} + \frac{1}{2}\frac{N}{g_0}\left(\sigma - \frac{1}{2}\frac{g_0}{N}\phi^a\phi^a\right)^2. \tag{2.3}$$

This added term has no effects on the dynamics of the theory. This is easy to see from the viewpoint of functional integration. The functional integral over σ is a trivial Gaussian integral; its only effect is to multiply the generating functional of the theory by an irrelevant constant. It is also easy to see from the viewpoint of canonical quantization. The Euler–Lagrange equation for σ is

$$\sigma = \frac{1}{2}\frac{g_0}{N}\phi^a\phi^a. \tag{2.4}$$

This involves no time derivatives; it is not a true equation of motion, but an equation of constraint, like the Euler–Lagrange equation for the fourth component of a massive vector field. When we construct the Hamiltonian, σ must be eliminated from the Lagrangian, using Eq. (2.4); this cancels the added term.

However, although the dynamics defined by our new Lagrangian are the same as those defined by the old one, the Feynman rules are different. By elementary algebra

$$\mathcal{L} = \frac{1}{2}\partial_\mu\phi^a\partial^\mu\phi^a - \frac{1}{2}\mu_0^2\phi^a\phi^a + \frac{1}{2}\frac{N}{g_0}\sigma^2 - \frac{1}{2}\sigma\phi^a\phi^a. \tag{2.5}$$

Thus, in the new formalism, the only non-trivial interaction is the $\phi\phi\sigma$ coupling. All factors of $1/N$ come from the σ propagator (ig_0/N). Every line on a closed ϕ loop must always carry the same index, and this index must always be summed over; thus, we need not write explicit indices on ϕ loops, and every closed ϕ loop always gives a factor of N.

Fig. 3 shows the graphs of Fig. 1 in our new formalism. (The dashed line is the σ propagator.) Counting powers of $1/N$ is now much easier than before, but things can be made easier yet. Let us imagine analyzing a general Feynman graph as follows: First, let us strip away all the ϕ lines that end on external lines, that is to say, that are not part of closed loops. This yields a graph that has only external σ lines. Second, let us do all the momentum integrals over the closed ϕ loops. Every ϕ loop thus becomes a (non-local) interaction between the σ fields that terminate on that loop. We thus generate a graph with only σ lines; it can be thought of as a graph in an effective field theory whose Feynman rules are derived from an effective action, $S_{\text{eff}}(\sigma)$.

There are two ways of describing S_{eff}, in terms of Feynman graphs or in terms of functional integrals. The description in terms of graphs is shown in Fig. 4. The first graph gives the term linear in σ; the second graph gives a term quadratic in σ, which must be added to the third graph, the quadratic term already present in Eq. (2.5); the fourth graph gives the cubic term; etc. In terms of functional integrals, the quantum theory is defined by integrating the exponential of iS, the classical action, over all configurations of all fields in the theory. The effective action is obtained by integrating over the ϕs only:

$$e^{iS_{\text{eff}}(\sigma)} = \int \prod_a [d\phi^a] e^{iS(\phi^a,\sigma)}. \tag{2.6}$$

Fig. 3

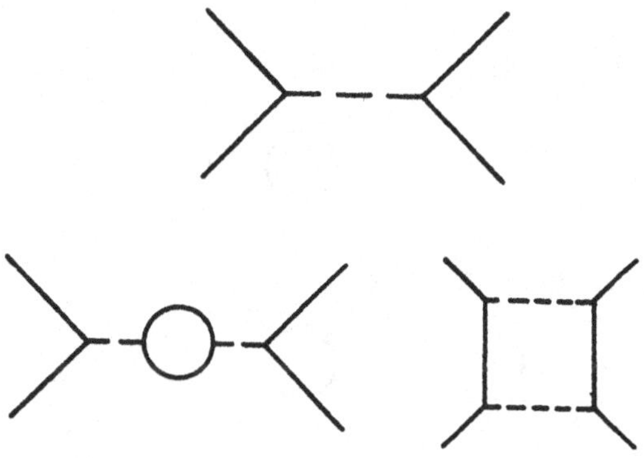

S is quadratic in the ϕs, so the integral is a Gaussian one, and can be done in closed form. Of course, 'closed form' is a hoax; the answer is a functional determinant that, for general σ, can be evaluated only by doing the Feynman graphs of Fig. 4.

Whichever way we describe S_{eff}, one thing about it is obvious; every term in it is proportional to N:

$$S_{\text{eff}}(\sigma, N) = N S_{\text{eff}}(\sigma, 1). \tag{2.7}$$

This makes counting powers of N very easy. Consider a graph in our effective field theory with E external lines, I internal lines, V vertices, and L independent loop integrations. These quantities are not independent. For a connected graph,

$$L = I - V + 1. \tag{2.8}$$

That is to say, we have one integration momentum for each internal line, but we also have one delta-function for each vertex; each delta-function cancels one momentum, except for one delta-function that is left over for overall momentum conservation. The power of N associated with a graph can be expressed in terms of these quantities. The propagator is obtained by inverting the quadratic part of the Lagrangian; thus each external and internal line carries a factor of $1/N$. (We put propagators on the external σ lines because we want eventually to attach them to external ϕ lines.) On the other hand, each vertex carries a factor of N. Thus the net power of N associated with a given graph is

$$N^{V-I-E} = N^{-E-L+1} \tag{2.9}$$

by Eq. (2.8). Thus, the smallest power of $1/N$ is obtained from graphs with no loops (tree graphs) and with the minimum number of external lines required to connect the external ϕ lines. In the case we began by studying,

Fig. 4

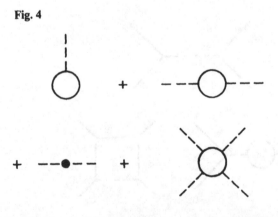

meson–meson scattering, two external σ lines are required, and thus the leading power is $1/N$.

This is no great surprise, of course; we hardly needed all this formalism to get this piddling result. However, we are almost in a position to compute meson–meson scattering to $O(1/N)$ in closed form. I say 'almost' because S_{eff} has the awkward feature of containing a term linear in ϕ. In the presence of such a term, there are an infinite number of tree graphs with two external lines; all one has to do is build a tree graph of arbitrary complexity, and then terminate all but two of its external lines on linear vertices. The cure for this problem is well known. We define a new, shifted field,

$$\sigma' \equiv \sigma - \sigma_0, \tag{2.10}$$

where σ_0 is a constant chosen such that $\sigma = \sigma_0$ is a stationary point of S_{eff},

$$\left.\frac{\delta S_{\text{eff}}}{\delta \sigma'}\right|_{\sigma' = 0} = 0. \tag{2.11}$$

(I will shortly show that σ_0 exists.) In terms of σ', there are no linear vertices.

The easiest way to construct $S_{\text{eff}}(\sigma')$ is to express $\mathscr{L}$ in terms of σ'. From Eq. (2.5),

$$\mathscr{L} = \frac{1}{2} \partial_\mu \phi^a \partial^\mu \phi^a - \frac{1}{2} \mu_1^2 \phi^a \phi^a$$
$$+ \frac{1}{2} \frac{N}{g_0} \sigma'^2 - \frac{1}{2} \sigma' \phi^a \phi^a + \frac{N}{g_0} \sigma_0 \sigma' \tag{2.12}$$

plus an irrelevant constant, where

$$\mu_1^2 \equiv \mu_0^2 + \sigma_0. \tag{2.13}$$

μ_1 is the ϕ mass, to leading (zeroth) order in $1/N$; it will be convenient to use it as an independent parameter of the theory instead of μ_0. The graphical construction of $S_{\text{eff}}(\sigma')$ is now the same as that shown in Fig. 4, with two exceptions. (1) The internal ϕ lines now carry a mass μ_1 rather than μ_0. (2) There is an additional linear vertex, coming from the last term in Eq. (2.12). We use this to cancel the linear vertex from the first graph in Fig. 4, thus at one stroke fixing σ_0 and eliminating it from all future computations.

We are now in a position to compute whatever we want. For example, let me sketch out the computation of ϕ–ϕ scattering, to leading order. There are only three graphs that can contribute, shown in Fig. 5. I have put a shaded blob on the σ' propagator to remind you that it is not just ig_0/N, but the full propagator obtained from inverting the quadratic

Fig. 5

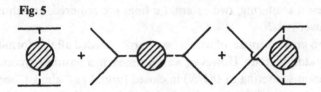

term in S_{eff}, the sum of the second and third graphs in Fig. 4. In momentum space,

$$D^{-1}(p) = -N\left[ig_0^{-1} + \int \frac{d^d k}{(2\pi)^d} \frac{1}{(k^2 - \mu_1^2 + i\varepsilon)} \frac{1}{([p+k]^2 - \mu_1^2 + i\varepsilon)} \right].$$
(2.14)

where d is the number of space-time dimensions. Note that for $d=4$, the integral is logarithmically divergent, but that its divergence can be absorbed in the bare coupling constant, g_0.

Of course, this is just the beginning. We could evaluate this integral, study the properties of the scattering amplitude it defines, investigate the interesting case in which we choose μ_1^2 to be negative (spontaneous symmetry breakdown?), worry about higher-order corrections, etc. However, as I warned you at the beginning, I am going to stop the discussion of ϕ^4 theory half-way, and go on to investigate other models. If you want to find out more about this theory, you will have to go to the literature[1] (or work it out yourself – it's not that hard).

2.2   The Gross–Neveu model[2]

The Gross–Neveu model is a renormalizable field theory that admits a $1/N$ expansion and displays asymptotic freedom, dynamical symmetry breakdown, and dimensional transmutation. This is the good news; the bad news is that it is a field theory in two space-time dimensions.

The fundamental dynamical variables of the theory are a set of Dirac fields, ψ^a, $a = 1 \ldots N$. In two dimensions, Dirac fields have only two components and the Dirac matrices are 2×2 matrices. In standard representation,

$$\gamma^0 = \sigma_z, \ \gamma^1 = i\sigma_y, \ \gamma_5 = \gamma^0 \gamma^1 = \sigma_x,$$
(2.15)

where the σs are the Pauli spin matrices. In all other ways, conventions are the same as in four dimensions.

The model is defined by

$$\mathcal{L} = \bar{\psi}^a i \partial_\mu \gamma^\mu \psi^a + \frac{g_0}{N} (\bar{\psi}^a \psi^a)^2.$$
(2.16)

We see from the first term that ψ^a has dimensions of (length)$^{-1/2}$, so g_0 is dimensionless and the interaction should be renormalizable. A mass term is excluded by the discrete chiral symmetry,

$$\psi^a \to \gamma_5 \psi^a, \quad \bar{\psi}^a \to -\bar{\psi}^a \gamma_5. \tag{2.17}$$

As we shall see, it is this symmetry (not the continuous U(N) symmetry) which suffers spontaneous breakdown.

The construction of the $1/N$ approximation is a duplicate of that for ϕ^4 theory. First we add a term to $\mathscr{L}$ that has no effect on the physics, involving an auxiliary field, σ,

$$\mathscr{L} \to \mathscr{L} - \frac{N}{2g_0}\left(\sigma - \frac{g_0}{N}\,\bar{\psi}^a\psi^a\right)^2$$

$$= \bar{\psi}^a i\partial_\mu\gamma^\mu\psi^a - \frac{N}{2g_0}\,\sigma^2 + \sigma\bar{\psi}^a\psi^a. \tag{2.18}$$

Next, we integrate over ψ loops to obtain $S_{\text{eff}}(\sigma)$. This is shown diagrammatically in Fig. 6. We only have even powers of σ because the trace of an odd number of Dirac matrices vanishes (alternatively, because σ changes sign under the discrete symmetry (2.17)).

At first glance, it would seem that there is no need for the third step in the analysis, shifting the σ field; S_{eff} is even in σ, and thus $\sigma = 0$ is automatically a stationary point. But this is begging the question; if we are interested in spontaneous breakdown of the discrete symmetry, the issue is precisely whether there are stationary points other than $\sigma = 0$.

Fortunately, to settle this issue we do not need to compute S_{eff} for general σ, merely for constant σ. In this case, we can put the universe in a box of spatial extent L and temporal extent T, and define

$$-V(\sigma) = \lim_{L,T \to \infty} S_{\text{eff}}(\sigma)/LT. \tag{2.19}$$

Each stationary point of V is a possible starting point for a $1/N$ expansion, defines a possible vacuum state of the theory to leading order in $1/N$.

Fig. 6

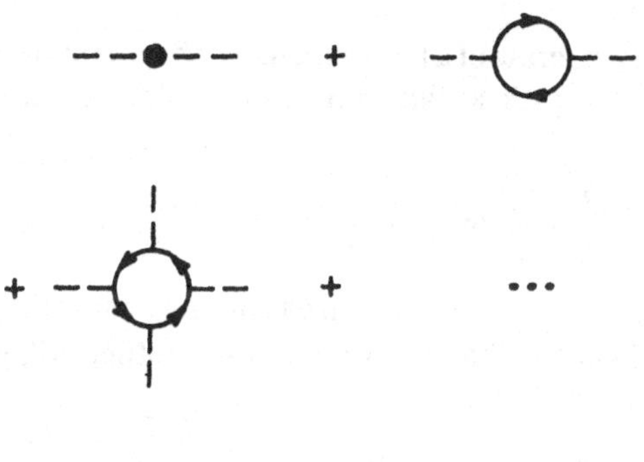

The energy densities of these vacua are easily computed. If we denote a vacuum energy density by $\mathscr{E}$, and if we denoted by Σ the sum of all connected vacuum-to-vacuum Feynman graphs, then a general formula of time-dependent perturbation theory states that

$$-i\mathscr{E} = \lim_{L,T\to\infty} \Sigma/LT. \tag{2.20}$$

To lowest (minus first) order in $1/N$, Σ is given by the sum of all connected tree graphs with no external lines. This set consists of precisely one graph, with only one vertex, the term in S_{eff} that contains no powers of the shifted field. That is to say, V at the stationary point *is* the vacuum energy density (to leading order); if there are several stationary points, only those of minimum V are true vacua.[3]

Diagrammatically, V is given by the sum of Feynman diagrams in Fig. 6, with all external lines carrying zero two-momentum, and with the momentum-conserving delta-functions left off. (These give the factor of LT.) The summation of these graphs has been done countless times in the literature. I will bore you by doing it once more:

$$-iV = -i\frac{N}{2g_0}\sigma^2 - \sum_{n=1}^{\infty} \frac{N}{2n}\,\text{Tr} \int \frac{d^2p}{(2\pi)^2}\left(\frac{-\not{p}\sigma}{p^2+i\varepsilon}\right)^{2n}. \tag{2.21}$$

The terms in this series have the following origins. (a) From Eq. (2.18), every vertex carries a factor of i and every propagator a factor of $i\not{p}/(p^2+i\varepsilon)$. (b) There is an N for the N Fermi fields and a (-1) for the Fermi loop. (c) Cyclic permutation of the external lines recreates the same graph; thus the $1/2n!$ in Dyson's formula is incompletely cancelled and we have a left-over factor of $1/2n$. It is trivial to do the trace, sum the series, and rotate the integration to Euclidean two-momentum, p_E. We thus obtain

$$V = N\left[\frac{\sigma^2}{2g_0} - \int \frac{d^2p_E}{(2\pi)^2}\ln\left(1+\frac{\sigma^2}{p_E^2}\right)\right]. \tag{2.22}$$

The momentum integral is ultraviolet divergent; we cut it off by restricting the integral to $p_E^2 \leqslant \Lambda^2$, with Λ some large number. We find

$$V = N\left[\frac{\sigma^2}{2g_0} + \frac{1}{4\pi}\sigma^2\left(\ln\frac{\sigma^2}{\Lambda^2} - 1\right)\right] \tag{2.23}$$

We wish to rewrite this in terms of a (conveniently defined) renormalized coupling constant, g. I will pick an arbitrary renormalization mass, M, and define g by

$$\frac{1}{g} \equiv N^{-1}\frac{d^2V}{d\sigma^2}\bigg|_M = \frac{1}{g_0} + \frac{1}{2\pi}\ln\frac{M^2}{\Lambda^2} + \frac{1}{\pi}. \tag{2.24}$$

Note that g is $g_0 + O(g_0^2)$, as a good renormalized coupling constant should be. If you do not like my choice of M and want to use another, M', you are free to do so. Your coupling constant is connected to mine by

$$\frac{1}{g'} = \frac{1}{g} + \frac{1}{2\pi} \ln \frac{M'^2}{M^2}.$$ (2.25)

V is now given by

$$V = N\left[\frac{\sigma^2}{2g} + \frac{1}{4\pi}\sigma^2\left(\ln\frac{\sigma^2}{M^2} - 3\right)\right].$$ (2.26)

Two of the announced properties of the model are now manifest. Firstly, it is renormalizable, at least in the order to which we are working; Eq. (2.26) is totally free of cutoff-dependence. Secondly, the theory is asymptotically free. This can be seen in the usual two equivalent ways. (1) In Eq. (2.24), if we hold g and M fixed, and let Λ go to infinity, g_0 goes to zero. In a fixed theory, the bare coupling constant vanishes for infinite cutoff. (2) In Eq. (2.25), if we hold g and M fixed, and let M' go to infinity, g' goes to zero. In a fixed theory, the renormalized coupling constant vanishes for infinite renormalization mass.

We can now search for spontaneous symmetry breakdown.

$$\frac{dV}{d\sigma} = N\left[\frac{\sigma}{g} + \frac{\sigma}{2\pi}\left(\ln\frac{\sigma^2}{M^2} - 2\right)\right].$$ (2.27)

This vanishes at

$$\sigma^2 = \sigma_0^2 \equiv M^2 \exp\left(2 - \frac{2\pi}{g}\right).$$ (2.28)

At this point

$$V = -N\sigma_0^2/4\pi.$$ (2.29)

This is negative, that is to say, less than $V(0)$. The third announced property of the model is now manifest. The discrete chiral symmetry suffers spontaneous breakdown, and the massless fermions acquire a mass. To leading order, this mass is just σ_0.

I stated earlier that one could choose the renormalization mass, M, arbitrarily; a change in M could always be compensated for by an appropriate change in g. I will now use this freedom to choose M to be σ_0. By Eq. (2.28) this fixes g:

$$g = \pi.$$ (2.30)

This is the fourth announced property, dimensional transmutation. We began with a theory that apparently depended on only one continuous parameter, g_0. We have arrived at a theory that depends on only one continuous parameter, σ_0. The surprise is that we began with a dimensionless parameter, on which we would expect observable quantities to depend in a complicated way, while we arrived at a dimensionful parameter, on which observable quantities must depend in a trivial way, given by dimensional analysis.

Of course, dimensional transmutation is an inevitable feature of any renormalizable field theory depending only on a single dimensionless coupling constant. Renormalization trades the single bare coupling constant, g_0, for the pair (g, M), the renormalized coupling constant and the renormalization point. But this is a redundant pair; we still have only a one-parameter theory; the (g, M) plane is the union of curves such that any two points on the same curve define the same theory. One way of parametrizing these curves is by that value of M at which they pass through some fixed value of g, like $1/2$ or π. Our one-parameter family of theories are now labeled by a single parameter, and it is a mass. Nevertheless, even though we expect dimensional transmutation to occur in very general circumstances, it is still pleasant to have a model in which we can explicitly see it happening.

2.3   The CP^{N-1} model[4]

Like the Gross–Neveu model, the CP^{N-1} model is a two-dimensional renormalizable field theory which displays dimensional transmutation. Also like the Gross–Neveu model, the theory contains a set of particles that are massless in perturbation theory but which acquire a mass in the leading $1/N$ approximation. However, this is not due to spontaneous symmetry breakdown, but to its reverse. The particles are massless because they are the Goldstone bosons of a spontaneously broken symmetry, and they acquire a mass because the symmetry is dynamically restored. This should be no surprise. In two dimensions there can be no spontaneous breakdown of a symmetry associated with a local conserved current; if the $1/N$ approximation had not predicted symmetry restoration, we would have known it was a lie.

More interestingly, the massive particles are confined; there is a linear potential between particle and antiparticle which prevents the components of a pair from being separated indefinitely. Of course, a linear potential is not as difficult to achieve in two space-time dimensions as in four; on a line, the classical electric force between oppositely charged particles is independent of distance. However, in this case the linear potential arises in a theory without any fundamental gauge fields; this is astonishing.

The CP^{N-1} model is a generalization of the nonlinear sigma model. I will first remind you of this latter theory and then go on to describe the generalization.

The linear sigma model is a theory of N scalar fields, assembled into an

N-vector, ϕ, with dynamics defined by

$$\mathcal{L} = \frac{1}{2}\partial_\mu\phi\cdot\partial^\mu\phi - \frac{\lambda}{8}(\phi\cdot\phi - a^2)^2, \tag{2.31}$$

where λ and a are positive numbers. This theory is SO(N)-invariant, but, at least in perturbation theory, the symmetry spontaneously breaks down to SO($N-1$); the ground states of the theory are constant fields lying on the $(N-1)$-dimensional sphere,

$$\phi\cdot\phi = a^2. \tag{2.32}$$

The nonlinear sigma model is the formal limit of this theory as λ goes to infinity. The fields in general, not just in their ground state, are restricted to obey Eq. (2.32). The Lagrange density then simplifies to

$$\mathcal{L} = \tfrac{1}{2}\partial_\mu\phi\cdot\phi\partial^\mu\phi. \tag{2.33}$$

Of course, this simple form does not mean that the dynamics is simple. The N components of ϕ are not independent, and Eq. (2.33) in fact describes a highly complicated nonlinear theory, as would be manifest were we to write ϕ as a function of $N-1$ independent variables (say, angles on the sphere). The nonlinear model can be thought of as a stripped-down version of the linear model, with only the Goldstone bosons retained.

The role of the parameter a can be clarified by rescaling the fields,

$$\phi \to a\phi. \tag{2.34}$$

Under this transformation, the constraint becomes

$$\phi\cdot\phi = 1, \tag{2.35}$$

while

$$\mathcal{L} = \frac{a^2}{2}\partial_\mu\phi\cdot\partial^\mu\phi. \tag{2.36}$$

From this we see that $1/a$ is a coupling constant. (See the discussion of powers of N following Eq. (2.7).) This makes sense; if a were infinite, we would not be able to tell the difference between the sphere defined by Eq. (2.32) and ordinary flat space, for which Eq. (2.33) would define a free field theory.

I emphasize that the passage to the nonlinear model is purely a formal limit. For example, the linear model is renormalizable in four dimensions or less, while the nonlinear model is renormalizable only in two dimensions (where scalar fields are dimensionless) or less. We have thrown away some important physics (at least at short distances) by throwing away the non-Goldstone modes. Nevertheless, once we have the nonlinear model, we can certainly study it as a theory in its own right. Indeed, I could have constructed the model directly, as a field theory where the field variables lie in a nonlinear space; I chose to build it from the linear model only for reasons of pedagogy.

The CP^{N-1} model can likewise be constructed directly as a field theory in a nonlinear space. However, again for reasons of pedagogy, I will obtain it as the formal limit of a linear theory.

The linear theory is a theory of $N^2 - 1$ scalar fields, assembled into an $N \times N$ traceless Hermitian matrix, ϕ, with dynamics defined by

$$\mathscr{L} = \tfrac{1}{2}\text{Tr}\, \partial_\mu \phi \partial^\mu \phi - \lambda\, \text{Tr}\, P(\phi), \tag{2.37}$$

where λ is a positive number and P is some polynomial in ϕ. This theory is invariant under SU(N):

$$\phi \to U\phi U^\dagger,\ U \in \text{SU}(N). \tag{2.38}$$

It is possible to choose P such that the minima of Tr P are matrices with $N - 1$ equal eigenvalues and one unequal eigenvalue; SU(N) then breaks down spontaneously to SU($N-1$)$\otimes$U(1). In equations, the ground states of the theory are constant fields of the form

$$\phi = g_0^{-1}[N^{\frac{1}{2}}zz^\dagger - N^{-\frac{1}{2}}I], \tag{2.39}$$

where z is an N-dimensional column vector of unit length,

$$z^\dagger z = 1, \tag{2.40}$$

and g_0 is some parameter derived from $P(\phi)$. I will assume that P has N dependence such that g_0 remains fixed as N goes to infinity. As we shall see shortly, this is necessary to get a $1/N$ expansion.

The CP^{N-1} model is the formal limit of this theory as λ goes to infinity. The fields in general, not just in their ground state, are restricted to obey Eq. (2.39). The Lagrange density then simplifies to

$$\begin{aligned}\mathscr{L} &= \tfrac{1}{2}\text{Tr}\, \partial_\mu \phi \partial^\mu \phi \\ &= (N/g_0^2)(\partial_\mu z^\dagger\, \partial^\mu z - j_\mu j^\mu),\end{aligned} \tag{2.41}$$

where

$$j_\mu \equiv (2\text{i})^{-1}[z^\dagger \partial_\mu z - (\partial_\mu z^\dagger)z]. \tag{2.42}$$

It is convenient to rescale z,

$$z \to g_0 N^{-\frac{1}{2}}z. \tag{2.43}$$

The Lagrange density then becomes

$$\mathscr{L} = \partial_\mu z^\dagger\, \partial^\mu z - g_0^2 N^{-1} j_\mu j^\mu, \tag{2.44}$$

while the constraint equation is

$$z^\dagger z = N/g_0^2. \tag{2.45}$$

We see that we have (in perturbation theory) a theory of massless particles with short-range interactions between them. The theory is slightly more complex than the nonlinear sigma model; there are not only interactions induced by the constraint but also explicit interactions in the Lagrange density.

It is interesting to count the number of particles in the theory. At first glance, it looks like we have N complex fields with one real constraint, yielding $2N-1$ real fields. On the other hand, if we count Goldstone bosons, we would expect the number of real fields to be

$$\dim \mathrm{SU}(N) - \dim \mathrm{U}(N-1) = N^2 - 1 - (N-1)^2 = 2N - 2. \quad (2.46)$$

The second count is the correct one. The first count ignored the fact that the transformation

$$z \rightarrow e^{i\theta} z, \quad (2.47)$$

does nothing to ϕ; the overall phase of z is not a dynamical variable and should not have been counted. The manifold in which our fields lie is the set of complex N-vectors with fixed length, *and* with N-vectors differing only by a multiplicative phase factor identified. This is complex projective $N-1$ space, $\mathbb{C}P^{N-1}$.

I will now solve the model to leading order in $1/N$, in two space-time dimensions, where it is renormalizable. The first step is standard, eliminating the quartic interaction by introducing an auxiliary field. Since the quartic interaction is of the form vector times vector, the auxiliary field must be a vector field. Thus, we change $\mathscr{L}$ by

$$\mathscr{L} \rightarrow \mathscr{L} + g_0^2 N^{-1} (j_\mu + g_0^{-2} N A_\mu)^2$$
$$= \partial_\mu z^\dagger \partial^\mu z + 2 j^\mu A_\mu + g_0^{-2} N A_\mu A^\mu. \quad (2.48)$$

Using the constraint, Eq. (2.45), this can be rewritten in the amusing form,

$$\mathscr{L} = (\partial_\mu - i A_\mu) z^\dagger (\partial_\mu + i A_\mu) z. \quad (2.49)$$

This looks like a piece of a gauge field theory, with the gauge transformation of the fields, Eq. (2.47), canceled by the gauge transformation of the vector potential,

$$A_\mu \rightarrow A_\mu - \partial_\mu \theta. \quad (2.50)$$

Of course, the 'gauge invariance' is a hoax, just a reflection of the fact that we are describing the theory in terms of highly redundant variables.[5]

The next step is to get the constraint into the Lagrange density. We do this with another auxiliary field, σ,

$$\mathscr{L} = (\partial_\mu - i A_\mu) z^\dagger (\partial_\mu + i A_\mu) z - \sigma [z^\dagger z - g_0^{-2} N]. \quad (2.51)$$

The new field is a Lagrange multiplier; its Euler–Lagrange equation is the constraint. Equivalently, performing the functional integral over σ yields a delta-function at each point of space-time which enforces the constraint. We now have a Lagrange density that is quadratic in z, so we can proceed as before to integrate out the internal z loops and obtain an effective action, a functional of σ and A_μ. The first few terms in the graphical expansion of S_{eff} are shown in Fig. 7; the directed lines are zs, the dashed lines σs, and the wiggly lines As. We see that we have pure σ terms (the first two lines), pure A terms (the third line) and mixed terms (the fourth line); however, all the terms are proportional to N, just as before.

Also just as before, to eliminate the linear terms, we must shift to a stationary point of $V(\sigma)$. The computation of V is essentially a rerun of that for the Gross–Neveu model. The relevant graphs are those in the first two lines of Fig. 7; the only difference in the computation is that the Fermi minus sign is missing and that we now have σ where before we had σ^2. Thus,

$$V = -N\left[\frac{\sigma}{g_0^2} + \frac{\sigma}{4\pi}\left(\ln\frac{\sigma}{\Lambda^2} - 1\right)\right], \tag{2.52}$$

where Λ is the cutoff. To renormalize this, I pick an arbitrary renormalization mass, M, and define the renormalized coupling constant, g, by

$$\frac{1}{g^2} \equiv -N^{-1}\frac{dV}{d\sigma}\bigg|_{M^2} = \frac{1}{g_0^2} + \frac{1}{4\pi}\ln\frac{M^2}{\Lambda^2}. \tag{2.53}$$

Note that g is $g_0 + O(g_0^2)$, as it should be. We can now rewrite V as

$$V = -N\left[\frac{\sigma}{g^2} + \frac{\sigma}{4\pi}\left(\ln\frac{\sigma}{M^2} - 1\right)\right]. \tag{2.54}$$

As before, the theory is renormalizable; all reference to the cutoff has

Fig. 7

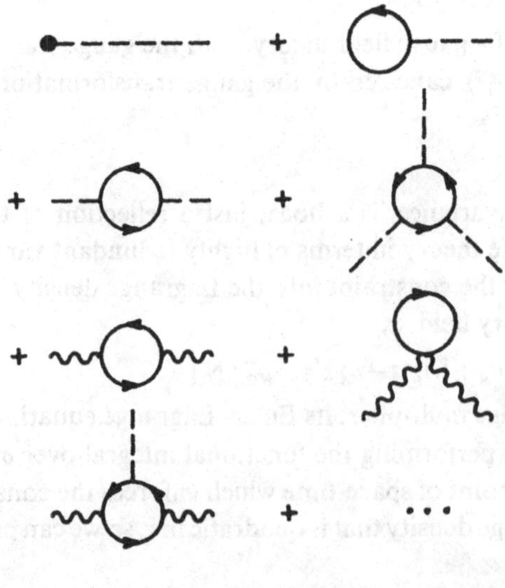

disappeared. As before, it is asymptotically free; for fixed g and M, g_0 vanishes as Λ goes to infinity.

We can now search for stationary points.

$$\frac{dV}{d\sigma} = -N\left[\frac{1}{g^2} + \frac{1}{4\pi}\left(\ln\frac{\sigma}{M^2}\right)\right].$$ (2.55)

This has a unique zero, at

$$\sigma = \sigma_0 \equiv M^2 \exp\left(-4\pi/g^2\right).$$ (2.56)

In terms of σ_0,

$$V = -\frac{N\sigma}{4\pi}\left(\ln\frac{\sigma}{\sigma_0} - 1\right).$$ (2.57)

Once again, dimensional transmutation has occurred; the dimensionless coupling constant, g, has disappeared from the theory, to be replaced by the dimensionful parameter σ_0.

From Eq. (2.51), σ_0 is the squared mass of the z-particles, to lowest (zeroth) order in $1/N$. Something remarkable has happened. We started out, in our linear model, with a set of fields transforming according to the adjoint representation of SU(N). By sending λ to infinity, we replaced these by a set of Goldstone bosons which transformed nonlinearly under the action of the group. These have now turned into an ordinary set of massive mesons, transforming linearly, according to the fundamental representation of the group. We are used to making mesons out of quarks; here we have made (bosonic) quarks out of mesons, fundamental representations out of adjoint ones.

Something even more remarkable happens when we study the long-range force between a pair of zs. By the same arguments as were given in ϕ^4 theory, to leading order the force is given by graphs like those shown in Fig. 5. To compute these graphs, we need the σ–σ propagator, the σ–A_μ propagator, and the A_μ–A_ν propagator. By current conservation and Lorentz invariance, the σ–A_μ propagator vanishes. At zero momentum transfer, the σ–σ propagator is given by

$$D_{\sigma\sigma}(0) = -i\left[\frac{d^2V}{d\sigma^2}\bigg|_{\sigma_0}\right]^{-1} = \frac{4\pi i}{N}.$$ (2.58)

This is not infinite; thus σ exchange gives no long-range force.

The A_μ–A_ν propagator is a different story. The term in S_{eff} quadratic in A obtained from the graphs on the third line of Fig. 7. These are just the standard second-order photon self-energy graphs; to compute their sum is a trivial exercise in Feynman-graph technology. The answer, in momentum space, is

$$- \frac{iN}{4\pi} [g_{\mu\nu}p^2 - p^\mu p^\nu] \int_0^1 \frac{dx(1-2x)^2}{\sigma_0^2 - p^2 x(1-x) - i\varepsilon}. \tag{2.59}$$

If we are only interested in long-range forces, that is to say, in small momenta, we may neglect the p^2 in the integrand. We thus obtain

$$- \frac{iN}{12\pi\sigma_0^2} [g_{\mu\nu}p^2 - p_\mu p_\nu]. \tag{2.60}$$

This corresponds to a term in the effective action of the form

$$S_{\text{eff}} = - \frac{N}{48\pi\sigma_0^2} \int d^2x (\partial_\mu A_\nu - \partial_\nu A_\mu)^2. \tag{2.61}$$

Aside from a trivial normalization, this is the action for the free electromagnetic field. This is the most astonishing feature of the model; a genuine gauge field has been dynamically generated, produced as a result of radiative corrections in a theory that perturbatively has only short-range interactions. The gauge field now produces a linear potential that confines the zs. Not only does the theory have (bosonic) quarks, it has confined quarks.

There is much more that can be said about the CP^{N-1} model. For example, the classical theory admits instantons (just as in chromodynamics), and there is no infrared cutoff on instanton sizes (again as in chromodynamics). Thus the model can be used as a laboratory for instanton physics (= arena for bloody controversies). Unfortunately, I do not have the time to go into any of this here, and must once again refer you to the literature.[6]

3   Adjoint representations, or, chromodynamics

3.1   The double-line representation and the dominance of planar graphs[7]

The $1/N$ expansion is vastly more difficult for SU(N) gauge theories than for any of the theories of Section 2. The source of the difficulty has nothing to do with the traditional problems of chromodynamics, the intricacies of gauge invariance or the uncontrollable infrared divergences. It is just that we have to deal with fields that transform according to the adjoint representation rather than the vector representation, objects that carry two group indices rather than one. ϕ^4 theory is just as difficult if ϕ is in the adjoint representation.

The dynamical variables of the theory are a set of Dirac fields, ψ^a, and a set of gauge fields, $A^a_{\mu b}$, where a and b run from 1 to N. The Dirac fields can be thought of as elements of an SU(N) column vector, the gauge fields as those of a traceless Hermitian matrix,

$$A^a_{\mu b} = A^{b\dagger}_{\mu a}, \qquad A^a_{\mu a} = 0. \tag{3.1}$$

From the gauge fields we define

$$F^a_{\mu\nu b} = \partial_\mu A^a_{\nu b} + i A^a_{\mu c} A^c_{\nu b} - (\mu \leftrightarrow \nu). \tag{3.2}$$

The dynamics of the theory is defined by

$$\mathcal{L} = \frac{N}{g^2} \left[-\tfrac{1}{4} F^a_{\mu\nu b} F^{\mu\nu b}_a + \bar{\psi}_a (i\partial_\mu + A^a_{\mu b}) \gamma^\mu \psi^b - m\bar{\psi}_a \psi^a \right], \tag{3.3}$$

where g and m are real numbers. For $N = 3$, this is the Lagrange density of quantum chromodynamics; the Dirac fields are quarks and the gauge fields gluons. I will retain this nomenclature for general N.

Remarks. (1) I have put the coupling constant in front of the total Lagrangian. Of course, by rescaling the fields, as we did in Section 2.3, we can remove it from the quadratic terms in Eq. (3.3) and put it in its conventional position, a factor of $g/N^{\frac{1}{2}}$ multiplying the cubic terms and one of g^2/N multiplying the quartic terms. (2) Equation (3.3) is incomplete; I have left out gauge-fixing terms, ghost couplings, renormalization counterterms, and the possibility of more than one flavor. I have done this to keep my equations as simple as possible. Practically all of my analysis will be purely combinatoric, hardly dependent at all on the detailed form of the interactions; thus, the extension to include all these neglected effects will be trivial. (3) I have taken advantage of my knowledge of how things are going to turn out to put the factor of N in the right place from the very beginning. As we shall see shortly, it is the theory defined by Eq. (3.3) that admits a non-trivial $1/N$ expansion, and not, for example, the one with an N^2 in place of the N.

To take proper account of factors of $1/N$, we must keep proper track of the indices within a Feynman graph. Let us begin our analysis with the propagators. The quark propagator is

$$\overline{\psi^a(x)\bar{\psi}_b}(y) = \delta^a_b S(x-y), \tag{3.4}$$

where S is the propagator for a single Dirac field. Thus there is no trouble following indices along a quark line; the index at the beginning is the same as the index at the end. The gluon propagator is

$$\overline{A^a_{\mu b}(x) A^c_{\nu d}}(y) = \left(\delta^a_d \delta^c_b - \frac{1}{N} \delta^a_b \delta^c_d \right) D_{\mu\nu}(x-y), \tag{3.5}$$

where $D_{\mu\nu}$ is the propagator for a single gauge field. The term proportional to $1/N$ is there because the gluon field is traceless; it would not be present if our gauge group were U(N) rather than SU(N). However, precisely because this term is proportional to $1/N$, we can drop it, even for SU(N), if we are only interested in the leading order in $1/N$. (Of course, we must remember we have dropped it if we want to compute subleading orders. See Appendix 3.) There is now no problem following indices along a gluon line; the index pair at the beginning is the same as the index pair at the end. As far as the index structure goes, a gluon propagates like a quark–antiquark pair.

This observation is at the root of the ingenious double-line representation of 't Hooft. This is an alternative way of drawing Feynman graphs in which we draw one line for each index rather than one line for each virtual particle. Thus a quark propagator is represented by a single index line, because a quark carries only one index, but a gluon propagator is represented by two index lines, a double line, because a gluon carries two indices. Fig. 8 is a translation dictionary from the old single-line representation on the left to the new double-line representation on the right. The figure shows the translations of propagators, vertices, and a typical vacuum-to-vacuum graph. The great advantage of the double-line representation is

Fig. 8

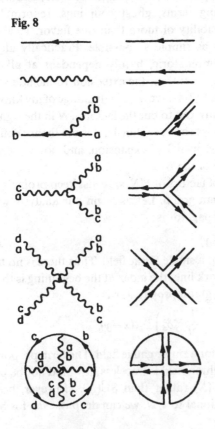

immediately obvious: we do not have to clutter our graphs with little letters to show where the indices go; to follow the indices all we have to do is follow the arrows. Phrased in another way, to each double-line graph there corresponds a single-line graph with indices assigned to the lines. Thus, if there is more than one way of assigning indices to the lines in a given single-line graph, there will be more than one double-line graph associated with it.

I will now show that the power of $1/N$ carried by a double-line graph is determined by certain topological properties of the graph. For simplicity, I will begin by restricting myself to vacuum-to-vacuum graphs, graphs with no external lines. I will later extend the analysis as needed.

Because the graph has no external lines, every index line must close to make an index loop. Let us imagine each index loop to be the perimeter of a polygon. The double-line graph can then be read as a prescription for fitting together these polygons. To be more precise, we identify one edge of one polygon with one edge of another if they both lie on the same double line (gluon propagator). In this way, we construct a two-dimensional surface.

We can give an orientation to each polygon by the direction of the arrows around its perimeter and the right-hand rule. Because the two halves of a double line are always oppositely directed, this orientation is consistent as we travel about the surface; we have constructed an oriented surface. Thus we can get spheres or toruses, but not Klein bottles. (If we were doing the parallel analysis for SO(N) rather than SU(N), quark and antiquark would transform equivalently, our lines would not carry arrows, and we could get Klein bottles.)

It is easy to count the power of N associated with this surface. Let the surface have V vertices, E edges, and F faces. Every vertex is an interaction vertex of a Feynman graph, and carries a factor of N, by Eq. (3.3). Every edge is a propagator, either quark or gluon, and carries a factor of $1/N$, again by Eq. (3.3). Every face is an index loop, and thus yields a factor of N when we sum over all possible values of the index. Thus the graph is proportional to

$$N^{F-E+V} \equiv N^{\chi}. \tag{3.6}$$

χ is the Euler characteristic. It is a famous topological invariant, and can be computed in quite another way. Every two-dimensional oriented surface is topologically equivalent to a sphere with some number of holes cut out of it and some number of handles stuck on to it. For example, a torus is a sphere with one handle; a disc is a sphere with one hole; a cylinder (without end caps) is a sphere with two holes; a loving cup is a sphere with one hole and two handles; etc. Let H be the number of handles and B (for boundary) be the number of holes. Then the Euler characteristic is given by

$$\chi = 2 - 2H - B. \tag{3.7}$$

(If you are not familiar with the Euler characteristic, a quick and dirty proof of this formula is given in Appendix 1.)

Thus the leading connected vacuum-to-vacuum graphs are proportional to N^2, and the associated surface has the topology of a sphere. What does this mean in terms of our original single-line graphs? The boundary of a hole is a loop of unpaired index lines, that is to say, a quark loop; thus, the leading graphs involve only gluons. (This is not a deep result. It takes just as many powers of the coupling constant to make a quark pair as to make a gluon pair, but there are N times more gluons than quarks to sum over.) Let us remove one randomly selected face from our spherical surface and project the remainder of the surface onto a plane. We thus obtain a planar graph. If we collapse the double lines to single lines, the graph remains planar. Conversely, given a planar graph made up only of gluon lines, we can always associate a double-line graph of the desired type with it. A planar graph divides the portion of the plane it occupies into regions. All we need do is draw a clockwise index line just inside the boundary of each of these regions and a counterclockwise index line just outside the boundary of the whole graph.

This analysis will shortly become important to us, so I summarize it in the following: *First Result – The leading connected vacuum-to-vacuum graphs are of order N^2. They are planar graphs made up only of gluons.*

We might be interested in the leading vacuum-to-vacuum graphs that have a nontrivial dependence on quark parameters, like quark masses or the number of flavors. These graphs must involve at least one quark loop; we see from Eq. (3.7) that the leading graphs have only one quark loop (= one hole) and no handles. Thus one of these graphs defines a spherical surface with one face removed. We can project this onto a plane, just as we did before. The only difference from the preceding case is that the outer boundary of the resultant planar graph is the perimeter of the hole, the quark loop.

Thus we obtain: *Second Result – The leading connected vacuum-to-vacuum graphs with quark lines are of order N. They are planar graphs with only one quark loop; the loop forms the boundary of the graph.* Thus the first graph in Fig. 9 is leading, but the second is not, even though it is planar.

As we shall see immediately, a very large amount of meson phenomenology is implicit in these two results.

Fig. 9

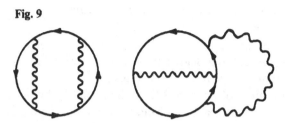

3.2   *Topology and phenomenology*
    The naive quark model deals with mesons (particles made of a quark and an antiquark) and baryons (particles made of three quarks). Somewhat more sophisticated models worry about glueballs (particles made of gluons) and exotics, in particular exotic mesons (particles made of more than one quark–antiquark pair). Because it takes N quarks to make a color singlet, baryons are a special problem for the $1/N$ expansion, and I will postpone their study to Section 3.4. However, we do have all the machinery needed to study mesons, glueballs, and exotic mesons.
    Our method will be to study the states made by applying certain gauge-invariant local operators to the vacuum. We will restrict ourselves to monomials in ψ, $\bar{\psi}$, $F_{\mu\nu}$, and their covariant derivatives, and, further, to monomials that cannot be written as the product of two gauge-invariant monomials of lower degree. (Practically everything I say will be valid even for non-local non-polynomial gauge-invariant operators, like $\bar{\psi}_a(x)U^a_b\psi^b(y)$, where U is the ordered exponential integral of the gauge fields over the line from x to y; as far as counting powers of N goes, the key point is that the operator cannot be decomposed into a product of gauge-invariant operators.)
    To begin with, let us study quark bilinears, operators involving one ψ and one $\bar{\psi}$. Let $B_1 \ldots B_n$ be a string of such bilinears, each at some point, and let $\langle B_1 \ldots B_n \rangle_C$ be the connected Green's function for this string. If we modify the action of our theory by an additional term,

$$S \to S + N \sum_i b_i B_i, \tag{3.8}$$

where the bs are numbers, and if W is the sum of connected vacuum-to-vacuum graphs, then

$$\langle B_1 \ldots B_n \rangle_C = (iN)^{-n} \left. \frac{\partial^n W}{\partial b_1 \ldots \partial b_n} \right|_{b_i = 0}. \tag{3.9}$$

    The reason for doing things this way is that all the analysis of Section 3.1 applies to the action (3.8) without a word of alteration; every interaction vertex carries a factor of N, and, in the double-line representation, every interaction vertex becomes a vertex of a polyhedron. (Note that this last point would not be true if one of the Bs was a product of two gauge-invariant monomials of lower degree. In that case we would get two polyhedral vertices from a single interaction vertex.)

Thus, from Result 2 of Section 3.1, we immediately know that the leading graphs are planar graphs with one quark loop, the boundary of the graph. Of course, all the bilinears must appear as insertions on the quark loop. Fig. 10 shows a leading graph for a three-bilinear Green's function; the bilinears are indicated by crosses. (We see that B_3 is linear in gluon fields as well as bilinear in quark fields.) We also know from Result 2 that the contribution of these graphs to W is proportional to N; thus, to leading order,

$$\langle B_1 \ldots B_n \rangle_C \propto N^{(1-n)}. \tag{3.10}$$

We will also have use for gauge-invariant local operators made up exclusively of gauge fields. I will denote such operators by G_i. The Green's function for a mixed string of Bs and Gs can be studied by the same method as before. Once again the leading graphs are planar graphs with only one quark loop, the boundary of the graph. Of course, the Gs can appear as insertions anywhere in the graph. Again, the contribution to W is proportional to N; thus, to leading order

$$\langle B_1 \ldots B_n G_1 \ldots G_m \rangle_C \propto N^{(1-n-m)}. \tag{3.11}$$

The situation is slightly different for a string made up of Gs alone. Here Result 1 of Section 3.1 applies; the leading graphs are planar graphs with no quark loops, and their contribution to W is proportional to N^2. Thus, to leading order,

$$\langle G_1 \ldots G_m \rangle_C \propto N^{(2-m)}. \tag{3.12}$$

We can derive much interesting physics from these equations if we make one assumption, that chromodynamics confines for arbitrarily

Fig. 10

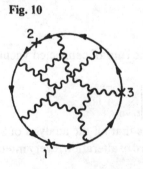

large N, that all states made by applying strings of gauge-invariant operators to the vacuum are states composed of SU(N)-singlet particles. I stress this is pure assumption. Our main reason for believing in confinement is that we can see quarks within hadrons but we cannot liberate them; this is an experimental reason, not a theoretical one, and experiment exists only for $N = 3$. Nevertheless, the assumption is not unreasonable. To the small extent to which we do have a theoretical understanding of confinement (for example, from strong-coupling lattice gauge theories), there does not seem to be anything special about small N. But the best reason for assuming large-N chromodynamics confines is that, if it does not confine, it bears no resemblance to reality, and the $1/N$ expansion is hopeless. Thus, we might as well make the assumption and see where it leads us.

I will define a meson to be a one-particle state made by applying a quark bilinear to the vacuum. From Eq. (3.10), our bilinears are not properly normalized to create mesons with N-independent amplitudes. Therefore, we renormalize them, and define

$$B_i' = N^{\frac{1}{2}} B_i. \tag{3.13}$$

For these,

$$\langle B_1' \dots B_n' \rangle \propto N^{(2-n)/2}. \tag{3.14}$$

Meson scattering amplitudes are obtained from these Green's functions by the reduction formula; thus a scattering amplitude with n legs is proportional to $N^{(2-n)/2}$. For large N, mesons interact weakly; $N^{-\frac{1}{2}}$ sets the scale of meson interactions just as e sets the scale of the interactions of electrons and photons.

Indeed, the parallel is exact. In quantum electrodynamics, in lowest non-vanishing order in perturbation theory, the tree approximation, an n-field Green's function is proportional to $e^{(n-2)}$. As far as dependence on N is concerned, it is as if our bilinears were linear functions of fundamental fields in some field theory with coupling constant proportional to $N^{-\frac{1}{2}}$, as if the leading order in $1/N$ were the tree approximation in this theory.

Of course, the tree approximation is characterized by more than just its dependence on the coupling constant. Green's functions in the tree approximation have very simple analytic structures in momentum space; their only singularities are poles. I will now argue that Eq. (3.14) implies that the same holds for our Green's functions.

I will begin with the two-point function. To show that the only singularities are poles is to show that a bilinear applied to the vacuum produces only single-meson states (to leading order). The proof is by contradiction. For example, let us assume a renormalized bilinear produces a pair of singlet particles (A, B), with an amplitude of order unity. We can then construct the sequence of events shown in Fig. 11. This is not a Feynman diagram, but a drawing of events in space-time (time runs upward). Initially a bilinear produces a pair; the components separate until they are each reflected by a bilinear; when they come back to the same point, they are absorbed by yet another bilinear. Every vertex in the figure is obtained from the assumed initial vertex by crossing; thus all the vertices are of order unity and the process produces a physical-region singularity of order unity in the four-bilinear connected Green's function. But this is impossible; Eq. (3.14) tells us that this Green's function is proportional to $1/N$. This argument generalizes instantly from a pair to a multi-particle state; all we need to do is declare that each line in the figure represents a cluster of particles.

The argument also generalizes to higher Green's functions. For example, let us consider a four-point function, and let us assume it has a two-particle cut, in a two-bilinear subenergy, in leading (first) order in $1/N$. Then there must be a connected amplitude for two bilinears to produce a pair, proportional to $N^{-\frac{1}{2}}$. If we reproduce the reasoning of the preceding paragraph, with the cross in Fig. 11 now denoting the double bilinear, we deduce that there is a singularity in an eight-bilinear Green's function proportional to $1/N^2$. But this is impossible; this Green's function is proportional to $1/N^3$. And so on.

There is one way in which the large-N theory does not resemble a conventional weakly coupled field theory; there is an infinite number of mesons. We know this is so because asymptotic freedom is not spoiled in the large-N limit; thus two-point functions must behave logarithmically for large spacelike momenta. This cannot be achieved if the only singularities of these functions are poles and if all poles lie within some bounded

Fig. 11

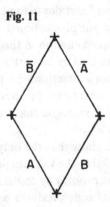

region. Thus there must be an infinite number of mesons of ever-increasing mass. We can sharpen this argument a bit: because we can build bilinears of any spin, there must be an infinite number of mesons of each spin. Such an infinite tower of stable mesons makes sense only if all mesonic S-matrix elements vanish in the large-N limit; otherwise, a heavy meson could decay into light ones. As we have seen, they do indeed vanish.

All of our mesonic analysis can be extended trivially to glueballs, particles made by applying gluonic operators to the vacuum. By Eq. (3.12), the Gs are already properly normalized and need no renormalization. From the same equation, a glueball scattering amplitude with n legs is proportional to $N^{(2-n)}$. Thus glueballs interact even more weakly than mesons; $N^{-\frac{1}{2}}$ in meson dynamics is replaced by N^{-1} in glueball dynamics.

If glueballs interact more weakly than mesons, they are not mesons. This assertion can be checked by an independent line of argument. From Eq. (3.11),

$$\langle B'_1 \ldots B'_n G_1 \ldots G_m \rangle_C \propto N^{(1-m-\frac{1}{2}n)}. \tag{3.15}$$

Thus, glueball–meson mixing vanishes like $N^{-\frac{1}{2}}$ as N becomes infinite. Also, in meson–meson scattering, glueball production is suppressed; to replace a final-state meson by a glueball costs a factor of $N^{-\frac{1}{2}}$ in amplitude.

Up to now, flavor has been irrelevant to our discussion. This will not be the case for our next (and last) two topics, the validity of Zweig's rule and the existence of exotic mesons. With only one flavor of quark, we would be hard pressed to distinguish Zweig-allowed from Zweig-forbidden processes, or to tell exotic mesons from ordinary ones.

The usual statement of Zweig's rule is that for any mesonic scattering graph, it is impossible to divide the meson legs into two sets unconnected by quark lines. We already have this, in the form of the statement that in leading graphs all bilinears must appear as insertions on a single quark loop. Graphs for Zweig-forbidden processes must involve at least two quark loops, and thus are down in amplitude by at least one factor of $1/N$. Thus we have all the usual consequences of the rule: in the limit of strict SU(3), mesons must fall into nonets, an $s\bar{s}$ meson cannot decay into a final state free of strange quarks, etc.

To show the nonexistence of exotics requires a little more work. Exotic mesons, if they existed, would be states created from the vacuum by the application of local gauge-invariant quark quadrilinears. Every such object is the sum of products of local gauge-invariant bilinears. (That is to say, the only way to make an SU(N) scalar is to take the inner product of each quark column vector with an antiquark row vector, perhaps with an intervening SU(N) matrix made of $F_{\mu\nu}$s and covariant derivatives.) With no loss of generality we can study the states made by a single product,

$$Q(x) = B'_1(x)B'_2(x), \qquad (3.16)$$

where, for purposes of this argument, I have restored explicit space-time dependence. For simplicity, I will assume we have chosen the flavors of these operators so that B_1, B_2, and Q all have vanishing vacuum expectation values. Then,

$$\begin{aligned}\langle Q^\dagger(x)Q(y)\rangle &= \langle B'^\dagger_1(x)B'_1(y)\rangle\langle B'^\dagger_2(x)B'_2(y)\rangle \\ &+ \langle B'^\dagger_1(x)B'_2(y)\rangle\langle B'^\dagger_2(x)B'_1(y)\rangle \\ &+ \langle B'^\dagger_2(x)B'^\dagger_2(x)B'_1(y)B'_2(y)\rangle_c.\end{aligned} \qquad (3.17)$$

The first two terms on the right are of order one, while the third is of order $1/N$, and thus should be dropped in the large-N limit. But the first two terms simply describe the independent propagation of two mesons from x to y. In the large-N limit, quadrilinears make meson pairs and nothing else.

3.3   The 't Hooft model[8]

The arguments of the preceding section have been powerful but abstract; it would be nice to have a concrete example in which we could see them at work. Such an example is provided by the 't Hooft model, large-N chromodynamics in two space-time dimensions. The model is almost exactly soluble; the simplest Green's functions can be found in closed form (in an appropriate gauge), and, although the computation of the particle spectrum requires numerical analysis, is is of a sort that can be carried out on a pocket calculator. The model also serves to eliminate a worrisome possibility, that the arguments of Section 3.2 are internally inconsistent, that confinement cannot exist for arbitrarily large N.

Of course, it is no surprise to find confinement in a two-dimensional gauge theory. As a warm-up for the 't Hooft model, let me remind you how confinement occurs in an even simpler theory, two-dimensional quantum electrodynamics. This theory is defined by

$$\mathscr{L} = \tfrac{1}{2}(F_{01})^2 + \bar\psi(i\partial_\mu\gamma^\mu - eA_\mu\gamma^\mu - m)\psi, \qquad (3.18)$$

where

$$F_{01} = \partial_0 A_1 - \partial_1 A_0. \qquad (3.19)$$

As always in the analysis of a gauge theory, the first step is to pick a gauge. I will choose axial gauge,

$$A_1 = 0. \qquad (3.20)$$

In this gauge,

$$\mathscr{L} = \tfrac{1}{2}(\partial_1 A_0)^2 + \bar{\psi}(\mathrm{i}\partial_\mu \gamma^\mu - eA_0 \gamma^0 - m)\psi. \tag{3.21}$$

No time derivatives of A appear in this equation; A is not a dynamical variable at all, but a constrained variable, one that must be eliminated from the theory before we can write it in canonical form.

The equation that determines A is

$$\partial_1^2 A_0 = -e\psi^\dagger \psi \equiv -ej^0. \tag{3.22}$$

The general solution of this is

$$A_0(x^0, x^1) = -\frac{e}{2} \int \mathrm{d}y^1 |x^1 - y^1| j^0(x^0, y^1) + Bx^1 + C, \tag{3.23}$$

where B and C are constants. C is irrelevant; it can always be eliminated by a gauge transformation; for simplicity, I will set it to zero. B is relevant; non-zero B corresponds to the existence of a constant background electric field, such as would be caused by classical charges at spatial infinity. In the Abelian case that occupies us at the moment such a background field has a real effect on the physics, and an interesting one. In the non-Abelian case we are heading for, it turns out that the corresponding object has no physical effect whatsoever. This is a fascinating byway, but it is a byway, and I do not want to spend time on it here. Thus, I will assume for this investigation that there is no background field, and set B to zero.

Thus, we can eliminate A_0 and write the Lagrangian as

$$L = L_{0f} + \frac{e^2}{4} \int \mathrm{d}x^1 \, \mathrm{d}y^1 \, j_0(x^0, x^1)|x^1 - y^1|j_0(x^0, y^1), \tag{3.24}$$

where L_{0f} is the free fermion Lagrangian. A linear potential has appeared between charges; confinement is manifest, at least for small coupling, where we can trust perturbation theory.

It will be convenient for our later work to express the current–current interaction in Eq. (3.24) as the effect of exchange of a photon propagator,

$$D_{\mu\nu}(k) = -\frac{\mathrm{i}}{2} \delta_{\mu 0}\delta_{\nu 0} \int \mathrm{d}^2 x \, \mathrm{e}^{\mathrm{i}k \cdot x}|x^1|\delta(x^0)$$

$$= \mathrm{i}\delta_{\mu 0}\delta_{\nu 0} \frac{\mathrm{P}}{(k_1)^2}, \tag{3.25}$$

where P is the principal-value symbol,

$$\mathrm{P}\frac{1}{z^2} = \frac{1}{2}\left[\frac{1}{(z+\mathrm{i}\varepsilon)^2} + \frac{1}{(z-\mathrm{i}\varepsilon)^2}\right]. \tag{3.26}$$

Of course, we could have obtained the momentum-space propagator directly from the Lagrange density, Eq. (3.21), by standard methods; I detoured through position space to justify the somewhat unusual principal-value prescription at the pole.

It requires no work to generalize this chromodynamics. The nonlinear terms in F_{01} are proportional to the product of A_0 and A_1. Thus they vanish in axial gauge, and with them vanishes one of the characteristic complications of chromodynamics, the self-coupling of the gauge field. Thus the only difference between chromodynamics and electrodynamics is a sprinkling of indices here and there; following the derivation that led to Eq. (3.24), we find

$$L = L_{0f} + \frac{g^2}{N} \int dx' \, dy' \, j_{0a}^b(x^0, x^1) |x^1 - y^1| j_{0b}^a(x^0, y^1), \tag{3.27}$$

where

$$j_{0a}^b = \psi_a^\dagger \psi^b - \frac{\delta_a^b}{N} \psi_c^\dagger \psi^c, \tag{3.28}$$

and I have rescaled the fields to put the coupling constant in its conventional location.

The elimination of gluon self-coupling drastically diminishes the number of graphs that contribute in the large-N limit. Fig. 12 shows the set of graphs that contribute to a Green's function for two quark bilinears; the shaded blobs represent quark propagators. The simple structure of the graphs arises because a gluon line that connects the upper quark line to the lower quark line forms an impassable barrier. No gluon line can cross it without interaction, because this would violate planarity; no gluon line can cross it with interaction, because there are no interactions. The same simplicity of structure appears in Fig. 13, the equations for the

Fig. 12

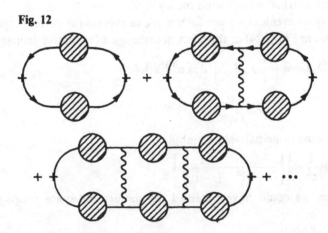

Fig. 13

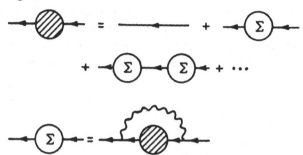

quark self-energy, Σ. The first gluon line to leave the quark must be the last to return, for it forms an impassable barrier.

Things are still not as simple as they could be. For one thing, we still have to keep track of the two components of the quark field; Fig. 13 defines a matrix equation. For another, we are working in a non-covariant gauge, so we do not have the advantages of manifest Lorentz invariance. Both these problems can be eliminated if we switch from axial gauge to light-cone gauge.

Light-cone coordinates are defined by

$$x^{\pm} = (x^0 \pm x^1)/\sqrt{2}. \tag{3.29}$$

In these coordinates,

$$g^{+-} = g^{-+} = g_{+-} = g_{-+} = 1. \tag{3.30}$$

All other components of the metric tensor vanish. Note that this implies a peculiar Dirac algebra,

$$(\gamma^+)^2 = (\gamma^-)^2 = 0, \ \{\gamma^+, \gamma^-\} = 2. \tag{3.31}$$

Light-cone gauge is defined by

$$A_- = A^+ = 0. \tag{3.32}$$

This condition is Lorentz-invariant (but not parity-invariant). A rerun of the derivation of Eq. (3.25) leads to the photon propagator,

$$\mathbf{D}_{\mu\nu}(k) = i\delta_{\mu+}\delta_{\nu+}\frac{\mathbf{P}}{(k_-)^2}. \tag{3.33}$$

At every interaction vertex, only the matrix γ^+ appears. Thus, if we consider a quark line joining two interaction vertices, most of its matrix structure is annihilated; from Eq. (3.31),

$$\gamma^+ \left\{ \begin{array}{c} 1 \\ \gamma^+ \\ \gamma^- \end{array} \right\} \gamma^+ = 2\gamma^+ \left\{ \begin{array}{c} 0 \\ 0 \\ 1 \end{array} \right\}. \tag{3.34}$$

382   1/N

Hence, all the Lorentz-index structure of our graphs is trivial. The photon propagator has only one non-zero component, and the interaction vertex is always proportional to a single Dirac matrix, as is the only surviving part of the quark propagator. There is no point in keeping track of these unvarying structures; we might as well drop them and make the substitutions

$$D_{\mu\nu}\rightarrow i\,\frac{P}{(k_-)^2}, \tag{3.35a}$$

$$-\frac{ig\gamma^+}{N^{\frac{1}{2}}}\rightarrow-\frac{2ig}{N^{\frac{1}{2}}}, \tag{3.35b}$$

and

$$i\frac{p_+\gamma^++p_-\gamma^-+m}{2p_+p_--m^2+i\varepsilon}\rightarrow\frac{ip_-}{2p_+p_--m^2+i\varepsilon}. \tag{3.35c}$$

The internal-index structure of our graphs is also trivial, by our earlier analysis; for each graph there is a unique way of distributing the internal indices, and the net effect of this distribution is to cancel the factor of $1/N^{\frac{1}{2}}$ in Eq. (3.35b).

Thus the first part of Fig. 13, the equation for the quark propagator, $S(p)$, becomes

$$S(p)=\frac{ip_-}{2p_+p_--m^2-p_-\Sigma(p)+i\varepsilon}. \tag{3.36}$$

while the second part of Fig. 13, the equation for the quark self-energy, Σ, becomes

$$-i\Sigma=-i4g^2\int\frac{dk_+\,dk_-}{(2\pi)^2}\,S(p-k)\,\frac{P}{(k_-)^2}. \tag{3.37}$$

If, in this equation, we make the shift of integration variables,

$$p_+-k_+\rightarrow-k_+, \tag{3.38}$$

we eliminate all reference to p_+. Σ is a function of p_- only; by Lorentz invariance, it must be a constant multiple of $1/p_-$. Thus,

$$m^2-p_-\Sigma\equiv M^2, \tag{3.39}$$

is a constant. To leading order in $1/N$, the sole effect of the interaction is to replace the bare quark mass, m, by the 'renormalized quark mass', M. (The quotation marks are to remind you that we are working with gauge-dependent entities, entities which do not necessarily have any physical meaning. I will return to this point shortly.) It is now straightforward to evaluate M^2. The details of the computation are in Appendix 2; the answer is

$$M^2 = m^2 - (g^2/\pi). \tag{3.40}$$

Now that we have the quark propagator, we can evaluate the sum of ladder graphs in Fig. 12, and discover the spectrum of meson states. I will only state the answer here (again, the details are in Appendix 2). There is a bound state of mass μ for every eigenvalue of the integral equation

$$\mu^2\phi(x) = \left(\frac{M^2}{x} + \frac{M^2}{1-x}\right)\phi(x) - \frac{g^2}{\pi}\int_0^1 dy \frac{P}{(x-y)^2}\phi(y), \tag{3.41}$$

where ϕ is a function defined on the interval $[0, 1]$ and vanishing at the end points of the interval.

Although it has been derived from field theory, Eq. (3.41) may be read as an equation in particle mechanics; to be more precise, it is a two-particle time-independent light-cone Schrödinger equation.

In the normal Schrödinger formalism, the state of the system is given at fixed x^0, and dynamics is defined by the operator that generates x^0-translations, P_0. This commutes with the generator of x^1-translations, P_1, and thus we can simplify dynamical problems by going to an eigenspace of P_1. In the light-cone Schrödinger formalism, the state of the system is given at fixed x^+, and dynamics is defined by the operator that generates x^+-translations, P_+. This commutes with the generator of x^--translations, P_-, and thus we can simplify dynamical problems by going to an eigenspace of P_-. For example, for a single free particle of mass M,

$$2P_+ = M^2/P_-, \tag{3.42}$$

and dynamics is totally diagonal in a P_- basis, just as it is totally diagonal in the normal formalism in a P_1 basis. An important difference between the two formalisms is that while the spectrum of P_1 is the entire real line, that of P_- is the positive half-line only.

For a two-particle system, it is convenient to work in an eigenspace of total P_- with eigenvalue one. Thus, if we denote the P_- operator for one of the particles by x, that of the other is $(1-x)$. Since each P_- must be positive, x must lie between 0 and 1. For two non-interacting particles of equal mass,

$$2P_+ = \frac{M^2}{x} + \frac{M^2}{1-x}. \tag{3.43}$$

Because P_- is one, the eigenvalues of this operator are the squared masses of the two-particle system. In terms of a momentum-space

Schrödinger wave function, the eigenvalue equation is

$$\mu^2 \phi(x) = \left(\frac{M^2}{x} + \frac{M^2}{1-x} \right) \phi(x). \tag{3.44}$$

This is almost Eq. (3.41); all that is missing is the last term. But such a convolution integral in a momentum-space Schrödinger equation is a familiar object; it corresponds to an ordinary potential back in position space, in the case at hand, to a linear potential. Once we strip away the heavy disguise of the light-cone momentum-space formalism, Eq. (3.41) is revealed to be the simplest meson model of all, two quarks interacting through a linear potential.

We would expect such a system to have a purely discrete spectrum. The easiest way to see that this is the case is to reinterpret Eq. (3.41), to think of x as a position operator and the conjugate variable as a momentum operator, p. The operator version of Eq. (3.41) then becomes

$$2P_+ = \frac{M^2}{x} + \frac{M^2}{1-x} + g^2 |p|. \tag{3.45}$$

Aside from a trivial multiplicative constant, this is the ordinary Hamiltonian for a mass-zero particle moving in a potential, *and* restricted to the box [0, 1]. It is this last condition that guarantees that the spectrum is purely discrete, that in fact our space of states does not contain any particles that correspond to two free quarks.

We can also use our reinterpretation to get a quantitative idea of the meson spectrum. For a particle moving in a potential and restricted to a box, we would expect the potential to be irrelevant for sufficiently high excited states, and the eigenstates to be those of a free particle in a box,

$$\phi_n = \sin \pi n x, \, n = 1, 2, \dots, \tag{3.46}$$

with associated eigenvalues

$$\mu_n^2 = g^2 \pi n. \tag{3.47}$$

An easy perturbative calculation shows that this expectation is correct, at least for large n; the corrections to Eq. (3.47) are $O(\log n / n)$.

Of course, Eq. (3.47) is no good for the low-lying mesons. Indeed, since M^2 can be negative for sufficiently large g/m, one might fear that for large g the low-lying spectrum might go crazy. This fear is groundless. From Eq. (3.41) and the identity

$$\int_0^1 \frac{dy \, P}{(x-y)^2} = -\left[\frac{1}{x} + \frac{1}{1-x} \right], \tag{3.48}$$

it follows that

$$
\mu^2 \int_0^1 |\phi|^2 \, dx = m^2 \int_0^1 |\phi|^2 \left(\frac{1}{x} + \frac{1}{1-x} \right) dx
$$
$$
+ \frac{g^2}{2\pi} \int_0^1 dx \int_0^1 dy \, \frac{|\phi(x) - \phi(y)|^2}{(x-y)^2}. \qquad (3.49)
$$

Thus if m^2 is positive, μ^2 is positive, no matter how large g^2 is; tachyonic quarks do not make tachyonic mesons.

To go on requires numerical analysis, so I will stop our investigation here. However, before I leave the model altogether, I would like to make two points.

(1) Everything worked out as we expected it to. Confinement and the large-N limit are not in contradiction.

(2) The structure of the quark propagator tells us nothing about confinement. Here we have a reasonable theory solved in a reasonable approximation; the quark propagator is that of a free particle (sometimes a tachyon); nevertheless, the theory contains no free quarks and no tachyons.

This point needs expansion. Why should the quark propagator be irrelevant? The first answer is that it is a gauge-variant object, and thus not an observable, but this is insufficient; the same might be said about the electron propagator in electrodynamics, and we know the location of the pole here tells us the electron mass, very much an observable quantity. However, it is observable only because it governs the singularity structure of gauge-invariant Green's functions.

For example, Fig. 14 shows a Feynman graph that contributes to a six-current Green's function in electrodynamics; the dots denote the

Fig. 14

currents. This graph has a singularity that corresponds to a reading of it as a process going on in space-time (time runs upward). The initial current creates an electron–positron pair, which are then bounced about by widely separated external fields until they recombine. The location of the corresponding singularity is obviously governed by the electron mass. (This is an idealization of an actual measurement of the mass of a charged particle; the external fields are idealized bending magnets or counters.)

Of course, Fig. 14 is just a lowest-order graph, and we must be sure that higher-order corrections cannot destroy the singularity. We need not worry about propagator corrections, where a photon returns to the same electron line from which it emerged; these merely renormalize the electron mass. Nor need we worry about vertex corrections, where the photon goes from one side of a dot to another; these merely renormalize the strength of the external field. Nor need we worry about corrections where the photon connects electron lines separated by two or more dots; because all the external currents are widely separated, the photon ends are necessarily widely separated, and the electrodynamic interaction is negligible at large distances.

Oh.

We have reached the crux of the matter. The argument that the location of the pole in the quark propagator is an observable quantity rests upon the assumption that there is no confinement. Those who look for confinement in the singularities of the quark propagator are like the man who settled in Casablanca for the waters. They have been misinformed.

3.4   Witten's theory of baryons[9]

Baryons present a special problem for the $1/N$ expansion. The method we used in mesodynamics, the study of fixed Feynman graphs in the large-N limit, cannot be applied here. It takes N quarks (in a totally antisymmetric color state) to make a baryon, and thus we must study graphs with more and more quark lines as N grows larger and larger. This does not mean that there is no hope, that baryons do not obey simple scaling laws. As Witten discovered, the proper procedure is to break the problem into two parts, to first use graphical methods to study n-quark forces in the large-N limit, and then to use other methods to study the effects of these forces on an N-body state.

Defining an n-body force is a delicate matter in a quantum field theory. For a two-body force, we must use the Bethe–Salpeter equation; for higher values of n, Faddeev equations. Fortunately, for our purposes, we need worry about none of these niceties. All we want to do is count powers of N, and for this all we need to know is that the n-body interaction kernel is obtained by summing up some family of graphs with n quark lines entering and n quark lines leaving. We can always imagine obtaining such graphs by breaking open n internal quark lines in a vacuum-to-vacuum graph. By result 2 of Section 3.1, the leading vacuum-to-vacuum graphs with internal quark lines are the planar graphs bounded by a single quark loop; these are proportional to N. We want an interaction that will be effective in a totally antisymmetric color state; thus each quark line should carry a different color index, and breaking the quark lines costs us a factor of N^{-n} from lost index sums. Hence, the n-quark interaction is proportional to N^{1-n}.

This completes the first part of the analysis. We must now study the effects of these interactions. From this point on I will assume that the states we are studying are non-relativistic, so we can use ordinary particle mechanics and treat the interactions as ordinary n-body potentials. I am embarrassed by the necessity of this assumption. Detailed dynamical assumptions should not be needed just to count powers of $1/N$; it should be possible to do the whole analysis in an elegant relativistic formalism. Unfortunately, I have not been able to find such a formalism, so we will just have to plug along with particle mechanics.

We wish to study a bound state made up of a very large number of particles interacting very weakly. This is the traditional domain of the Hartree approximation: each particle is treated as moving independently of the others in a common potential, which in turn is determined self-consistently from the motions of all the particles. Let me remind you of the justification for this approximation by estimating the sources of error. Firstly, the approximation neglects the fact that as each particle moves, it changes the state of the other particles, and thus the potential which it feels. This effect is proportional to the square of the interaction strength, and is thus a negligible correction to the Hartree potential (directly proportional to the interaction strength) if the interaction is weak. Secondly, the approximation neglects the fact that each particle feels not the potential caused by all the particles, but the potential caused by all the particles but itself. The correction this makes to the Hartree potential is inversely proportional to the number of particles and is negligible if this number is large.

The Hartree approximation leads to many-particle energy eigenfunctions that are products of single-particle wave functions. In our case, the many-particle wave function is antisymmetric in color, and thus symmetric in the remaining quark variables: space, spin, and flavor. Thus, if we factor out the color part of the wave function, the quarks act like identical bosons; in the baryon ground state, all the quarks will be in the same state, the ground state of the Hartree potential.

Let me make this more quantitative. For notational simplicity, I will ignore flavor and spin. The baryon Hamiltonian is then

$$H = \frac{1}{2m} \sum_a |\mathbf{p}_a^2| + \frac{1}{2N} \sum_{a \neq b} V^{(2)}(\mathbf{r}_a, \mathbf{r}_b)$$

$$+ \frac{1}{6N^2} \sum_{a \neq b \neq c} V^{(3)}(\mathbf{r}_a, \mathbf{r}_b, \mathbf{r}_c) + \cdots, \tag{3.50}$$

where the Vs are functions independent of N. The approximate ground state wave function is of the form

$$\psi(\mathbf{r}_1 \cdots \mathbf{r}_N) = \prod_a \phi(\mathbf{r}_a) \tag{3.51}$$

We find the best choice of ϕ by the variational method. We compute

$$\langle \psi | H | \psi \rangle = N \left[\frac{1}{2m} \int d^3\mathbf{r} |\nabla \phi|^2 \right.$$

$$+ \frac{1}{2} \int d^3\mathbf{r}_1 \, d^3\mathbf{r}_2 \, V^{(2)}(\mathbf{r}_1, \mathbf{r}_2) |\phi(r_1)\phi(r_2)|^2$$

$$+ \frac{1}{6} \int d^3\mathbf{r}_1 \, d^3\mathbf{r}_2 \, d^3\mathbf{r}_3 \, V^{(3)}(\mathbf{r}_1, \mathbf{r}_2, \mathbf{r}_3) |\phi(r_1)\phi(r_2)\phi(r_3)|^2$$

$$\left. + \cdots \right], \tag{3.52}$$

plus terms of $O(1)$, which we neglect. We now minimize this as a function of ϕ, subject to the constraint

$$\int d^3\mathbf{r} \, |\phi|^2 = 1. \tag{3.53}$$

We find

$$\left[-\frac{\nabla^2}{2m} + V(\mathbf{r}) \right] \phi = \varepsilon \phi, \tag{3.54}$$

where ε is the Lagrange multiplier associated with the constraint, and V is the Hartree potential,

$$V = \int d^3\mathbf{r}_1 \, V^{(2)}(\mathbf{r}, \mathbf{r}_1) |\phi(r_1)|^2$$

$$+ \frac{1}{2} \int d^3\mathbf{r}_1 \, d^3\mathbf{r}_2 \, V^{(3)}(\mathbf{r}, \mathbf{r}_1, \mathbf{r}_2) |\phi(r_1)\phi(r_2)|^2 + \cdots. \tag{3.55}$$

Equations (3.54) and (3.55) define the Hartree approximation; note that they are independent of N. This is a consequence of the fact that every term in the expression for the total energy, Eq. (3.52), is proportional to N; the n-quark interaction is $O(N^{1-n})$, but there are $O(N^n)$ distinct n-quark clusters. As we shall see, this lucky cancellation of powers of N leads to simple scaling laws for baryon physics.

Let me begin with the static properties of baryons. ϕ is independent of N; thus so is the shape of the ground-state baryon, as measured, for example, by its charge or mass distribution. This is just like the situation for a meson. In contrast to a meson, though, the energy of the baryon grows with N. This is essential to the resolution of what would otherwise be a problem for the approximation, that the ground state is not a momentum eigenstate. Precisely because the Hamiltonian is translationally invariant, any spatial translation of a Hartree ground state is also a Hartree ground state. Thus we can form linear combinations of these states that are momentum eigenstates; more carefully phrased, the problem is not that the energy eigenstates are not momentum eigenstates but that the momentum eigenstates are degenerate in energy. But this is as it should be, if the mass of the states is $O(N)$; in the non-relativistic approximation, for example, the momentum dependence of the energy is proportional to (momentum)^2/mass; this is $O(1/N)$, and should not be seen in leading order.

Low-lying excited baryons are obtained by placing a few quarks in excited energy eigenstates of Eq. (3.54). Note that, to leading order, there is no need to change V; the change in the Hartree potential caused by exciting only a few quarks is an effect of $O(1/N)$. Thus, for example, if ϕ_1 is the first excited state of Eq. (3.54) and ε_1 is the associated eigenvalue, the first excited baryon is given by

$$\psi_1 = \frac{1}{N^{\frac{1}{2}}} \sum_a \phi_1(\mathbf{r}_a) \prod_{b \neq a} \phi(\mathbf{r}_b). \tag{3.56}$$

Its energy exceeds that of the ground-state baryon by $\varepsilon_1 - \varepsilon$. Thus, although the baryon spectrum begins high, the spacing between successive baryons is $O(1)$, just like the spacing of mesons.

So much for the static properties of baryons. Now let me turn to their interactions with mesons. As explained in Section 3.2, mesons are created and annihilated by quark bilinears, like

$$B = N^{-\frac{1}{2}} \bar{\psi}^a \psi^a. \tag{3.57}$$

Here the ψs are quark fields, normalized such that they obey canonical commutation relations, that is to say, such that in the non-relativistic approximation they create and annihilate quarks with amplitudes of order unity. Let us consider the matrix element of such a bilinear between two ground-state baryons. Each of the N terms in Eq. (3.57) can annihilate and recreate a quark; thus the baryon–meson–baryon vertex is $O(N/N^{\frac{1}{2}})$ $= O(N^{\frac{1}{2}})$. Of course, because the baryon energy is not changed, the meson must carry energy zero, and cannot be real. However, it can be a virtual meson, for example, one exchanged between baryon and meson in a meson–baryon scattering graph. Because the trilinear meson vertex is $O(1/N^{\frac{1}{2}})$, this contribution to meson–baryon scattering is $O(1)$.

Indeed, in general the meson–baryon scattering amplitude is $O(1)$. To prove this, let us consider the one-baryon expectation value of the time-ordered product of two quark bilinears. There are two classes of terms that contribute to this expression. (1) One of the bilinears can annihilate and recreate a quark; the other can do the same. In this case, the two quarks can be of different colors; the sum over colors gives a factor of N^2, and the total contribution is proportional to N. However, since the intermediate state is the ground-state baryon, the bilinears must each carry energy zero, and this contribution vanishes on the meson mass shell. (2) One of the bilinears can annihilate a ground-state quark and replace it with an excited quark; the other can then reverse the process. In this case, all the quarks must be of the same color, so the sum over colors only gives a factor of N, and the total contribution is $O(1)$. However, this contribution does not vanish on the meson mass shell.

We can also study the matrix element of a bilinear between a ground-state baryon and an excited baryon, like the one given by Eq. (3.56). Here each term in Eq. (3.57) must match up with the corresponding term in Eq. (3.56); thus, as before, we only get a factor of N from the color sum, and the amplitude for mesonic decay of an excited baryon is $O(1)$. By similar reasoning, the amplitude for the process meson + baryon → meson + excited baryon is $O(N^{-\frac{1}{2}})$.

Because the meson–baryon scattering amplitude is $O(1)$, the contribution of continuum states to the absorptive part of the amplitude is of the same order as the contribution of excited baryons. This is in striking contrast to the situation for meson–meson scattering. If, in the real world, meson–baryon scattering is well approximated by a sum over narrow resonances, the explanation of this phenomenon does not lie in the $1/N$ expansion.

Baryon–baryon scattering has a special feature because the baryon mass increases with N; if one studies scattering at fixed center-of-mass energy and momentum transfer, one soon finds oneself below threshold.

The solution is to study scattering at fixed center-of-mass velocity and scattering angle. This implies that momentum transfer grows linearly with N, so it is not profitable to study the scattering process in terms of one-meson exchange, or, indeed, in terms of exchange of any finite number of mesons. The proper strategy is to directly compute the baryon–baryon interaction. When we first worked through the Hartree approximation, we found the interaction of one quark in a baryon with all the rest of the baryon was $O(1)$. For a two-baryon system, by the same reasoning, the interaction of a quark in one baryon with the entirety of the other baryon is $O(1)$. Thus the baryon–baryon interaction is proportional to N, just like the baryon energies, and N factors neatly out of the baryon–baryon scattering equation. Baryon–baryon scattering, at fixed velocity and scattering angle is $O(1)$.

Let me summarize our results. The meson spectrum is independent of N, and meson–meson scattering is $O(1/N)$. The baryon spectrum begins at an energy proportional to N, but, after it begins, the spacing of baryons is independent of N. The sizes and shapes of the baryons are also independent of N, as are the amplitudes for meson–baryon scattering and baryon–baryon scattering (at fixed velocities and angles).

As Witten has pointed out, we have heard this tune before. Certain classical field theories admit finite-energy time-independent solutions of the field equations, like the soliton in the sine–Gordon equation and the monopoles that arise in many unified electroweak and grand unified theories.[10] These lumps of energy become particles in the quantum versions of these theories, and, for small coupling, it is possible to study the properties of these particles. They are, word for word, the properties of the baryons enunciated in the preceding paragraph, with $1/N$ replaced by the small coupling constant (e^2 in the electroweak theories). This is a tantalizing parallelism; it strongly suggests that there should be some way of formulating the $1/N$ expansion such that the baryons appear directly as lumps. Unfortunately, at this moment, I know of no such formulation.

3.5   The master field

It would be good to know the leading term in the $1/N$ expansion of chromodynamics. We have been able to go far without this knowledge, but with it we could go much farther, and much faster. A direct approach, an attempt to compute and sum all planar graphs, is hopeless; some indirect method is needed. In this section, I will explain an indirect method recently proposed by Witten.[11] Witten's program has not yet been brought to completion. Nevertheless, I think it is worth talking about even in its incomplete state; it may well succeed, and, even if it fails, it involves such novel insights that it may inspire you to discover some other, better method.

I will first describe the method for pure gauge theory, and then go on to explain how to assimilate quarks. I think the clearest way to explain things is by drawing a parallel with the classical limit. Feynman's path integral formula tells us that Green's functions for a quantum theory are obtained by integrating over all possible classical motions. However, as $\hbar$ goes to zero, the measure in function space becomes more and more sharply concentrated about the solution to the classical equations of motion; in the limit of vanishing $\hbar$, all quantities are given by their values at the classical solution. A very similar statement applies to the large-N limit. There is a classical gauge-field configuration, which I will call the master field, such that the large-N limits of all gauge-invariant Green's functions are given by their values at the master field. I emphasize that once things are evaluated at the master field, there are no further steps; in particular, no integrations, functional or other, need to be done.

I will first comment on the method and then give the proof of the existence of the wonder-working master field.

Comments. (1) The master field is a field for the large-N limit of chromo-dynamics, that is to say, it is a gauge field for gauge group $U(\infty)$. (2) The master field is not unique; because we are interested only in gauge-invariant quantities, any gauge transform of a master field is also a master field. However, this is the end of the non-uniqueness; two gauge-inequivalent fields assign different values to some gauge-invariant quantity, and thus cannot both be master fields. A purist would thus speak not of 'the master field' but of 'the master orbit of the gauge group'. (3) For the classical limit, we not only know that everything is dominated by a single field configuration, we have an algorithm for finding it, solution of the classical equations of motion. This is the missing element in Witten's program. We know that the master field exists but we have no algorithm for finding it. (4) Nevertheless, we can say some things about the master field. We expect large-N Green's functions to be translationally invariant; thus the master field should be translationally invariant also. That is to say, we expect that, in an appropriate gauge, A_μ should be independent of space-time. (Note that this does not mean that the master field is trivial, that $F_{\mu\nu}$ vanishes; the components of A_μ need not commute.) Thus, to find the master field we need only find four matrices. True, these are infinity-by-infinity matrices, so this is not necessarily an easy task. Nevertheless, this is a remarkable reformulation of the problem of summing all planar graphs.

Proof. We wish to show that for large N the measure in function space becomes concentrated on a single orbit of the gauge group. This is equivalent to showing that the probability of finding any gauge-invariant quantity away from its expectation value goes to zero as N goes to infinity. Because all gauge-invariant quantities are sums of products of the Gs of Section 3.2, it suffices to prove the proposition for an arbitrary G. Of course, before we begin, we must normalize G such that its expectation value has a large-N limit, that is to say, is of order unity. Thus we define

$$G' = G/N. \tag{3.58}$$

Now let us estimate the probability of G' departing from $\langle G' \rangle$ by computing the variance:

$$\begin{aligned}
\langle (G' - \langle G' \rangle)^2 \rangle &= \langle G'G' \rangle - \langle G' \rangle \langle G' \rangle \\
&= \langle G'G' \rangle_c \\
&= O(1/N^2),
\end{aligned} \tag{3.59}$$

by Eq. (3.12).   Q.E.D.

All of this has been for pure gauge field theory, but the method can readily be extended to the computation of Green's functions involving quark bilinears. For simplicity, let me assume that we are only interested in the bilinear $\bar{\psi}\psi$. We can compute Green's functions for strings of these operators by giving the quarks a space-time dependent mass, $m(x)$, computing the vacuum-to-vacuum amplitude, and then functionally differentiating this with respect to m. We know that in the large-N limit, the dominant vacuum-to-vacuum graphs are those with one quark loop. In a given external gauge field, the sum of all these graphs is given by a famous expression,

$$\text{Tr} \ln [i\partial\!\!\!/ - A\!\!\!/ - m]. \tag{3.60}$$

But this is a gauge-invariant function of gauge fields only; thus, in the large-N limit, when we integrate over gauge fields, it is given by its value with A_μ replaced by the master field.

You may find these arguments a bit too slick and abstract, and yearn for a concrete example in which one can explicitly find the master field. Fortunately, such an example exists. As I have said, the master field has not been found for four-dimensional chromodynamics. It has not even been found for the vastly simpler 't Hooft model, two-dimensional chromodynamics. However, it has been found for zero-dimensional chromodynamics. This is quite a come-down from field theory; instead of functional integrals over matrix-valued fields we have ordinary integrals over ordinary matrices, and the master field is just a master matrix. Nevertheless, even though the dynamics has been trivialized, the combinatorics retains much of its four-dimensional horror, and the integrals evaluated easily with the master matrix would be nightmares if attempted by the summation of planar graphs. These matrix integrals were first evaluated by this method in a brilliant paper by the Saclay group.[12] I will follow their analysis closely in what follows.

We wish to evaluate integrals of functions of an $N \times N$ Hermitian matrix, H. To begin, we must define integration over H:

$$dH = \prod_{a,b} dH_{ab}, \ a, b = 1 \ldots N, \tag{3.61}$$

where integration over complex variables is defined in the usual way,

$$dH_{ab}dH_{ba} = d(\text{Re } H_{ab})d(\text{Im } H_{ab}). \tag{3.62}$$

This measure is invariant under zero-dimensional gauge transformations,

$$H \rightarrow U^\dagger HU, \tag{3.63}$$

where U is a unitary matrix. We wish to study the zero-dimensional version of the chromodynamic formula for the expectation value of a gauge-invariant operator,

$$\langle \text{Tr } g \rangle \equiv \frac{\int dH \, e^{-S(H)} \, \text{Tr } g(H)}{\int dH \, e^{-S(H)}}, \tag{3.64}$$

where g is some function, and

$$S(H) = N \, \text{Tr } f(H), \tag{3.65}$$

for some function f. (Note that the positioning of the factor of N, and thus the combinatoric analysis, is the same as in four-dimensional chromodynamics.)

We want to evaluate (3.64) for large N. We begin by writing H in canonical form,

$$H = U^\dagger DU, \tag{3.66}$$

where D is a diagonal matrix, with eigenvalues $\lambda_1 \ldots \lambda_N$. We can rewrite the integration measure in terms of the λs and U. By gauge invariance, it must be of the form

$$dH = \left(\prod_a d\lambda_a \right) h(\lambda_1 \ldots \lambda_N) dU \tag{3.67}$$

where dU is the invariant measure on $U(N)$ and h is a function we shall find immediately. To determine h, we compare the two sides of Eq. (3.67) in the neighborhood of the identity in $U(N)$. Here,

$$U = 1 + i\varepsilon, \tag{3.68}$$

where ε is a Hermitian matrix. At this point, Eq. (3.66) becomes

$$H = D - i[\varepsilon, D], \tag{3.69}$$

or, written out in components,

$$H_{ab} = \lambda_a \delta_{ab} + i\varepsilon_{ab}(\lambda_a - \lambda_b), \tag{3.70}$$

where there is no sum on the repeated indices. Hence,

$$dH = \left(\prod_a d\lambda_a \right) \prod_{a \neq b} i(\lambda_a - \lambda_b) d\varepsilon_{ab}. \tag{3.71}$$

Aside from a possible (but irrelevant) multiplicative constant, the term involving ε is dU. Thus,

$$h = \prod_{a \neq b} (\lambda_a - \lambda_b). \tag{3.72}$$

The integration over $U(N)$ factors out of Eq. (3.64). Thus we are left with

$$\langle \mathrm{Tr}\, g \rangle = \frac{\displaystyle \int \prod_a d\lambda_a \sum_b g(\lambda_b) e^{-S_{\mathrm{eff}}}}{\displaystyle \int \prod_a d\lambda_a\, e^{-S_{\mathrm{eff}}}}, \tag{3.73}$$

where

$$S_{\mathrm{eff}} = N \sum_a f(\lambda_a) - \sum_{a \neq b} \ln|\lambda_a - \lambda_b|. \tag{3.74}$$

If we count both explicit factors of N and factors of N arising from the number of terms in a sum, we see that both terms in S_{eff} are $O(N^2)$. This can be made more apparent by introducing

$$\rho(\lambda) \equiv \frac{1}{N} \sum_a \delta(\lambda - \lambda_a). \tag{3.75}$$

The factor of N has been introduced so ρ obeys an N-independent normalization condition,

$$\int d\lambda\, \rho(\lambda) = 1. \tag{3.76}$$

The integral of ρ over any interval gives the fraction of the total number of eigenvalues that lie on that interval; ρ is the fractional density of eigenvalues. For any finite N, ρ is a spiky sum of delta-functions; however, as we shall see shortly, it has a continuous limit as N goes to infinity. In terms of ρ,

$$S_{\text{eff}} = N^2 \left[\int d\lambda \, \rho(\lambda) f(\lambda) - \int d\lambda \, d\lambda' \, \rho(\lambda) \rho(\lambda') \ln |\lambda - \lambda'| \right]. \qquad (3.77)$$

All factors of N have now been made explicit, and the character of the large-N limit is now clear. The integral is dominated by the ρ which minimizes S_{eff}; to leading order,

$$\langle \text{Tr } g \rangle = N \int d\lambda \, \rho(\lambda) g(\lambda). \qquad (3.78)$$

This can be thought of as $\text{Tr } g(H)$, where H is a master matrix, a matrix whose density of eigenvalues is given by the minimizing ρ. We can find the minimizing ρ by searching for the stationary points of Eq. (3.77),

$$f(\lambda) - 2 \int d\lambda' \, \rho(\lambda') \ln |\lambda - \lambda'| = \text{constant}, \qquad (3.79)$$

where the constant is the Lagrange multiplier associated with the constraint equation, (3.76). It is convenient to eliminate the constant by differentiating with respect to λ. We find

$$f'(\lambda) - 2 \int d\lambda' \, \rho(\lambda') \frac{\text{P}}{\lambda - \lambda'} = 0. \qquad (3.80)$$

This is the equation that must be solved to find the master matrix. There is an important technical point: because ρ is restricted to be positive, Eq. (3.80) holds only within the support of ρ, the region where ρ is non-zero. The easiest way to see this is to enforce positivity by writing ρ as σ^2; $\delta\rho$ is then $2\sigma\delta\sigma$, and deriving Eq. (3.80) outside the support of ρ involves an illegitimate division by zero.

I could go on to solve Eq. (3.80) for special choices of f, but I would prefer to stop here; if you want more, you can find it in the literature.[12] The point has been made. There is nothing wrong with the general arguments; for zero-dimensional chromodynamics, the master field exists, and we have found an algorithm for constructing it, Eq. (3.80). The unsolved problem is to find the appropriate generalization of Eq. (3.80) to four dimensions.

3.6   Retrospect and prospect
Where are we?

For mesons, things are wonderful. Only an enthusiast who has spent too much time studying dual resonance models and too little time studying reality would claim that the properties we found in Section 3.2 form an accurate portrait of the mesons. They form a caricature. But it is a recognizable caricature; we look upon it and cry, 'These are the mesons!'

I know of no method other than the $1/N$ expansion in which the lineaments of the mesons emerge so clearly and unambiguously from chromodynamics.

For the baryons, things are not so good. Witten's theory is an analytical triumph but a phenomenological disaster. It is true that in significant ways baryon phenomenology is qualitatively different from meson phenomenology. To take one famous example, duality plus no exotics works wonderfully for mesons, but leads to contradictions for baryons. Baryons are different from mesons, but not as different as they are in the $1/N$ expansion. Baryons are not much heavier than mesons, and baryon resonances are not much broader than meson ones. If our picture of the mesons is a good caricature, our picture of the baryons is a bad one.

There are two possibilities. One is that the $1/N$ expansion is a better approximation for mesons than for baryons. This statement is not as silly as it seems. A reasonable person might agree that a quark and an antiquark is a quark and an antiquark, pretty much the same no matter how many colors there are, while a three-quark baryon is very different from a hundred-quark baryon, no matter how we adjust our coupling constants. The second possibility is that the $1/N$ expansion is terrible for both mesons and baryons. After all, most of our meson phenomenology was derived only from the dominance of graphs with a single quark loop. Although this is certainly a consequence of the $1/N$ expansion, it is not inconceivable that it might also be a consequence of some other principle altogether, and thus be valid even though the $1/N$ expansion is not.

Of course, we would know which of these possibilities is correct, and know much more, if we had an explicit expression for the leading approximation. As Witten has stressed, to seek such an expression is not ridiculously ambitious. One is not searching for a portrait of nature in all its fine shadings, not all phase shifts at all energies, but just for a recognizable caricature of mesodynamics, a table of resonance masses and couplings.

In these lectures I have discussed one attempt to find such an explicit expression. This particular attempt may succeed or fail, but, in any event, I feel future progress in this field rests upon constructing the leading approximation. It is amazing how far we have gone while avoiding this problem, but I do not think we can go much farther without solving it.

Appendix 1. The Euler characteristic

Given a surface composed of polygons, with F faces, E edges, and V vertices, the Euler characteristic is defined by

$$\chi = F - E + V. \tag{A.1.1}$$

In this appendix I give (very sloppy) proofs of two propositions. (1) The

Euler characteristic is a topological invariant. (2) If our polygonal surface is topologically equivalent to a sphere with B holes cut out of it and H handles stuck on to it, then

$$\chi = 2 - 2H - B. \tag{A.1.2}$$

Proof of (1). One can convince oneself that there are only three fundamental ways one can change a polygonal surface continuously. (i) One can distort the surface without changing either F, E, or V. Of course, this does not change χ. (ii) One can shrink an edge to a point. This eliminates one edge, merges two vertices, causing a net loss of one vertex, and does not change the number of faces. Thus χ does not change. The same argument applies to the reverse process. (iii) One can shrink a face to a point. If the face is a polygon with n sides, this procedure eliminates one face, n edges, and $n-1$ vertices. Again χ does not change. The same argument applies to the reverse process. Processes like conversion of a face to an edge can be obtained as (iii) followed by the reverse of (ii) and do not require independent analysis.

Proof of (2). The argument goes in stages. First we prove the equation for a sphere, then for a sphere with holes, and finally for a sphere with holes and handles. (i) A sphere can be constructed by taking two n-sided polygons and identifying their perimeters. We thus obtain a surface with n edges, n vertices, and two faces, for which

$$\chi = 2. \tag{A.1.3}$$

(ii) If we have a polygonal surface that is topologically equivalent to a sphere, we cut holes out of it by removing faces. Thus, each hole reduces χ by one, and

$$\chi = 2 - B. \tag{A.1.4}$$

(iii) To make a handle, we cut two holes that are both n-sided polygons (reducing χ by two) and then identify the perimeters of the two polygons (reducing both E and V by n and not changing χ at all). Thus,

$$\chi = 2 - 2H - B. \tag{A.1.5}$$

Appendix 2. The 't Hooft equations

This appendix gives the computations promised in Section 3.3. I will begin with the quark self-energy. We need a preliminary identity:

$$\int \frac{dx}{x \pm i\varepsilon} = \int dx \left[\frac{P}{x} \mp i\pi\delta(x) \right] = \mp i\pi. \tag{A.2.1}$$

Hence,

$$\int \frac{dp_+ p_-}{2p_+ p_- - a + i\varepsilon} = -i\frac{\pi}{2}\operatorname{sgn} p_-, \qquad (A.2.2)$$

and Eq. (3.37) becomes

$$\Sigma = \frac{g^2}{2\pi} \int dk_- \operatorname{sgn}(p_- - k_-) \frac{P}{(k_-)^2}$$

$$= -\frac{g^2}{\pi p_-}. \qquad (A.2.3)$$

I now turn to the eigenvalue equation for meson masses. If the Green's function of Fig. 12 has a meson pole, then standard arguments lead to the Bethe–Salpeter equation shown in Fig. 15. Here all momenta are oriented to the right, and the shaded blob on the right is the matrix element of the time-ordered product of two quark fields between the vacuum and the meson state. If we denote the Fourier transform of this matrix element by ψ, then the Bethe–Salpeter equation is

$$\psi(p, q) = -4g^2 iS(p-q)S(-q) \int \frac{d^2k}{(2\pi)^2} \frac{P}{(k_- - q_-)^2} \psi(p, k). \quad (A.2.4)$$

If we define

$$\phi(p, q_-) = \int dq_+ \psi(p, q), \qquad (A.2.5)$$

then

$$\phi(p, q_-) = -\frac{ig^2}{\pi^2} \int dq_+ S(p-q)S(-q) \int dk_- \frac{P}{(k_- - q_-)^2} \phi(p, k_-). $$

$$(A.2.6)$$

The q_+ integral is an integral over known rational functions and can be done explicitly:

$$\int dq_+ S(p-q)S(-q) \equiv I(p, q_-)$$

$$= \int dq_+ \frac{1}{2(p+q)_+ - [(M^2 - i\varepsilon)/(p-q)_-]} \frac{1}{2q_+ - [(M^2 - i\varepsilon)/q_-]}. $$

$$(A.2.7)$$

Fig. 15

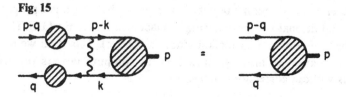

If q_- is outside the interval $[0, p_-]$, the two poles of the integrand are on the same side of the real axis, and the integral vanishes. This implies that ϕ vanishes outside this interval. For q_- within the interval, the integral may be done trivially by closing the contour; the result is

$$I = -\pi i/[2p_+ - M^2/q_- - M^2/(p_- - q_-)].$$ (A.2.8)

Equation (A.2.6) thus becomes

$$[2p_+ - M^2/q_- - M^2/(p_- - q_-)]\phi(p, q_-)$$
$$= -\frac{g^2}{\pi} \int_0^{p_-} dk_- \frac{P}{(k_- - q_-)^2} \phi(p, k_-).$$ (A.2.9)

If we make the substitutions,

$$2p_+ = \mu^2/p_-, \quad q_- = xp_-, \quad k_- = yp_-,$$ (A.2.10)

this becomes Eq. (3.41).

Appendix 3. U(N) as an approximation to SU(N)

In Section 3.1, I dropped the second term in the gluon propagator, Eq. (3.5). To drop this term is to approximate SU(N) gauge theory by U(N) gauge theory; to restore this term is to correct the approximation by introducing a correction gluon, a negative-norm color-singlet gauge meson that cancels the positive-norm color-singlet gauge meson of the U(N) theory. In this appendix I compute the powers of $1/N$ associated with graphs containing internal correction gluons.

As in Section 3, I will restrict myself to vacuum-to-vacuum graphs; from these, all Green's functions can be obtained by functional differentiation. If all we had in our theory were gauge fields, we would have made no error and we would need no corrections; in a U(N) gauge theory, the U(1) gauge meson is completely decoupled from the SU(N) gauge mesons. Thus we need only consider graphs where all correction gluons terminate on quarks. Because the correction gluon carries no color indices, we can always imagine constructing such a graph by adding correction gluons to graphs without them. How many powers of $1/N$ do we introduce by this process? For each quark line, we introduce one extra propagator and one extra vertex; the $1/N$ for the propagator cancels the N from the vertex. However, the correction gluon itself carries $1/N$ because it is a propagator and an additional $1/N$ from the explicit factor of $1/N$ in Eq. (3.5). Thus the net effect of adding a correction gluon is $O(1/N^2)$. This is all we need to know, because we already know how to compute powers of $1/N$ for graphs without correction gluons.

As an example, let me compute the power of $1/N$ associated with the leading connected vacuum-to-vacuum graphs containing a correction gluon. We must add the correction gluon to graphs that contain quark loops. The leading graphs are those in which the correction gluon connects two disconnected graphs each of which contains a quark loop; these are $O(N \times N \times 1/N^2) = O(1)$. However, these graphs are trivial; they serve only to exactly cancel the corresponding gluon-exchange graphs in the uncorrected theory. The leading non-trivial graphs are those in which a correction gluon is added to a connected graph containing a quark loop; these are $O(N \times 1/N^2) = O(1/N)$.

Notes and references

1. To my knowledge, the first to observe that ϕ^4 theory became simple for large N was K. Wilson, *Phys. Rev.* **D7**, 2911 (1973). Wilson drew on ideas developed in statistical mechanics, especially Stanley's work on the spherical model (H. E. Stanley, *Phys. Rev.* **176**, 718 (1968)). The auxiliary-field method used here was also developed in statistical mechanics (but not in the context of large-N expansions), by R. L. Statonovich, *Doklady Akad. Nauk. S.S.S.R.* **115**, 1097 (1957). The treatment of ϕ^4 theory given here follows that of S. Coleman, R. Jackiw, and H. D. Politzer, *Phys. Rev.* **D10**, 2491 (1974). Some points left confused in this paper are clarified in L. Abbott, J. Kang, and H. Schnitzer, *Phys. Rev.* **D13**, 2212 (1976).
2. D. Gross and A. Neveu, *Phys. Rev.* **D10**, 3235 (1974).
3. This note is for the cognoscenti only; it is written in shorthand. You may know that you can define an effective potential, much like V, in theories with fundamental scalar fields, and that, in such theories, V can be interpreted as an energy density for general arguments, not just stationary points. This is important; it implies that if V is unbounded below the theory is sick, no matter how nicely behaved V is at its stationary points. *There is no such interpretation of the effective potential for composite fields.* For example, in the case at hand, if we add a source term to the Lagrange density,

$$\mathscr{L} \to \mathscr{L} + J(x)\sigma,$$

this corresponds to adding a term to the Hamiltonian density,

$$\mathscr{H} \to \mathscr{H} - J\phi^a\phi^a + \frac{g_0}{2N} J^2.$$

The J^2 term has no analog for a fundamental scalar field, and destroys the standard energy arguments (except at stationary points of V, where it vanishes).
4. The model was devised by H. Eichenherr, *Nucl. Phys.* **B146**, 215 (1978) and V. Golo and A. Perelomov, *Phys. Lett.* **79B**, 112 (1978). The $1/N$ expansion is worked out in A. D'Adda, M. Lüscher, and P. Di Vecchia, *Nucl. Phys.* **B146**, 63 (1978), and **B152**, 125 (1979).
5. This is a bit too slick. After all, in electrodynamics, gauge invariance could also be thought of as a mere reflection of the presence of a redundant variable, the longitudinal part of the vector potential. The real difference is slightly more subtle. In electrodynamics, it is not possible to eliminate the redundant variables and still have a local theory, as is shown by the Bohm–Aharonov effect; in the case at hand, it is possible to do just this, as is shown by our

derivation of the model. As we shall see, this distinction will disappear when we sum up the radiative corrections.

6. E. Witten, *Nucl. Phys.* B149, 285 (1979); A. Jevicki, *Phys. Rev.* D20, 3331 (1979); I. Affleck, *Phys. Lett.* 92B, 149 (1980); *Nucl. Phys.* B162, 461 (1980); B171, 420 (1980).

7. (This note covers both this subsection and the next.) The large-N expansion for chromodynamics was invented by G. 't Hooft, *Nucl. Phys.* B72, 461 (1974). There are numerous parallels and connections with the topological expansions of S-matrix theory; see G. Veneziano, *Nucl. Phys.* B117, 519 (1976) and G. Chew and C. Rosenzweig, *Phys. Rep.* 41C, 263 (1978).

8. G. 't Hooft, *Nucl. Phys.* B75, 461 (1974). This paper has spawned a large literature, with two branches. (1) Papers which investigate the model in more detail (for example, by computing form factors) and/or use the model to gain insight into four-dimensional chromodynamics. Some examples: C. G. Callan, N. Coote, and D. J. Gross, *Phys. Rev.* D13, 1649 (1976); M. B. Einhorn, *Phys. Rev.* D14, 3451 (1976); R. Brower, J. Ellis, M. Schmidt, and J. Weis, *Nucl. Phys.* B128, 131, 175 (1977). (2) Papers which attempt to clean up 't Hooft's original derivations and/or to find inconsistencies in the model. To my mind, the paper of this kind that does things best and settles all the problems is I. Bars and M. B. Green, *Phys. Rev.* D17, 537 (1978).

9. E. Witten, *Nucl. Phys.* B160, 57 (1979).

10. For a review of lumps, see Chapter 6 in this volume.

11. The proposal was made in a lecture given at Harvard in the Spring of 1979. E. Witten in *Recent Developments in Gauge Theories*, 1979 Cargese Lectures, edited by G. 't Hooft *et al.* Plenum (1980).

12. E. Brezin, C. Itzykson, G. Parisi, and J. B. Zuber, *Comm. Math. Phys.* 59, 35 (1978).